Solid Phase Microextraction

Gangfeng Ouyang · Ruifen Jiang
Editors

Solid Phase Microextraction

Recent Developments and Applications

Springer

Editors
Gangfeng Ouyang
School of Chemistry
Sun Yat-sen University
Guangzhou, Guangdong
China

Ruifen Jiang
School of Environment
Jinan University
Guangzhou, Guangdong
China

ISBN 978-3-662-53596-7 ISBN 978-3-662-53598-1 (eBook)
DOI 10.1007/978-3-662-53598-1

Library of Congress Control Number: 2016953638

Printed on acid-free paper

This Springer imprint is published by Springer Nature
The registered company is Springer-Verlag GmbH Germany
The registered company address is: Heidelberger Platz 3, 14197 Berlin, Germany

Preface

Solid-phase microextraction (SPME) is a sample preparation technique introduced by Dr. Janusz Pawliszyn, professor of University of Waterloo, in 1990. Compared with traditional sample preparation techniques, SPME combines sampling, enrichment, separation, and introduction into one step and becomes a simple, fast, efficient, and green sample preparation technique. Since its inception, SPME has been widely used in environmental, pharmaceutical, food, and aroma analyses, as well as the studies of chemistry, materials, biology, medicine, etc. In recent 10 years, the hot research areas of SPME include developing novel fiber coatings based on nano or porous materials, detecting free concentration of analyte in complex sample matrices, and nondestructive sampling of living systems using in vivo SPME. The purpose of this book is to summarize the applications of SPME in recent years.

In this book, 10 representative chapters cover the recent development of SPME in coating preparation, environmental analysis, in vivo sampling, etc. Chapter 1 briefly introduces the fundamental knowledge of SPME technique. In Chap. 2, the progresses of the new SPME coating preparation in recent years are summarized. Then, from Chaps. 3 to 7, the applications of SPME for environmental samples, including aerosol particulate in air, complex water sample, and soil/sediment, are reviewed. The main feature of SPME for sensing the freely dissolved organic compounds in complex environmental samples using active and passive SPME sampler is highlighted in these chapters. In addition, SPME combined with derivatization techniques to overcome the challenge of polar compounds' analysis in environmental samples is also summarized and discussed. In the last three chapters, the most distinguished application of in vivo SPME technique is reviewed and discussed, including tracing volatile organic compounds' emission from plant and analyzing the metabolites of pharmaceuticals.

I believe this book will be useful to the researchers in the fields of chemistry, materials, environment, etc., especially who are doing works relative to SPME as well as the newcomers who want to employ the SPME techniques. I am especially grateful to the authors and my colleagues. This book cannot be done without their outstanding contributions.

Guangzhou, China

Gangfeng Ouyang
Ruifen Jiang

Contents

Contributors

Guosheng Chen MOE Key Laboratory of Aquatic Product Safety/KLGHEI of Environment and Energy Chemistry, School of Chemistry, Sun Yat-Sen University, Guangzhou, China

Anna Gracka Faculty of Food Science and Nutrition, Poznań University of Life Sciences, Poznań, Poland

Siming Huang MOE Key Laboratory of Aquatic Product Safety/KLGHEI of Environment and Energy Chemistry, School of Chemistry, Sun Yat-Sen University, Guangzhou, People's Republic of China

Henryk Jeleń Faculty of Food Science and Nutrition, Poznań University of Life Sciences, Poznań, Poland

Ruifen Jiang School of Environment, Guangzhou Key Laboratory of Environmental Exposure and Health, and Guangdong Key Laboratory of Environmental Pollution and Health, Jinan University, Guangzhou, People's Republic of China

Xiang Li Shanghai Key Laboratory of Atmospheric Particle Pollution and Prevention (LAP3), Department of Environmental Science and Engineering, Fudan University, Shanghai, China

Shuqin Liu MOE Key Laboratory of Aquatic Product Safety/KLGHEI of Environment and Energy Chemistry, School of Chemistry, Sun Yat-Sen University, Guangzhou, People's Republic of China

Tiangang Luan MOE Key Laboratory of Aquatic Product Safety, School of Marine Sciences, Sun Yat-Sen University, Guangzhou, China

Małgorzata Majcher Faculty of Food Science and Nutrition, Poznań University of Life Sciences, Poznań, Poland

Gangfeng Ouyang MOE Key Laboratory of Aquatic Product Safety/KLGHEI of Environment and Energy Chemistry, School of Chemistry, Sun Yat-Sen University, Guangzhou, People's Republic of China

Dajana Vuckovic Department of Chemistry and Biochemistry and PERFORM Centre, Concordia University, Montreal, Canada

Jianqiao Xu MOE Key Laboratory of Aquatic Product Safety/KLGHEI of Environment and Energy Chemistry, School of Chemistry, Sun Yat-Sen University, Guangzhou, China

Lihua Yang MOE Key Laboratory of Aquatic Product Safety, School of Marine Sciences, Sun Yat-Sen University, Guangzhou, China

Juan Zheng MOE Key Laboratory of Aquatic Product Safety/KLGHEI of Environment and Energy Chemistry, School of Chemistry, Sun Yat-Sen University, Guangzhou, PR, China

Fang Zhu MOE Key Laboratory of Aquatic Product Safety/KLGHEI of Environment and Energy Chemistry, School of Chemistry, Sun Yat-Sen University, Guangzhou, China

Chapter 1
Introduction of Solid-Phase Microextraction

Shuqin Liu and Gangfeng Ouyang

Abstract The simplification of tedious sample preparation procedures and the integration of sampling and convenient introduction of extracted components to analytical instruments are significant challenges in contemporary analytical chemistry. As a solvent-free and green equilibrium-based sample preparation technique developed by Janusz Pawliszyn in 1990, solid-phase microextraction (SPME) requires only a small volume of adsorbent as the extraction phase and can integrate sampling, isolation, and enrichment into one step. The flexibility of SPME in design, geometry, and calibration strategies makes it a prior choice for the applications in various fields, especially in environmental and life sciences. In this chapter, the fundamentals and practical information about SPME are introduced and discussed, which is supposed to be helpful for readers to explore the following chapters.

Keywords Sample preparation · Solid-phase microextraction (SPME) · SPME devices and coatings · Principle of SPME · Automated SPME systems · Calibration of SPME · SPME method development · Applications of SPME

1.1 Sample Preparation in Analytical Process

Typically, the analytical process for complex samples always contains several steps, including sampling, sample pretreatment, separation, identification and quantitation, data processing, and decision making [1–3]. Each step may cause significant effects on the accuracy of the final result.

The sampling step means to choose the representative sampling points and the appropriate sampling method. In most cases, the complex samples are not able to be

S. Liu · G. Ouyang (✉)
MOE Key Laboratory of Aquatic Product Safety/KLGHEI of Environment and Energy Chemistry, School of Chemistry, Sun Yat-Sen University, Guangzhou 510275, People's Republic of China
e-mail: cesoygf@mail.sysu.edu.cn

G. Ouyang and R. Jiang (eds.), *Solid Phase Microextraction*,
DOI 10.1007/978-3-662-53598-1_1

handled by most analytical instruments directly. So the sample pretreatment step is conducted to separate the analytes from the complex matrix and obtain a suitable concentration of the target analytes. While conducting the separation step, it means to divide the isolated mixture which contains the target analytes into single components by means of chromatography or electrophoretic techniques. The identification and quantitation step can confirm the unknown analytes and determine its amount by retention time and peak area combined with selective detection. Nowadays, gas chromatography or liquid chromatography coupled with mass spectrometers which can provide more specific information is widely used to improve the accuracy of the detection. Finally, the data processing step can provide the analytical results of the target analytes which support the following decision making. It is noted that the analytical steps mentioned above follow one from another, which means the latter step cannot begin before the former one has finished. Hence, the slowest step determines the final speed of the whole analytical procedure and the errors of each step will affect the accuracy of the whole process.

The miniaturization and integration of the analytical instrument have brought great improvement in analytical science. However, the automation of the analytical process and the on-site analysis are still hot research topics in the instrumentation technology, which will lead to more rapid, accurate, and precise results. Presently, the gas chromatography/mass spectrometers (GC/MS) and the liquid chromatography/mass spectrometers (LC/MS) have successfully realized the automatic separation, quantitation, and data processing of the complex samples. Nevertheless, the sample pretreatment techniques always contain multi-steps and organic solvents consumption, which make it tough to integrate the sampling and sample pretreatment steps into automatic procedures. Therefore, almost 80% of the analysis time is spent on the sampling and the sample pretreatment steps.

As for complex samples, it is believed that it will not be possible to improve the extraction efficiency by the reasonable design and the optimization of the extraction process [1, 3]. Therefore, the development of sample pretreatment technique is indistinctive because it is considered as a problem of technique instead of science. The exhaustive techniques tend to completely remove the target analytes from the complex samples to the extraction phase, which do not require calibration during the transfer of the analytes. Consequently, although it will consume much more resources, exhaustive techniques are still more popular than the non-exhaustive techniques among the researchers and regulatory agencies.

1.2 Solid-Phase Microextraction (SPME)

The non-exhaustive microextraction techniques possess many excellent advantages. It will not change the chemical components and the concentrations of the analytes because only a very small amount of the target analytes is removed from the samples. Hence, using microextraction techniques will result in a more representative information and lead to a more accurate characterization of the analytical system or process compared with the exhaustive extraction [4–7].

Solid-phase microextraction (SPME) is a technique developed for the rapid pretreatment of laboratory samples and on-site samples [4–6]. Normally, a small amount of extraction phase dispersed on a supporting medium is exposed to the sample for a specific time before its desorption and analysis. Before equilibrium, the adsorption amount of the target analytes always increases with the increase in the adsorption time. But the distribution of the analytes between the sample and the extraction phase will get to a stable value after reaching the adsorption equilibrium, which will not be affected by the changes in the convection conditions. So it is favorable to quantitatively analyze the target analytes of the samples by conducting equilibrium extraction. In other cases, if the convection conditions remain constant, adopting pre-equilibrium extraction can also successfully receive the accurate quantitation results because the extraction amount is related to the adsorption time, which will notably shorten the analysis time. Up to now, several configurations of SPME have been developed, including fiber, stirrer, vessel walls, suspended particles, tube, and membrane. Among all these approaches, fiber SPME technique is the most widely used one [8, 9].

Taking advantage of equilibrium extraction and selective adsorption, the analytes are moved from the sample matrix onto the coating by conducting SPME procedures. In the first step, the coating is exposed to the sample to selectively extract the analytes by the strong affinity of the analytes and the extraction phase. And then, all of the components extracted by the coating are desorbed into an analytical instrument.

A degree of selectivity is required for every kind of sample pretreatment techniques, and it is impractical to lead all the compounds in the sample into the analytical instrument. An ideal sample pretreatment technique is supposed to remove the compounds incompatible with the instrument including the matrix compounds. It is also desirable to remove the unwanted compounds as many as possible to avoid the interference. Generally, the sensitivity, selectivity, and reproducibility of the SPME technique are mainly determined by the properties of the adsorbent coated on SPME device [10, 11]. Therefore, choosing a proper coating can simplify the analytical procedures and save plenty of time.

1.3 SPME Devices and Coatings

1.3.1 Historical Retrospect

In the original work of SPME, fused silica optical fibers, which were uncoated or coated with liquid or solid polymer, were dipped into the aqueous sample containing the target analytes and desorbed into a GC injector. It needs to open the injector during the insertion and movement of the fiber, which results in the loss of head pressure at the column [8].

Then, the combination of coated fiber into a microsyringe tremendously accelerates the development of SPME technique, resulting in the first SPME devices [5, 6]. Another SPME sampler is based on a piece of microtube with coating inside. Such a tube can be installed inside a needle, or it can just be the "needle" of a syringe. Heating and cooling the air in the upper part of the tube can push the liquid or gas samples into and out of the microtube, which will accelerate the mass transport of analytes from the sample to the coating and realize the active sampling.

Some other samplers have been reported too, including the coating interior of vessels, the coating exterior of the magnetic stirring bars, and even the pieces of poly(dimethyl)siloxane (PDMS) tubes and thin films. Although SPME devices are mainly used in laboratory presently, the recent research has paid more attention to the remote monitoring, clinical, and field environmental applications [9].

1.3.2 Commercial SPME Fiber Coatings and Devices

SPME commercial device was made in Supelco Co. in 1993, which consists of a holder and a fiber [6]. The SPME holder looks like a remolded syringe. The fiber is a fused silica or stainless steel fiber coated with polymer membranes with specific thickness which is protected by a hollow stainless steel tube, and the fiber can stretch freely into the tube. There are two kinds of holders: automatic holder and manual holder. The automatic holder is used for GC autosampler and high-performance liquid chromatography (HPLC). Correspondingly, there are automatic fibers without spring and manual fibers with spring to match the two kinds of holders [12].

The general procedures using commercial fibers contain extraction step and desorption step. In the extraction step, the SPME needle tubing impales the sampling vail and stretches into it. Then, the pushrod of the holder is pushed to expose the fiber coating to the sample to conduct headspace SPME or direct extraction until reaching the adsorption equilibrium. After that, the fiber is retracted and the SPME needle tubing is pulled out of the sampling vail. In the desorption step, the SPME needle tubing is desorbed in the GC injector or the SPME/HPLC interface for a specific period of time.

Choosing an appropriate SPME fiber coating is the key to achieve the favorable analysis results [6, 10]. There are four major criteria that are commonly used in selecting a proper fiber coating for a specific application and are as follows: (a) the molecular weight (MW) and size of the analytes, (b) the polarity of the analytes, (c) the concentration levels and range of the analytes, and (d) the complexity of the sample. Supelco Co. has listed the general rule for the selection of the SPME commercial fiber toward different analytes according to their molecular weight (MW), volatility, and polarity (Table 1.1).

Table 1.1 Selection of SPME commercial fiber

Analyst type	Recommended SPME fiber
Gas and low MW compounds (MW 30–225)	75 μm/85 μm carboxen/PDMS
Nonpolar and volatile compounds (MW 60–275)	100 μm PDMS
Volatile, amino, and nitro aromatic compounds (MW 50–300)	65 μm PDMS/DVB
Polar and semi-volatile compounds (MW 80–300)	85 μm PA
Nonpolar and semi-volatile compounds (MW 80–500)	30 μm PDMS
Nonpolar and high MW compounds (MW 125–600)	7 μm PDMS
Alcohols and polar compounds (MW 40–275)	60 μm PEG
Aromatic compounds (volatile and semi-volatile C_3–C_{20}) (MW 40–275)	50/30 μm DVB/carboxen/PDMS

Additionally, it is noted that the adsorption capacity of the adsorbent fibers is smaller than the absorbent fibers with the same thickness. There are some other guidelines for choosing an appropriate SPME fiber:

- Adsorbent fibers are better for analytes at low concentration levels.
- Adsorbent fibers have a smaller adsorption capacity, so it is necessary to determine the linear range for each analyte.
- Absorbent fibers are better for complex samples with different concentration levels and dirty analytes containing multiple unknown compounds.

1.3.3 Combination with Analytical Instruments

SPME can be coupled with various analytical instruments conveniently due to its solvent-free nature and the tiny size of the fiber or tube. Although in most instances, not all of the analytes are able to be extracted from the samples, all of the extracted analytes can be detected in the analytical instrument and achieve good analysis results. Besides, the sensitivity of the determination by using SPME technique is pretty high. Presently, GC, HPLC, matrix-assisted laser desorption/ionization (MSLDI), and some other analytical instruments have been applied to the coupling with SPME technique [5, 6].

GC has been the most widely used analytical instrument for the coupling of SPME. Only the analytes will be induced into the GC for analysis because the extraction phase is nonvolatile. Besides, it is no need to design complex injector to deal with the evaporation of the solvents. The standard GC injectors such as split or splitless injector can be easily applied to SPME injection by using a narrow liner tube whose inner diameter is slightly wider than the SPME needle tubing, which can increase the gas flow around the fiber and facilitate the effective desorption of the analytes.

For the analysis of nonvolatile and thermally unstable compounds, HPLC will be a better detector. Typically, the SPME/HPLC coupling device is made of a specialized desorption port and a six-port injection value. When the injection value is in the "load" mode, it allows the fiber to insert into the desorption chamber under ambient pressure. Then, the value is switched to "inject" mode to transfer the desorbed analytes to the column.

MSLDI has been used for the analysis of biomacromolecules coupling with SPME technique. The tip of the optical fiber is silanized and used to extract the target analytes of the sample. The laser energy is transferred to the sample by the optical fiber. An ion mobility spectrometer and a quadrupole/time-of-flight (QqTOF) mass spectrometer are used to detect the SPME/MSLDI signals. SPME/MSLDI coupled with QqTOF can simplify the pretreatment of the sample, which at the same time broadens the SPME technique to the analysis of macromolecules.

1.4 Principle of SPME

1.4.1 Introduction

There are three basic modes in SPME: direct immersion (DI), headspace (HS) extraction, and membrane-protected extraction [6, 11]. Figure 1.1 shows the differences among these extraction modes.

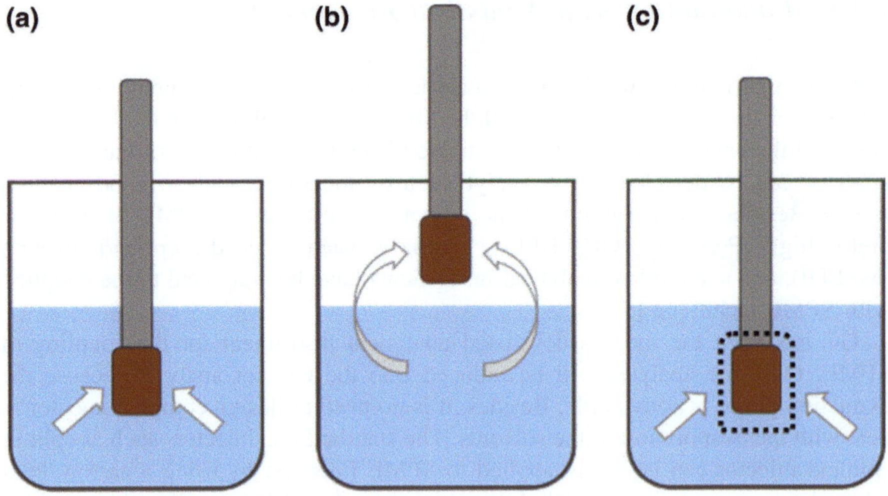

Fig. 1.1 SPME extraction modes: **a** direct extraction, **b** headspace extraction, and **c** membrane-protected extraction

In the direct extraction mode (Fig. 1.1a), the coated fiber is immersed into the sample solution and the analytes are transferred directly from the sample matrix to the coating. Agitation of the sample can accelerate the mass transfer of the analytes from the sample to the extraction phase, which in consequence accelerates the extraction process.

In the headspace extraction mode (Fig. 1.1b), the fiber is placed in the headspace above the sample matrix. In order to receive a favorable diffusion rate, the analytes here are required to be volatile. Headspace extraction can effectively avoid the interference of macromolecules.

For the analysis of nonvolatile compounds with the interference of macromolecules, both direct extraction and headspace extraction are inapplicable. On such occasions, membrane-protected extraction will be a best choice. With the protection of the selective-access membrane, the unwanted macromolecules can be hindered away and the coating will be kept away from the pollution.

1.4.2 SPME Principle

SPME is a technique based on the partition mechanism and the equilibrium established between the analytes and the sample matrix [5, 6]. Typically, the analyte concentration will reach equilibrium in the sample matrix and the fiber coating when the extraction is complete. If only two phases are considered in this case (e.g., the sample matrix and the coating), the equilibrium conditions can be described by Eq. 1.1 according to the law of mass conservation.

$$c_0 V_s = c_s^{\infty} V_s + c_f^{\infty} V_f \tag{1.1}$$

where c_0 is the initial concentration of the analyte, V_s and V_f are volumes of the sample and the fiber coating, and c_s^{∞} and c_f^{∞} are equilibrium concentrations of the fiber coating and the sample, respectively.

The distribution coefficient K_{fs} of the analyte between the fiber coating and the sample matrix is defined as follows:

$$K_{fs} = c_f^{\infty} / c_s^{\infty} \tag{1.2}$$

Equations 1.1 and 1.2 can be combined as follows:

$$c_f^{\infty} = c_0 K_{fs} V_s / (K_{fs} V_f + V_s) \tag{1.3}$$

And finally, the amount of analyte extracted by the fiber coating can be calculated as follows:

$$n = c_f^{\infty} V_f = c_0 K_{fs} V_s V_f / (K_{fs} V_f + V_s) \tag{1.4}$$

Equation 1.4 is the quantitative basis of SPME technique. It indicates that the amount of the extracted analyte (n) is in direct proportion to the initial concentration of the analyte (c_0) in the sample.

When the sample volume is very large, which means $V_s \gg K_{fs}V_f$, Eq. 1.4 can be simplified as follows:

$$n = K_{fs}V_f c_0 \quad (1.5)$$

Equation 1.5 makes it feasible for the application of SPME technique to the cases where sample volume is unknown. It means that when the fiber coating is directly exposed to air, water, or flowing blood, the amount of the extracted analyte will correspond directly to its concentration in the sample instead of the sample volume.

In general, the extraction system is very complex. And the discussion below will only consider three phases: the fiber coating, the gas phase or headspace, and a homogeneous matrix such as pure water or air. During the extraction process, the analyte will migrate among the three phases until the equilibrium is reached.

The amount of analyte extracted by the liquid polymeric coating is related to the overall equilibrium of the three-phase system. The total mass of an analyte should remain constant during the whole process, so we have:

$$c_0V_s = c_f^\infty V_f + c_h^\infty V_h + c_s^\infty V_s \quad (1.6)$$

where c_0 is the initial concentration of the analyte; V_s, V_f, and V_h are volumes of the sample, the fiber coating, and the headspace, respectively; and c_s^∞, c_f^∞, and c_h^∞ are equilibrium concentrations of the fiber coating, the sample, and the headspace, respectively.

If we define the distribution coefficient K_{fh} of the analyte between the fiber coating and the headspace as $K_{fh} = c_f^\infty/c_h^\infty$ and the distribution coefficient of the analyte between the headspace and sample matrix K_{hs} as $K_{hs} = c_h^\infty/c_s^\infty$, then the amount of analyte extracted by the fiber coating can be turned into:

$$n = K_{fh}K_{hs}V_f c_0 V_s/(K_{fh}K_{hs}V_f + K_{hs}V_h + V_s) \quad (1.7)$$

Also:

$$K_{fs} = K_{fh}K_{hs} \quad (1.8)$$

So, Eq. 1.7 can be simplified as follows:

$$n = K_{fs}V_f V_s c_0/(K_{fs}V_f + K_{hs}V_h + V_s) \quad (1.9)$$

The equation indicates that the presence of headspace does not influence the amount of the extracted analytes. As long as the volumes of the fiber coating, headspace, and sample are constant, it does not matter whether the coating is placed in the headspace or into the sample.

1.5 Automated SPME Systems

Automation of an analytical method provides a large number of advantages such as the reduction in analysis time, the simplification of the operations, and the improvement in the reproducibility. Nowadays, using commercially available autosampler systems with robotic arms, the automation of SPME can be easily achieved [12]. The autosampler contains sample preparation, pre-incubation, and extraction steps. Additionally, it also includes the fully automated desorption of the analyte and the fiber bake-out steps in a separated chamber between the runs which can avoid the carryover of the analyte. At present, the automation of SPME-GC and SPME-LC systems has been widely used [6].

1.6 Calibration of SPME

Since its appearance in 1990s, SPME has been widely used in the detection of various contaminants in air, water, soil, and sediment samples. A proper calibration method is required for the analysis of SPME results. Figure 1.2 shows several methods which have been used for the calibration of SPME.

1.6.1 Traditional Calibration Methods

Traditional calibration methods, such as external standard, internal standard, and standard addition methods, can be used for the quantitation of SPME [6, 13, 14]. They are more suitable for laboratory uses, and each of them has its own advantages and disadvantages.

Calibration in SPME

- Traditional calibration methods: External standard, internal standard, standard addition
- Equilibrium extraction: $n = c_0 K_{fs} V_s V_f / (K_{fs} V_f + V_s) \xrightarrow{V_s \gg K_{fs} V_f} n = K_{fs} V_f c_0$
- Exhaustive extraction: $n = V_s c_0$
- Pre-equilibrium extraction: $n = c_0 K_{fs} V_s V_f [1 - \exp(-a_R t)] / (K_{fs} V_f + V_s)$
- Diffusion-based calibration
  - Various empirical mass transfer correlations
  - Fick's First Law of diffusion
  - Kinetic calibration

Fig. 1.2 Several calibration methods in SPME

For external standard method, drawing a calibration curve involves a series of standard solutions at different concentrations to obtain the relationship between the peak responses and the concentrations of the standard solutions. Then, the analyte concentration in the unknown sample can be calculated by the calibration curve. This method does not need too much sample, but it is crucial to make sure all the extraction procedures and analysis conditions for each extraction are strictly invariant. A blank sample matrix is needed if there are matrix effects, and a standard gas mixture or a standard gas generator is needed for the gas samples.

In the internal standard calibration process, a compound whose properties are highly similar to the target analyte is added to the calibration solutions and the samples. That chosen compound should be easily isolated from the analyte and can simulate the treatment steps including equilibrium and non-equilibrium process of the analyte. The ratio of the peak area of the analyte to the internal standard for calibration solutions can be received by using the internal standard method. That ratio will be adopted to calibrate the real samples. The internal standard method is commonly used to compensate the matrix effects, the analyte loss during preparation steps, and the parameters with low reproducibility. However, it is not very easy to find out a proper internal standard compound for the complex samples.

The standard addition method involves adding different amount of target analytes to the sample matrix and analyzing the resulting mixtures. Then, a plot of the responses for the range of target analyte concentrations is developed, and the extrapolation of the response to zero defines the initial concentration of the analyte in the unspiked sample. It needs to prepare a large amount of samples in this method. Nonetheless, it can calibrate the matrix effects in the sample, and it is suitable for the analysis of small amount of complex samples.

1.6.2 Equilibrium Extraction

Equilibrium extraction has been widely used as the quantitation method in SPME technique [6]. When the extraction equilibrium is reached, the amount of analyte extracted by the fiber coating can be calculated using Eq. 1.4 $\left(n = c_f^\infty V_f = c_0 K_{fs} V_s V_f / (K_{fs} V_f + V_s)\right)$.

When the sample volume is very large, as in field sampling, it can be simplified as Eq. 1.5 ($n = K_{fs} V_f c_0$). The above equation indicates that the amount of the extracted analyte will correspond directly to its concentration in the sample matrix and has nothing to do with the sample volume. In practice, the fiber can be exposed directly to the specific environment without collecting a defined sample before the analysis, which will accelerate the whole analytical process and reduce the errors. For field sampling, if the distribution coefficient of the analytes between the fiber coating and the sample matrix is known, the concentration of the target analyte can be determined from the amount of extracted analytes under equilibrium condition.

1.6.3 Exhaustive Extraction

When the sample volume is extremely small and the distribution coefficient of the analyte between the fiber coating and the sample matrix is pretty large ($V_s \ll K_{fs}V_f$), then Eq. 1.4 can be simplified as follows:

$$n = V_s c_0 \quad (1.10)$$

Equation 1.10 indicates that all of the analytes in the sample matrix are extracted by the fiber coating. In this case, the concentration of the target analyte can be easily calculated from the sample volume and the amount of the analyte extracted by the coating.

Exhaustive extraction is not a usual case in SPME because of the strict demand of large distribution coefficient and small sample volume. However, utilizing special devices such as an internally cooling fiber device during the extraction can achieve exhaustive extraction. Another approach to achieve exhaustive extraction is multiple SPME. By repeating the extraction procedures of the same sample using the fiber coating for several times, the total amount of the analyte can be calculated from the several extraction results even if the analyte in the sample matrix is not extracted exhaustively [6].

1.6.4 Pre-equilibrium Extraction

A diffusion-based theoretical model for the mass transfer process of SPME has been proposed by Ai [15, 16].

$$n = c_0 K_{fs} V_s V_f [1 - \exp(-a_R t)] / (K_{fs} V_f + V_s) \quad (1.11)$$

The a_R here is a time constant related to the extraction phase, the headspace volume, the sample volume, the mass transfer coefficient, the distribution coefficient, and the surface area of the extraction phase. Equation 1.11 implies a linear relationship between the amount of analyte extracted and the initial concentration of the analyte in the same sample matrix. In other words, under the stable agitation condition and the constant extraction time and extraction temperature, quantitation of SPME can be achieved even if equilibrium extraction is not reached.

The equilibrium time for SPME depends on the agitation condition and the physical and chemical properties of the analytes and the fiber coating. When the equilibrium is reached, the amount of analyte extracted by the fiber coating will be the maximum, which results in the highest sensitivity. If the sensitivity is not the primary factor, the extraction can be stopped before equilibrium to shorten the extraction time. However, the calibration method can be used only when the time constant a_R is known [6].

1.6.5 Diffusion-Based Calibration

Recently, several diffusion-based calibration methods have been developed for the quantitation of SPME. These methods are mainly applied to the on-site sampling and in vivo sampling, including grab sampling and long-term monitoring. The calibration methods are mainly based on Fick's first law of diffusion, the interface model, the cross-flow model, and the kinetic processes of absorption/adsorption and desorption [6].

1.7 Method Development of SPME

Several procedures are needed to conduct during the development of SPME method, such as the selection of the fiber coating, the selection of the sampling mode, the optimization of the extraction conditions, the optimization of the sample matrix, and the selection of quantitation and calibration methods. The above-mentioned steps are usually carried out no matter what type of separation/detection system is used for the following analysis. Generally, SPME can be coupled with all kinds of commonly used separation and detection systems such as GC, HPLC, capillary electrophoresis (CE), or supercritical fluid chromatography (SFC) combined with conventional detectors. Different analytical system is chosen depending on the properties of the analyte especially on its volatility and distribution behavior.

1.7.1 Method Developing Procedures

The sensitivity, selectivity, and reproducibility of SPME are mainly determined by the properties of the adsorbent coated on SPME device. Therefore, the first step of developing a SPME method is to select a suitable fiber coating. Nowadays a variety of commercial fibers with different polarity, thickness, and coating length are available in the market, either in manual or autosampler versions. The general selection of the SPME commercial fiber toward different analytes according to their MW, volatility, and polarity has been listed in Sect. 1.3.2, which can be a useful reference. Besides, numerous novel fiber coatings with high sensitivity and selectivity have been proposed by the researchers for the extraction of the specific analytes [17, 18].

Secondly, as mentioned in Sect. 1.4.1, there are three sampling modes in SPME: DI, HS, and membrane-protected. Both DI and HS sampling modes are widely used in SPME-GC applications. DI-SPME is more suitable for gaseous or simple liquid sample, while HS-SPME is usually used for extraction of complex liquid sample or solid sample. For the dirty samples, only HS-SPME and membrane-protected

SPME can be used. For the analytes with highly volatility and non-polarity, HS mode is the best choice, while DI mode is suitable for the analysis of the less volatile and polar analytes. In general, the analytes for SPME-LC applications are not volatile enough to conduct the HS mode, so DI mode is mainly used in this case. Membrane-protected SPME is rarely used for SPME-LC applications due to its slow kinetics.

Thirdly, several extraction conditions including sample volume, agitation mode, and extraction time should be optimized. Depending on the analytes and the sampling mode, these extraction conditions can be optimized until the best extraction performance is obtained, which will at the same time improve the sensitivity of the method.

Fourthly, optimization of matrix conditions involves the optimization of the temperature, pH value, ionic strength, and the addition of an organic modifier. As a crucial step in SPME method development, optimization of matrix conditions can achieve the best sensitivity and method performance.

Fifthly, the calibration method is chosen according to the actual circumstances. Calibration is a process that relates the analytical signals measured to the analyte concentration. The details of the several calibration methods have been discussed in Sect. 1.6, and depending on the real demand, an appropriate calibration method can be adopted for the quantitation.

1.7.2 Method Validation

After all related parameters of SPME have been selected and optimized, the method should be validated to prove its suitability for the specific applications. Validation experiments are needed to evaluate the selectivity, sensitivity, linear range, limit of detection (LOD), limit of quantitation (LOQ), method precision, and method accuracy of the proposed method. Depending on the intended use of the proposed SPME method, it can be further validated by comparison with the results to those obtained from standard methods. Besides, it can also be validated by analyzing the certified reference materials or the participation in inter-laboratory studies.

1.8 Applications

1.8.1 Environmental Analysis by SPME

SPME techniques are very suitable for the analysis of environmental pollutants including volatile organic compounds (VOC), semi-volatile organic compounds (SVOC), pesticides, herbicides, metal organic compounds, and amines [19]. HS-SPME, DI-SPME, membrane-protected SPME, and in-tube SPME have been applied to the analysis of diverse environmental systems, such as air, water, soil, and sediment samples.

SPME air sampling can be conducted in the field or in the laboratory. The air samples can be collected by bags or SPME devices and brought to the laboratory before extracted by the SPME coating at DI mode or HS mode. DI, HS, and in-tube SPME have been applied to the analysis of water samples. Moreover, many novel coatings have been used for the determination of the pollutants in environmental water sample. Ultrasonic and microwave are sometimes used as the assisted devices during the analysis process. For soil and sediment samples, HS-SPME and DI-SPME are the most commonly used extraction modes, and the extraction always requires the assistance of ultrasonic, microwave, heat, or the cooling of the coating.

1.8.2 Drug Analysis by SPME

For drug applications, SPME was firstly used for the analysis of forensic drugs due to the non-polarity and volatility of the forensic drugs, which was convenient for the SPME-GC application coupled with PDMS-coated fiber [20]. However, with the development of solid sorbent coatings, especially the divinyl benzene (DVB)-related products, a broader range of drugs can be successfully analyzed. It opened a new avenue for the application to the semi-volatiles such as cocaine and benzodiazepines. Application to nonvolatile compounds appeared somewhat later, which might result from the limitation of the available extraction phases and the complex pretreatment for these drugs. However, these challenges have been mostly overcome due to the improvement in the extraction phases for nonvolatile compounds and the combination of SPME with LC and CE, which at the same time broadens the SPME applications to the biomedical and biological compounds [21].

1.8.3 Food and Fragrance Analysis by SPME

Increasing consumer demand for safety and nutritional excellence together with higher market competition underlines the importance of food analysis. Researches on food analysis are mainly focused on the nutritional value, food freshness, food additives or toxic components, changes occurred during the food processing, and the effects of processing or storage on food texture and microbiological quality.

Food fragrance is another factor that plays an important role in consumer acceptance, which is mainly composed of volatile and semi-volatile compounds. Fragrance can often reveal the freshness and processing and storage history of the food, which makes it a hot issue for many chemical and sensory studies [22, 23].

As an extremely complex matrix, several procedures are necessary for the pretreatment of the food before it is analyzed by GC, LC, or other instruments. In recent years, many food studies involving SPME have been proposed and reviewed.

SPME has been proved to be sensitive enough to satisfy the legislation requirements related to low determination and quantitation limits, as well as method accuracy and precision requirements.

1.8.4 In Vivo Application of SPME

In vivo analysis is a special application area where SPME is gaining ground because of its unique format and convenient device. In its most common configurations, a syringe-like (for headspace sampling) or a needle-like device (for direct sampling) is exposed directly to the living system under study. After a short sampling time, the device is directly introduced to GC for thermal desorption or desorbed using solvents for injection into LC–MS. As a matter of fact, some of the presented research could not be carried out using any other sampling and sample preparation methods, as they would probably cause severe damage to the living system or would demand its sacrifice [24–26].

In the past decades, SPME has been applied to analyze the compounds produced by microorganisms, insects, plants, animals, and humans. The target analytes investigated to date include environmental pollutants, pharmaceuticals, pheromones, metabolites, and proteins. Future feasibility and applicability of in vivo SPME techniques depends on the development and commercialization of devices and extraction phases suitable for different types of applications, which can make the techniques more convenient and more applicable [6, 24–26].

Acknowledgement We acknowledge financial support from the projects of NNSFC (Grants 21225731, 21377172, and 21477166) and the NSF of Guangdong Province (Grant S2013030013474).

References

1. Dean J (1998) Extraction methods for environmental analysis. Wiley, New York
2. Pawliszyn J (1995) New directions in sample preparation for analysis of organic compounds. Trends Anal Chem 14(3):113–122
3. Pawliszyn J (2002) Sampling and sample preparation for field and laboratory. Elsevier, Amsterdam
4. Arthur CL, Killam LM, Buchholz KD et al (1992) Automation and optimization of solid-phase microextraction. Anal Chem 64(17):1960–1966
5. Pawliszyn J (1997) Solid-phase microextraction: theory and practice. Wiley, New York
6. Pawliszyn J (2009) Handbook of solid phase microextraction. Chemical Industry Press of China, Beijing
7. Greenwood PA, Greenway GM (2002) Sample manipulation in micro total analytical systems. Trends Anal Chem 21(11):726–740
8. Arthur CL, Pawliszyn J (1990) Solid phase microextraction with thermal desorption using fused silica optical fibers. Anal Chem 62(19):2145–2148

9. Augusto F, Koziel J, Pawliszyn J (2001) Design and validation of portable SPME devices for rapid field air sampling and diffusion-based calibration. Anal Chem 73(3):481–486
10. Dietz C, Sanz J, Cámara C (2006) Recent developments in solid-phase microextraction coatings and related techniques. J Chromatogr A 1103(2):183–192
11. Pawliszyn J (2000) Theory of solid-phase microextraction. J Chromatogr Sci 38(7):270–278
12. O'Reilly J, Wang Q, Setkova L et al (2005) Automation of solid-phase microextraction. J Sep Sci 28(15):2010–2022
13. Martos PA, Pawliszyn J (1997) Calibration of solid phase microextraction for air analyses based on physical chemical properties of the coating. Anal Chem 69(2):206–215
14. Ouyang G, Pawliszyn J (2008) A critical review in calibration methods for solid-phase microextraction. Anal Chem Acta 627(2):184–197
15. Ai J (1997) Solid phase microextraction for quantitative analysis in nonequilibrium situations. Anal Chem 69(6):1230–1236
16. Ai J (1997) Headspace solid phase microextraction. Dynamics and quantitative analysis before reaching a partition equilibrium. Anal Chem 69(16):3260–3266
17. Spietelun A, Pilarczyk M, Kloskowski A et al (2010) Current trends in solid-phase microextraction (SPME) fibre coatings. Chem Soc Rev 39(11):4524–4537
18. Xu J, Zheng J, Tian J et al (2013) New materials in solid-phase microextraction. Trends Anal Chem 47:68–83
19. de Fatima Alpendurada M (2000) Solid-phase microextraction: a promising technique for sample preparation in environmental analysis. J Chromatogr A 889(1):3–14
20. Snow NH (2000) Solid-phase micro-extraction of drugs from biological matrices. J Chromatogr A 885(1):445–455
21. Mirnaghi FS, Pawliszyn J (2012) Reusable solid-phase microextraction coating for direct immersion whole-blood analysis and extracted blood spot sampling coupled with liquid chromatography–tandem mass spectrometry and direct analysis in real time–tandem mass spectrometry. Anal Chem 84(19):8301–8309
22. Steffen A, Pawliszyn J (1996) Analysis of flavor volatiles using headspace solid-phase microextraction. J Agric Food Chem 44(8):2187–2193
23. Kataoka H, Lord HL, Pawliszyn J (2000) Applications of solid-phase microextraction in food analysis. J Chromatogr A 880(1):35–62
24. Ouyang G, Vuckovic D, Pawliszyn J (2011) Nondestructive sampling of living systems using in vivo solid-phase microextraction. Chem Rev 111(4):2784–2814
25. Togunde OP, Lord H, Oakes KD et al (2013) Development and evaluation of a new in vivo solid-phase microextraction sampler. J Sep Sci 36(1):219–223
26. Bai Z, Pilote A, Sarker PK et al (2013) In vivo solid-phase microextraction with in vitro calibration: determination of off-flavor components in live fish. Anal Chem 85(4):2328–2332

Chapter 2
Development of Novel Solid-Phase Microextraction Fibers

Jianqiao Xu and Gangfeng Ouyang

Abstract The basic principles for the preparation of SPME fibers with satisfying selectivity, sensitivity, loading capacities and stabilities are summarized here in terms of the physicochemical properties of the coating materials and the supporting substrates of SPME fibers. While the main part of this chapter focuses on the advances in developing the most widely used coating materials, including ionic liquids, polymeric ionic liquids, carbonaceous materials, molecularly imprinted polymers, metal-organic frameworks, metals and metal oxides, conductive polymers, modified silica, as well as their composites, into SPME fibers. Meanwhile, the applications of novel SPME fibers in analyzing diverse analytes in different sample matrices are briefly reviewed, including the emerging applications of SPME in the extraction of bioactive compounds in living animals and plants.

Keywords SPME fiber coating · Ionic liquid · Polymeric ionic liquid · Graphene · Carbon nanotube · Molecularly imprinted polymer · Metal-organic framework · Metal and metal oxide · Conductive polymer · Modified silica

2.1 Introduction

The outstanding features of solid-phase microextraction (SPME), including rapid and simple operating procedures, slight disturbance to investigated systems, reduced consumption of solvents, and feasibility of automation, continuously inspire new explorations of SPME in various frontiers, such as metabolome detection [1], central nervous systems studies [2], pharmacokinetic studies [3], environmental analysis [4], and food analysis [5]. On the other hand, the diversity of the physicochemical properties of the target analytes and the complicity of the sample matrices require the development of new SPME devices to nourish the new

J. Xu · G. Ouyang (✉)
MOE Key Laboratory of Aquatic Product Safety/KLGHEI of Environment and Energy Chemistry, School of Chemistry, Sun Yat-Sen University, Guangzhou 510275, China
e-mail: cesoygf@mail.sysu.edu.cn

G. Ouyang and R. Jiang (eds.), *Solid Phase Microextraction*,
DOI 10.1007/978-3-662-53598-1_2

explorations. Since the conception of SPME, multiple configurations (Table 2.1) [6], extensive extraction materials [7], and numerous preparation methods for SPME devices [8] have been introduced for the implementation of novel SPME.

Different configurations are designed mainly with three benefits: (1) to be conveniently coupled to analytical instruments (Table 2.1) [4, 9–16], (2) to deal with different samples (e.g., the in-tube SPME and stir bar microextraction are mostly used in aqueous samples, and SPME fibers can be used in liquid, gaseous, and semisolid samples), and (3) to improve analytical sensitivity by increasing surface areas or surface area-to-volume ratios of extraction phases (such as thin-film and coated blade SPME) [12]. Nowadays, fiber is the most widely used SPME configuration, mainly for its wide applicability in gaseous, liquid, and semisolid samples. For this reason, the preparation of SPME fibers has been more extensively studied than the SPME devices in other configurations. In fact, the strategies of preparing SPME fibers can also be readily extrapolated to the preparation of SPME devices in other configurations [9, 10].

The development of novel SPME fibers aimed at promoting the sensitivity, selectivity, loading capacities, and mechanical/thermal/chemical stabilities of SPME fibers, and lowering the costs of preparing SPME fibers as well [7]. Figure 2.1 presents the principles for the preparation of SPME fibers with the desirable features. In general, the extraction performances of SPME fibers, i.e., sensitivity, selectivity, and loading capacities, are determined by the chemistry and physics of the extraction materials. On the other hand, the preparation method (Table 2.2) plays an important role in improving the stabilities and inter-fiber reproducibility, which, however, should be developed according to the coating materials and supporting substrates selected.

The development of new SPME fibers is recently one of the most active research field of this technique, a large quantity of works on this topic have been published in recent years. This chapter is to sketch the progresses in developing the most

Table 2.1 Direct coupling of SPME in different configurations to analytical instruments and automatic sample preparation platforms

Configuration	Instrument coupled to	References
Coated stainless steel mesh	Direct analysis in real time (DART)–MS/MS	[9]
Coated blade	Electrospray ionization (ESI)–MS/MS	[10]
	96-well automatic sample preparation platform	[11]
Thin-film SPME	GC	[12]
In-tube SPME	Direct analysis in real time (DART)–MS/MS	[13]
	Online LC	[14]
SPME fibers	GC	[4]
	Online LC	[15]
	Multiple MS	[16]

Many configurations of SPME samplers, such as thin-film SPME and SPME fibers, can be coupled to off-line LC, i.e., solvent desorption prior to LC analysis, which are not included in the direct coupling strategies

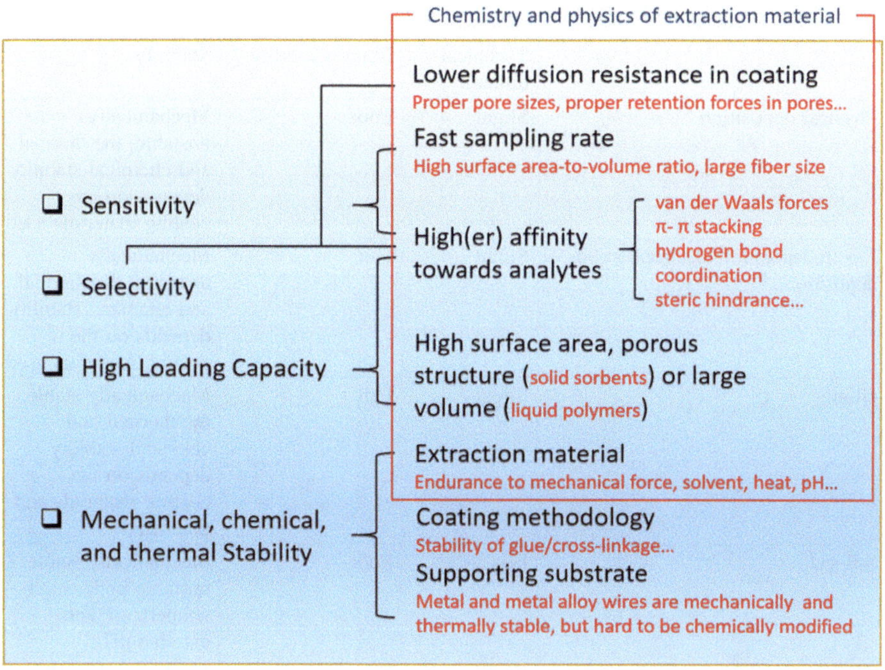

Fig. 2.1 Principles in chemical and physical perspectives for preparing SPME fibers with desirable features

widely used extraction materials into SPME fibers with selected works, and to provide examples to show the utilization of these fibers in practical uses with satisfactory extraction performances. Extraction materials, including ionic liquids (ILs) and polymeric ionic liquids (PILs), carbonaceous materials, molecularly imprinted polymers (MIPs), metal-organic frameworks (MOFs), metals and metal oxides, conductive polymers, and modified silica, are reviewed. In addition, a separate section for reviewing the latest progress of preparing fibers for direct immersion (DI) SPME in complex biological matrices is presented.

2.2 An Overview of Preparation of Fibers with Desirable Features

Figure 2.1 presents the principles from the perspectives of chemistry and physics for the preparation of fibers that fulfill the proposed requirements. Here, an overview of preparation of desirable fibers is provided with several special notes.

To achieve high sensitivity to target analytes, various interactions can be introduced for trapping the target analytes in fiber coatings, including not only the

Table 2.2 Comparison of different preparation methods of SPME fibers

	Technical difficulty	Reproducibility	Stability
Physical deposition	Slight	Poor	Mechanically unstable, the thermal and chemical stability depends on the coating materials used
Electrodeposition/electrophoretic deposition	Slight	Fair	Mechanically unstable, the thermal and chemical stability depends on the coating materials used
Gluing	Slight	Fair	Mechanically stable, the thermal and chemical stability depends on the coating materials and glue used
Sol–gel	Fair	Poor	Mechanically stable, unstable under high temperature and extreme pH
Chemical bonding	High	Fair	Mechanically stable, the thermal and chemical stability depends on the coating materials and cross-linkages used
Electrospinning	Fair	Good	Mechanically stable, the thermal and chemical stability depends on the coating materials used
Anodizing	Fair	Fair	Mechanically and thermally stable, the chemical stability depends on the coating materials used
In situ polymerization/hydrothermal growth	High	Poor	Fair mechanical stability, the thermal and chemical stability depends on the coating materials used

non-covalent interactions such as π–π stacking effect and dipole–dipole interaction, but also the covalent bonds. For example, when using boronate sorbent to extract glycoproteins, covalent bonds formed between the boronate groups of the sorbent and the glycosyl groups of the proteins [17]. It is also notable that thorough desorption of the analytes from fiber coatings is also important for improving sensitivity, which should be realizable under the conditions which the coating materials can endure. For example, the boronate ester bond can be broken under relative mild acidic pH [17]. In addition, increasing surface area-to-volume ratios of the coatings can accelerate the extraction kinetics; thus, a porous morphology can be expected to improve the sensitivity of SPME. In particular, improving the accessibility of the inner surfaces of nanomaterial-based fibers can further accelerate the sampling kinetics, as the inner surfaces in the nanomaterial-based coatings might be blocked due to the highly stacking structures [18]. Fibers with larger dimensions can extract more analytes than smaller fibers in the same time duration; thus, fibers with larger dimensions can potentially possess higher sensitivity [12].

High selectivity toward target analytes mainly depends on higher affinities to target analytes than other compounds coexisted in the sample matrices (thermodynamic factors). In addition, high selectivity might also originate from the differences in adsorption kinetics. For example, diffusion resistances in pores of MOF might differ for analytes due to the different interactions between the analytes and the ligands of MOF, which could result in different adsorption kinetics [19]. High selectivity is ideal for lowering backgrounds and interferences, and thus, it guarantees a satisfactory sensitivity of SPME. However, in contrast to the conventional applications, the applications of SPME in metabolomic studies require the fibers with wide metabolome coverage [1].

Larger extraction-phase volumes or more available binding sites in the extraction phases can lead to higher loading capacities. Simultaneously, the partition coefficient between the extraction phases and the sample matrices should be large for guaranteeing high loading capacities, which also requires that the extraction phases should possess high affinities toward the target analytes.

Immobilizing extraction materials on substrates is an essential step for preparing substrate-supported coating fibers. Table 2.2 presents a brief comparison among the most widely used preparation methods in terms of technical difficulty, reproducibility, and stability. Silica fibers and stainless steel wires (SSWs) are usually used as the supporting substrate. Silica fibers are fragile; however, their surfaces are easy to be modified with reactive functional groups to chemically bond the coatings (Fig. 2.2a). SSWs are popular substrates mainly for the outstanding flexibilities under mechanical forces. Except for the difficulty in chemically bonding the coatings on SSWs, other preparation methods can be readily realized with SSWs as the substrates, especially the wires can be used as electrodes or electrospinning receptors in the preparation of conductive polymer coatings and electrospun nanofibrous coatings [18, 20]. Actually, fiber coatings can be chemically bonded on SSWs by precoating noble metals [21–23], silicon [24] and polydopamine (PDA) [25], which can be further functionalized with reactive functional groups. Figure 2.2b summarizes the strategies of chemically bonding coatings on SSWs. The reactive functional groups can chemically bond coatings through specific covalent bonds. For example, amino groups can react with carboxyl groups [21, 26], vinyl groups can participate in polymerization [25], sulfhydryl groups can act as chain transfer agent in polymerization reactions [22] and can chemically bonding noble metal nanoparticles [23],

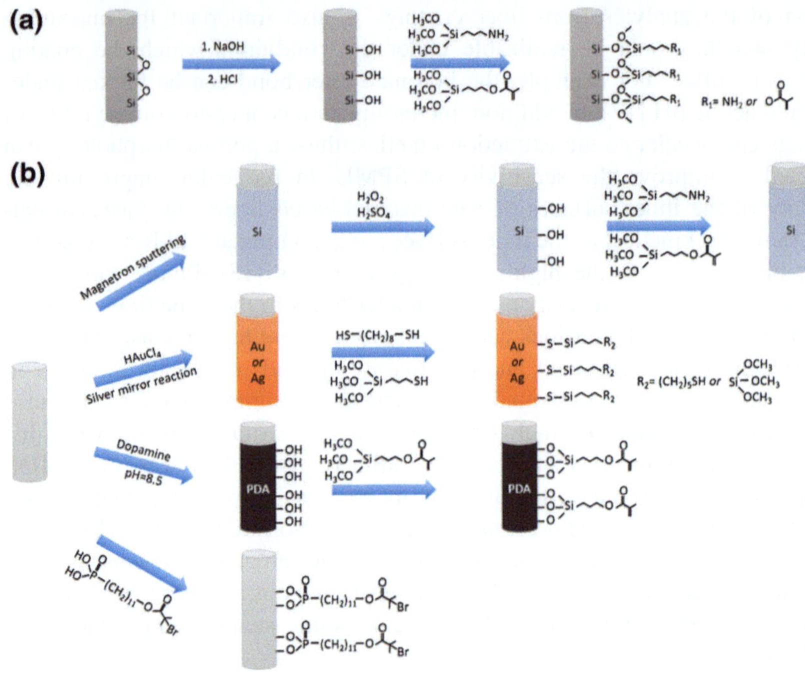

Fig. 2.2 **a** Modification of fused-silica fibers and **b** modification of SSWs for chemical bonding of coatings

methoxysilyl can react with other alkoxysilyls [21], and bromine atoms can also act as chain transfer agent in polymerization reactions [27]. Other metal and metal alloy wires, such as gold wires [28] and NiTi alloy wires [29], have also been used as supporting substrates. Substrateless fibers, i.e., monolithic fibers, were also reported in the literature [30, 31]. Polymeric monoliths are also flexible under mechanical forces. However, monolithic SPME fibers are not as much reported as substrate-supported ones, probably because not all the extraction materials can be facilely prepared into monoliths.

When the fibers are used for DI SPME in complex sample matrices, such as in biological samples and in sediments, besides the requirements aforementioned in Fig. 2.1, the fibers should be resistant to binding of proteins and humic substances on the coatings, which could hinder the surfaces of fibers, and shorten the lifespans of fibers [32]. In addition, SPME fibers used for in vivo sampling in biological tissues are also required to be biocompatible to avoid triggering rejection reactions [18].

2.3 ILs and PILs

ILs are now extensively studied for the applications in organic chemistry, electrochemistry, and analytical chemistry [33–35], due to the versatile and intriguing physicochemical properties including low volatility, thermal stability, electrical conductivity, tunable miscibility with solvents, tunable viscosity, and designable and anticipatable solvation properties [33]. For the perspective of providing high affinities to target analytes by introducing proper functional groups on the cations and by using anions with different Lewis basicity, ILs are recently explored for preparing task-specific SPME fiber coatings. Polymeric ionic liquids (PILs) are prepared by polymerization of alkenyl-functionalized ILs and have also been used as SPME fiber coatings.

IL was first used as a disposable SPME fiber coating by Liu et al. [36] by dipping a fused-silica fiber in the mixture of dichloromethane and 1-octyl-3-methylimidazolium hexafluorophosphate. Later, to load more ILs on fused-silica fibers, fused-silica fibers were etched or coated with a Nafion membrane prior to dipping in ILs [37, 38]. Cui et al. [39] also found etched SSWs could load more IL than non-etched SSWs. Even though Ho et al. [40] demonstrated their IL coating prepared on untreated fused-silica fiber with the dipping method could stand no less than eight extraction–desorption circles, these IL coatings were not stable enough, recoating was required during uses [36–40]. The first relatively long-term reusable IL-based SPME fiber was prepared by fixing IL-impregnated silicone elastomer on a bare fused-silica fiber [41], followed by Carda-Broch et al. [42] and Wanigasekara et al. [43] who glued IL-functionalized silica particles on fused-silica fibers, which loaded more IL on the fiber as well. IL was also chemically bonded on a fuse silica fiber by using 3-chloropropyltrimethoxysilane as a cross-linker [44]. In other works, reusable IL-based SPME fibers were also prepared by chemically bonding allyl-ILs to sol–gel substrates [45–49], or using 3-triethoxysilylpropyl-ILs as sol–gel precursors [49, 50]. Electrodeposition of aniline at the presence of ILs in aniline solution could prepare fibrous nanostructured polyaniline-IL composites, which were also stable for repeated uses [51, 52]. Abolghasemi et al. [53] prepared IL-based periodic mesoporous organosilica (PMO-IL) and glued PMO-IL on a SSW to prepare a SPME fiber. Other two studies used ILs as porogens in sol–gel processes, where ILs were actually not used as coating materials [54, 55]. However, Pena-Pereira et al. [56] used sol–gel technique to entrap IL in the gel, this SPME fiber coating was demonstrated to contain the high content of IL and, therefore, exhibited excellent EEs.

At the same time, PILs were also introduced as SPME fiber coatings through polymerization of allyl-, vinyl- or vinylbenzyl-ILs. Anderson and colleagues for the first time used PILs in SPME coatings [57]. In the pioneer works, PILs were firstly prepared in solutions, and then, the dipping method was used for coating sealed fused-silica capillaries with the PILs [57–63]. Later, on-fiber polymerization initiated by ultraviolet was also utilized for the preparation of a PIL coating [64]. To improve the stability of PIL fibers, silver- and gold-coated SSWs were used as the supporting substrates and subsequently functionalized with 3-mercaptopropyltriethoxysilane, and then with vinyltrimethoxysilane or 1-vinyl-3-(3-triethoxysilylpropyl)-4,5-dihydroimidazolium chloride, to functionalize the surface with reactive site for polymerization [21, 65, 66]. In another work of Feng et al. [67], 1,8-octanedithiol was also used to chemically bond the PIL coating on the silver-coated SSW. Anodized titanium wire [68, 69] and hydrogen peroxide-oxidized nitinol wire [70, 71] were also used as supporting substrates. In some cases, cross-linker with two vinylbenzylimidazolium ends or vinylimidazolium was used to improve the stabilities of PIL coatings [68–71]. Even though cation moieties of PILs were polymerized, many PIL coatings were still inapplicable for DI SPME, since the anion moieties might be exchanged with anions in sample matrices. Feng et al. [67] introduced a double-confined PIL, of which the cation and anion moieties were both polymerized, to be used for DI SPME (Fig. 2.3).

Of particular interest is that by tuning the structures of ILs and PILs, the affinities toward the target analytes could be possibly tuned. Amini et al. [72] prepared fibers based on four different ILs (1-butyl-3-methylimidazolium tetrafluoroborate, 1-octyl-3-methylimidazolium tetrafluoroborate, 1-octyl-3-methylimidazolium hexafluorophosphate, and 1-ethyl-3-methylimidazolium ethylsulfate) to extract methyl tertbutyl ether in gasoline samples. Both the cations and the anions affected the EEs. Cations with longer carbon chains exhibited higher EEs. And ethylsulfate exhibited a higher EE than tetrafluoroborate and hexafluorophosphate.

In another work [46], the SPME fiber coating based on a crown ether-functionalized IL, 1-allyl-3-(6′-oxo-benzo-15-crown-5-hexyl)imidazolium hexafluorophosphate, exhibited higher EEs over the 1-allyl-3-methylimidazolium hexafluorophosphate coating for extracting polar compounds (e.g., alcohols, phthalate esters, environmental phenolic estrogens, fatty acids, and aromatic amines).

Three different ILs with the same anion but different cations were used to analyze PAHs in aqueous samples with the headspace (HS) mode [37]. 1-methyl-3-octylimidazolium trifluoromethanesulfonate-coated fibers exhibited the highest EEs, followed by 1-benzyl-3-methylimidazolium trifluoromethanesulfonate-coated

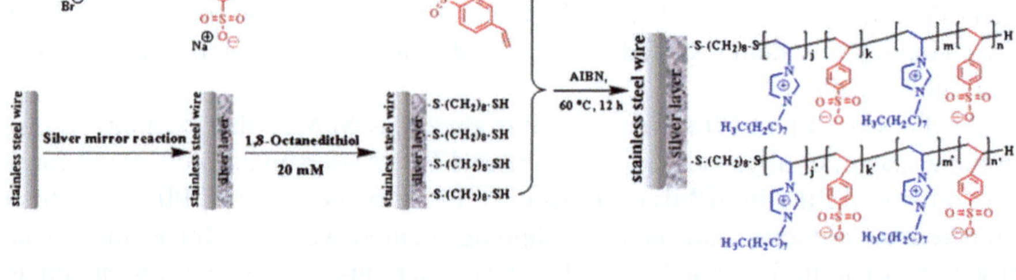

Fig. 2.3 Preparation of the double-confined PIL coating. Reprinted with the permission from Ref. [67]. Copyright 2012 Elsevier

fibers and 1-methyl-3-phenyl-propylimidazolium trifluoromethanesulfonate-coated fibers. The diffusion rates of analytes in fiber coatings other than the affinities toward analytes were thought to be the predominant factor that affected the EEs, which suggested a better understanding of the thermodynamics and kinetics of SPME was needed when ILs were used as fiber coatings.

Three PILs with the same anion, poly(1-vinyl-3-hexylimidazolium bis [(trifluoromethyl)sulfonyl]imide), poly(1-vinyl-3-dodecylimidazolium bis[(trifluoromethyl)sulfonyl]imide), and poly(1-vinyl-3-hexadecylimidazolium bis[(trifluoromethyl)sulfonyl]imide), were coated on fused-silica capillaries for comparison studies [57]. The PIL-based fibers could be satisfactorily used to extract fatty acid methyl esters from complex and hydrophilic matrices, while few differences in EEs among the fibers were observed.

In another work [63], the extraction performances of poly(1-vinyl-3-hexylimidazolium chloride) and poly(1-vinyl-3-hexylimidazolium bis[(trifluoromethyl)-sulfonyl]imide) toward volatile fatty acids and alcohols were compared to evaluate the counter-anion-tuned selectivity. Poly(1-vinyl-3-hexylimidazolium chloride) exhibited better extraction performance because of the higher hydrogen-bond basicity of Cl^-, which also demonstrated the importance of anions in determining the extraction properties.

Two PILs with different cations and anions were used as SPME coatings in the study by López-Darias et al. [61] to analyze complex coffee aromas. Despite of their small film thicknesses, the coating of poly(1-(4-vinylbenzyl)-3-hexadecylimidazolium bis[(trifluoromethyl)sulfonyl]imide) exhibited superior selectivity to aldehydes, while the poly(1-vinyl-3-hexylimidazolium chloride) coating showed good EEs to aromatic alcohols and organic acids possessing alkyl chains lower than pentadecyl.

Meng et al. [62] and López-Darias et al. [73] also reported remarkable selectivity toward aromatic compounds resulted from the aromatic groups in PILs in their works.

2.4 Carbonaceous Materials

Many carbonaceous materials have been developed into SPME fiber coatings; here, we will focus on the most studied ones in recent years, i.e., graphene and carbon nanotubes (CNTs). These two allotropes are much closed to each other on many aspects. Their basic constructing units are both hexatomic rings with sp^2 hybridized carbon atoms. They both possess excellent mechanical strength, thermal stabilities, and high specific surface areas. And both of them can form hydrogen bonds, π–π stacking, dispersion forces, and hydrophobic interactions with organic compounds to exhibit good affinities toward organic compounds. And they both can be functionalized after being oxidized.

Recent studies were also searching other carbon sources those could be facilely obtained for the preparation of SPME fibers, which provided intriguing protocols for preparing inexpensive SPME fibers. The relevant studies are also discussed below.

2.4.1 Graphene

Graphene has been used as an SPME fiber coating, mainly due to its high specific surface area, remarkable thermal and chemical stabilities, high mechanical strength, and high affinities to organic compounds. In addition, graphene can be readily prepared from the low-cost graphite by oxidation exfoliation, and followed by reduction with hydrazine, *p*-phenylene diamine, or ascorbic acid, or by using electrochemical method, which also facilitates its application as an extraction material in SPME fibers.

The first graphene-based SPME fiber was prepared by physical deposition of graphene on SSWs [74]. After that, graphene was also immobilized on substrates by gluing [75, 76], using sol–gel method [77–79], or codepositing in polypyrrole [80, 81]. The aggregation of graphene sheets during the preparation of graphene-based SPME fibers hindered the adoption of the aforementioned methods. Thus, in the following studies, graphene oxide (GO), which is less likely to aggregate in its suspension than graphene, was firstly chemically bonded to substrates, then reduced to graphene on the substrates [26, 82, 83]. Banitaba et al. [84] prepared a SPME fiber by electrochemically reducing GO on the surface of copper wire. Liu et al. [31] directly oxidized the surface of pencil lead and then reduced it for the preparation of a graphene fiber (Fig. 2.4). And Luo et al. [85] prepared another substrateless graphene fiber by hydrothermal reduction of graphene oxide in a capillary mold. In addition, GO, functionalized graphene, and the composites of graphene were also developed into SPME fiber coatings by using the similar methods aforementioned [86–90].

Up to now, almost all the custom-made graphene-based fibers were used to extract analytes with aromatic rings (Table 2.3). Still, in some works, the

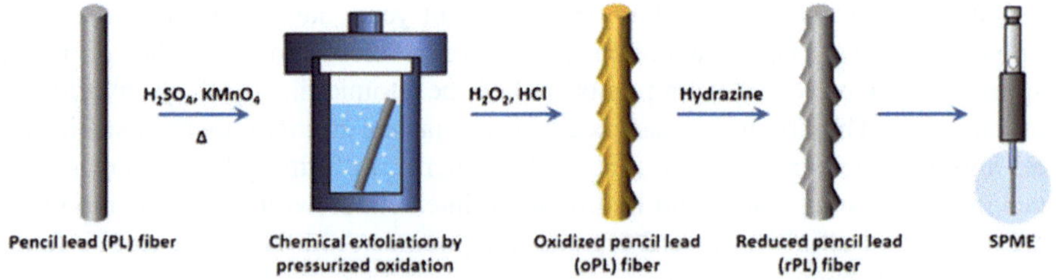

Fig. 2.4 Preparation of the graphenized pencil lead fibers. Reprinted with the permission from Ref. [31]. Copyright 2014 Elsevier

Table 2.3 The applications of graphene in the preparation of SPME fibers

References	Preparation method	Graphene format	Graphene source	Supporting substrate	Analyte	Sample matrix	SPME mode
[74]	Physical deposition	Graphene	Graphite	SSW	Pyrethroids	Pond water	DI
[75]	Silicone glue	MA-reduced GO	Graphite	SSW	Organochlorine pesticides	River water	MA-HS
[76]	Ultrafast gluing	Graphene	Graphite	SSW	Polycyclic aromatic hydrocarbons	Tea productions	HS
[77]	Sol–gel	GO	Graphite	Stainless steel plunger	Polybrominated diphenyl ethers	Canal water	DI
[78]	Sol–gel	Graphene	Commercial graphene	SSW	Organochlorine pesticides	Rainwater	DI
[79]	Sol–gel	Graphene	Not described	SSW	Halogenated aromatic hydrocarbons	Mineral water, lake water, and tap water	HS
[80]	Electrochemically polymerization of pyrrole and graphene	Graphene	Graphite	SSW	Phenols	Pond water	DI
[81]	Electrochemical polymerization	Graphene	Commercial graphene	Platinum wire	Tricyclic antidepressant	Plasma, milk, urine, hair	HS
[26]	Chemical bonding	On-fiber-reduced GO	Graphite	Silica fiber	Polycyclic aromatic hydrocarbons	River water and pond water	HS
[82]	Chemical bonding	On-fiber-reduced GO	Graphite	Titanium fiber	Phthalates	Bottled water and intravenous drip samples in plastic-packaging	DI

(continued)

Table 2.3 (continued)

References	Preparation method	Graphene format	Graphene source	Supporting substrate	Analyte	Sample matrix	SPME mode
[83]	Chemical bonding	On-fiber-reduced GO	Graphite	Silver-coated SSW	*n*-Alkanes	Rainwater and water solution of ethanol extract from a soil sample	DI
[84]	Electrochemical reduction	Electrochemically reduced GO	Graphite	Copper wire	Tricyclic antidepressants	plasma	HS-CF
[31]	In situ chemical exfoliation	Graphene	Pencil lead	None (monolith)	Bisphenols	Tap water and effluent of WWTP	DI
[85]	Hydrothermal method	GO	Graphite	None (monolith)	Organochlorine pesticides	Lake and pond water	HS
[86]	Sol–gel	Composite of dehydrated Zn $(OH)_2$ and reduced GO	Graphite	Fused-silica fiber	Sulfur volatiles	Chinese chive and garlic sprout	HS
[87]	Sol–gel	Poly(ethylene glycol) grafted graphene	Graphite	SSW	Volatile aromatic compounds	Tap, river, and mineral water	HS
[88]	Electrochemically polymerization of 3,4-ethylenedioxythiophene and GO	GO	Graphite	Gold wire	Polycyclic aromatic hydrocarbons	Seawater	DI, HS and HS-CF
[89]	Electrochemical polymerization	Sulfonated graphene	Graphite	SSW	Terpenes	Star anise and fennel	HS
[90]	Electrochemical polymerization of pyrrole and GO	GO	Commercial GO	Platinum wire	Phthalate esters	Boiling water from disposable drinking cups	DI

MA Microwave assisted, *CF* cold fiber

graphene-based fibers were also reported with high EEs to compounds without aromatic rings, such as some organochlorine pesticides [78] and *n*-alkanes [83], which might be owing to the hydrophobic interactions [83]. It is also notable that the composites of graphene with other materials could exhibit high EEs to some specific analytes due to the combination effects of each component of the composites. For example, the composite coating of graphene and ZnO exhibited high EEs to sulfur volatiles, where ZnO provided the coordination sites for binding the sulfur volatiles, and graphene helped to form a large specific surface area in the coating [86].

Due to the superior thermal conductivity of graphene and GO, the fibers of graphene and GO coated on the metal wires were also used for headspace-cold fiber-SPME [84, 88]. Cooling the fibers could increase the partition coefficients between the fibers and sample matrices, since adsorption of analytes on graphene coatings were exothermic processes, which could potentially promote the sensitivity of SPME.

2.4.2 CNTs

The methods for preparing CNT-based SPME fibers were very similar to those for preparing graphene-based ones. Up to now, physical deposition [91], gluing [92–99], sol–gel method [100–103], electrophoretic deposition [104, 105], depositions in conductive polymers [106–109], and chemical bonding [24, 27, 28, 110, 111] have also been adopted to immobilize CNTs on supporting substrates. Besides, CNTs and β-cyclodextrin-functionalized CNTs were also entrapped in pores of polypropylene hollow fibers for the preparation of novel SPME fibers (Fig. 2.5) [112, 113].

The CNT-based SPME fibers were used to analyze various analytes in a wide range of sample matrices via DI extraction mode or HS extraction mode, such as:

- Polybrominated diphenyl ethers in water (DI [91] and HS [100]) and milk (DI [91]),
- Phenols in water (DI [24, 27, 92, 96, 98, 104, 110, 111] and HS [101, 106]),
- Pesticides in water (HS [28, 93] and DI [108]), and tea samples (DI [95]),
- Butyltin compounds in seawater (HS) [94],
- *p*-Nitroaniline, aniline, *o*-dinitrobenzene in water (DI) [96],
- Aniline, 4-methylaniline, 4-methoxybenzoic acid, benzyl alcohol, benzoic acid, and 3-chlorobenzoic acid in water (DI) [97],
- Hydrocarbons in water (DI) and soil samples (HS) [99],
- Benzene series in water (HS) [27, 101, 103],
- Furan in food samples (HS) [102],
- F^-, Cl^-, Br^-, NO_3^-, and SO_4^{2-} in water (DI) [105],
- Phthalate esters in water (DI) [107],
- Polycyclic aromatic hydrocarbons in water (HS) [109],

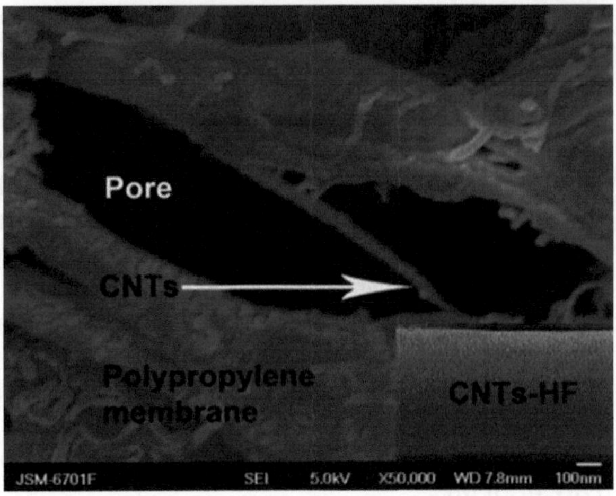

Fig. 2.5 Scanning electron microscopic image of carbon nanotube-reinforced hollow fiber. Reprinted with the permission from Ref. [112]. Copyright 2013 Elsevier

- Benzaldehyde, acetophenone, and 2-octanone in water (HS) [27],
- Strychnine and brucine in urine (DI) [112],
- Hormones of tomatoes (DI) [113].

Most analytes aforementioned contain aromatic rings. Due to the electrical conductivity of CNTs, CNT-based fiber, which were prepared by Nafion gluing [97] and electrophoretic deposition [105], were also used to extract organic ions [97] and inorganic anions in water [105]. When the CNT-based fibers were used to extract target analytes in complex sample matrices, the analytes were firstly extracted by solvents, and then, the solvents were vaporized and the residues were dissolved in water for DI SPME [91, 95].

2.4.3 Other Carbonaceous Materials

Gluing bamboo charcoal on SSWs for the preparation of SPME fiber was reported by Zhao et al. [114]. Bamboo charcoal is facilely available and environmentally friendly. And bamboo charcoal possesses porous structure and large specific surface area up to 300 $m^2\ g^{-1}$. The SPME fiber prepared with bamboo charcoal exhibited good sensitivities for the extraction of phthalate esters from environmental water samples, with the detection limits reached 4–23 ng L^{-1}.

Sun et al. [115] prepared a carbon nanoparticle-based fiber coating by simply deploying the SSWs in the flame of a common alcohol burner. After a very short time duration (60 s), carbon derived from incomplete combustion of alcohol deposited on the wire and presented a nanoscale brushy structure. The fiber exhibited high sensitivities to phthalate esters and polycyclic aromatic hydrocarbons, the detection limits were at parts-per-trillion levels. Similarly, Lin et al. [116] prepared mesoporous carbon coatings on metal wires by carbonizing ethanol with room-temperature atmospheric-pressure dielectric barrier discharge plasma. The detection limits of polycyclic aromatic hydrocarbons were also at parts-per-trillion levels.

Huang et al. [117] reported another SPME fiber by carbonizing polydopamine spontaneous polymerized on SSWs at a very mild alkaline pH. The fiber exhibited high sensitivities to organochlorine pesticides with the detection limits at parts-per-trillion levels.

2.5 MIPs

MIPs are widely used as artificial recognition systems for macromolecules, small organic molecules, metal ions, and even virtues and cells. Compared with biological recognition systems, MIPs also possess many other intriguing characteristics, such as mechanical/chemical/thermal stabilities, low cost, and ease of preparation. These features have also facilitated the application of MIPs in the preparation of SPME fibers (Table 2.4).

MIPs are usually synthesized through polymerization of functional monomers in the presence of templates. After removing the templates, the cavities formed in MIPs can trap the analytes of interest. Cross-linkers are also needed to form rigid polymer networks to keep the cavities for target molecules stable, and make the polymer robust both mechanically and thermally. Porogens are sometimes used to realize porous morphologies, which could facilitate the removal of the templates and make the cavities for the target analytes more accessible. The functional groups on the monomeric units of MIPs trap the target analytes via non-covalent interactions or covalent bonds. Non-covalent interactions, such as hydrogen bonds, are more common for trapping the target analytes. The covalent interactions are rare, because it is not easy to find functional monomer that can readily form covalent bonds with the template and target analyte, and avoid suffering from the difficulty of removing the template [118].

The works applied MIPs in SPME in recent years are selectively listed in Table 2.4. Hydrogen bonds were the most common interactions in these works. For examples, when the target compounds contained Lewis basic or Lewis acidic groups, methacrylic acid or acrylamide was often used as the functional monomer. The reason for the wide adoption of hydrogen bonds is that, compared with van der Waals forces, hydrogen bonds are much stronger and have specific directions, which would result in stronger affinities and better recognition abilities toward the target analytes. Still, π–π stacking effect was also adopted for recognizing target analytes [119, 120]. In addition, metal ion coordination, which plays an important role in biological recognition systems, was used in MIP coating for the recognition of target analyte as well [121].

Generally, MIP coatings were immobilized on silica fibers [119–132]. Prior to coating, silica fibers needed to be activated and silylated in ahead [121–130]. However, it was no easy to prepare uniform MIP coatings with the thickness well

Table 2.4 The applications of molecularly imprinted polymers in the preparation of SPME fibers

References	Functional monomer	Template	Cross-linker	Substrate	Target compounds	Sample matrix	SPME mode
[119]	Phenyltrimethoxysilane	BDE-209	Tetraethoxysilane	Activated silica fiber	Polybrominated diphenyl ethers	Municipal wastewaters	DI in samples with salinity adjusted
[120]	Calix[4]arene	Parathion methyl		Activated silica fiber	Organophosphorus pesticides	Apple and pineapple	HS
[121]	4-vinylpyridine	2,2′-dipyridine		Silylated silica fiber	2,2′-dipyridine	Tap water, laboratorywastewater, human urine samples	DI in pH adjusted samples
[122]	MAA	Sudan I	TRIM	Silylated silica fiber	Sudan dyes	Chilli tomato sauce and chilli pepper	DI in hexane solutions
[123]	MAA	Testosterone	TRIM	Silylated silica fiber	Anabolic steroids	Human urine	DI in samples
[124]	MAA	17β-estradiol	TRIM	Silylated silica fiber	Estrogens	Fishery samples	DI in hexane and acetonitrile solutions
[125]	MAA	Terbutylazine	TRIM	Silylated silica fiber (HF protected)	Triazine herbicides	Sludge water, watermelon, milk, and urine samples	DI in samples, milk was filtered
[126]	Acrylamide	Biphenol A	TRIM	Silylated silica fiber	Biphenol A	Tap water, human urine, milk	DI in tap water and diluted urine, and diluted ACN extract of milk
[127]	MAA	Propranolol	TRIM	Silylated silica fiber	β-blockers	Urine and plasma samples	DI in toluene solutions
[128]	Acrylamide	Tetracycline	TRIM	Silylated silica fiber	Tetracyclines	Chicken feed, chicken muscle, and milk	DI in toluene solutions
[129]	MAA	Prometryn	TRIM	Silylated silica fibers	Triazines	Soybean, corn, lettuce, and soil samples	DI in benzene solutions
[130]	MAA	Ofloxacin and sulfamethazine	TRIM	Silylated silica fiber	Antibiotic drugs	Pork liver and chicken	DI in dichloromethane solutions

(continued)

Table 2.4 (continued)

References	Functional monomer	Template	Cross-linker	Substrate	Target compounds	Sample matrix	SPME mode
[131]	Melamine and chloranil	Ascorbic acid		Polymethyl methacrylate-coated silica fiber	Ascorbic acid	Diluted aqueous, blood serum and pharmaceutical samples	DI in samples with pH adjusted
[132]	Polyethylene glycol	Diazinon		Activated silica fiber	Organophosphorus pesticides	Vegetables	HS of homogenized samples
[133]	MAA	Propazine	EDMA	Monolith	Triazines	Soil, potato, pea	DI in toluene solutions
[134]	MAA	Biphenol A	EDMA	Monolith	Biphenol A and analogues	Tap water, sewage, human urine	DI in aqueous matrices
[135]	MAA	Epinephrine	EDMA	Monolith	Catecholamines	Human urine and serum	DI in acetonitrile solutions
[136]	MAA	Epinephrine	EDMA	Monolith	Ephedrine and pseudoephedrine	Human urine and serum	DI in toluene solutions
[137]	MAA	Dibutyl phthalate	EDMA	Monolith	Phthalates	Reservoir and bottled water	DI in samples
[138]	MAA	Atrazine	EDMA	Monolith	Triazine herbicides	Tap water, rice, onion	DI in aqueous or methanol solutions
[139]	MAA	Ametryn	EDMA	Monolith	Triazine herbicides	Tap water, rice, maize, onion	DI in aqueous solutions
[140]	MAA	Thiabendazole	TRIM	Monolith (HF protected)	Thiabendazole	Orange juices	DI in samples
[25]	MAA	Ofloxacin	EDMA	Polydopamine-coated SSW	Ofloxacin	Milk	DI in diluted samples
[141]	MAA	Sudan I	TRIM	Silylated SSW	Sudan dyes	Hot chili powder and poultry feed samples	DI in hexane solutions
[142]	Pyrrole, thiophene, 3-methylthiophene	Linezolid		SSW	Linezolid	Synthetic body fluid and human plasma	DI in aqueous matrices

(continued)

Table 2.4 (continued)

References	Functional monomer	Template	Cross-linker	Substrate	Target compounds	Sample matrix	SPME mode
[143]	MAA	Estradiol	EDMA	Fe_3O_4@SiO_2 microspheres	Estrogens	Milk powder	DI in acetone solutions
[144]	ABBA	Estradiol		Fe_3O_4@ZIF-8 particles	Estrogens	Fish and pork	DI in hexane solutions
[145]	MAA	Dopamine	EDMA	Polypropylene hollow fiber supported	Dopamine	Human serum and dopamine hydrochloride injection	DI in aqueous solutions
[146]	MAA	Chlorogenic acid	Vinyltriethoxysilane	Polypropylene hollow fiber supported	Chlorogenic acid	Herbal product	DI in hydro-alcoholic extract of samples

MAA Methacrylic acid, *ABBA* 3-aminobenzeneboronicacid, *TRIM* trimethylopropane trimethacrylate, *EDMA* ethylene dimethacrylate controlled. To ensure the uniformity and control the thicknesses of the MIP coatings, the polymerization conditions, such as the temperature, the polymerization time, and the doses of functional monomer and cross-linker, all needed to be investigated [128]. It is notable that using a capillary as a mold has been an intriguing approach for controlling the diameter of the fiber [116].

Monolithic MIP fiber was first prepared by Turiel et al. [133] by in-mold polymerization (Fig. 2.6). Compared with the supported coatings on fused-silica fibers or SSWs, the extraction would need longer time to reach equilibrium for monolithic fibers, and it might be more difficult for removing the templates and eluting the adsorbed analytes from the fibers. However, monolithic fibers were much flexible and would not be easily broken during uses. In addition, monolith fibers could be readily prepared in batch with the diameters well controlled through in-mold polymerization, and there is no worry about the exfoliation of the coatings [133–140].

MIP coatings were also immobilized on polydopamine-coated SSWs [25], silylated SSWs [141], or electropolymerized on SSWs [142]; these fibers would be much more flexible than those with the silica fibers as the supporting substrates. Lan et al. [143, 144] immobilized MIPs coated magnetic particles on electromagnets, which were readily used for automatic SPME practices. MIPs were also supported by polypropylene hollow fibers to be used as SPME fibers [145, 146].

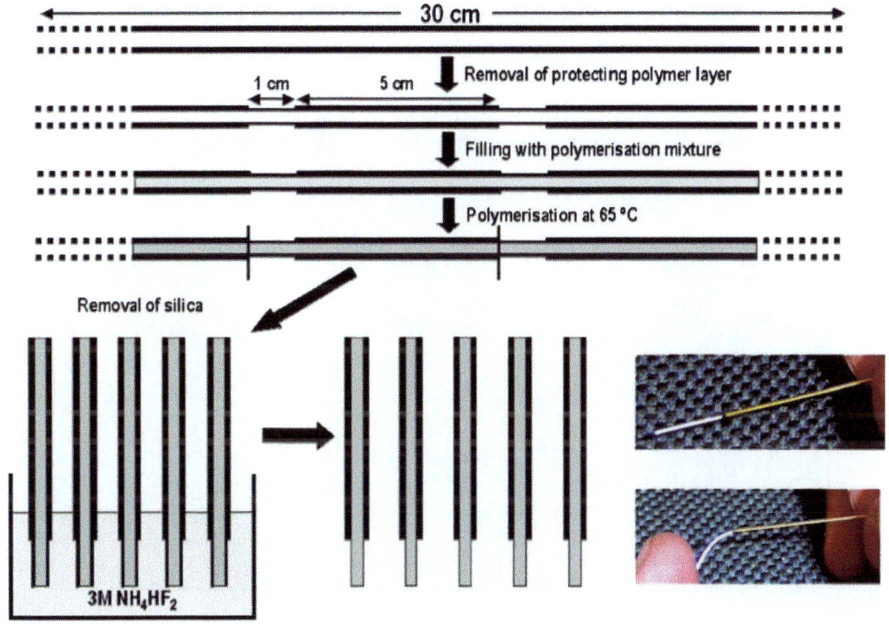

Fig. 2.6 Preparation of monolithic molecularly imprinted fibers. Reprinted with the permission from Ref. [133]. Copyright 2007 American Chemical Society

As shown in Table 2.4, in most cases, MIP fibers were not directly used in aqueous sample matrices. That is because in aqueous sample matrices, MIPs would act as reverse-phase sorbents that retain both the target analytes and other matrix components through non-specific interactions [125, 139]. In other words, the selectivity of MIPs would be ruined in aqueous sample matrices. Therefore, the analytes in the sample matrices were first extracted with organic solvents, and then, the MIP fibers were immersed in the organic solvents for the extraction of the analytes. In other studies [125, 139], to enable direct sampling in aqueous sample matrices, and to simplify the sample preparation procedures using MIP fibers, liquid–liquid extraction was coupled with SPME by inserting MIP fibers in polypropylene hollow fibers filled with organic solvents. This technique was named as liquid–liquid–solid microextraction by Hu et al. (Fig. 2.7) [125].

For biomolecule templates which slightly solubilize in organic solvents, an aqueous imprinting process might be preferred. To achieve this goal, Golsefidi et al. [146] developed an organic–inorganic hybrid MIP. In the preparation procedure, functional monomer and cross-linker were dissolved in organic phase, while the template was dissolved in the mixture of ethanol and water, and then, the two phases were mixed and solidified to form the MIP.

Metal complex imprinted polymer was also used as SPME fiber coatings, which could trap the analytes through coordination bonds [121]. Coordination bonds exhibit higher strength, specificity, directionality than hydrogen bond and van der Waals forces; therefore, the metal complex imprinted polymer could achieve higher selectivity in aqueous sample matrices. In addition, sol–gel method was also used for preparing MIP coatings, by incorporating calix[4] arene [120] and polyethylene glycol [132] for trapping the target analytes.

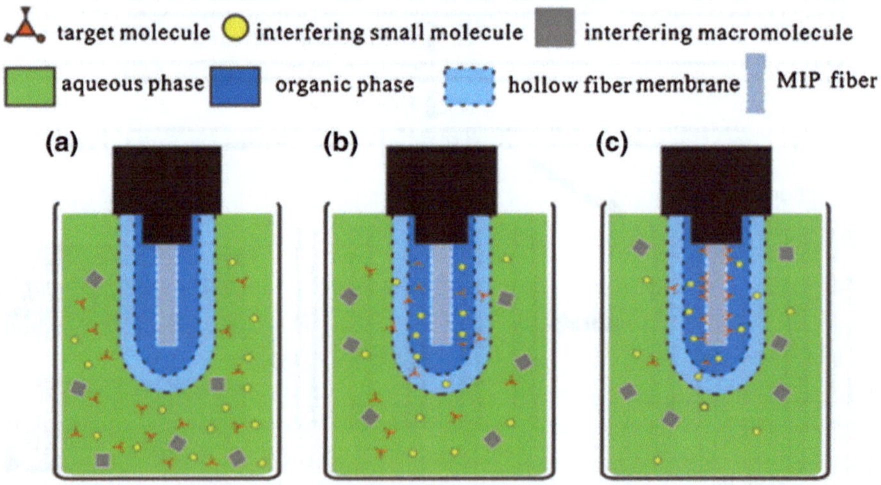

Fig. 2.7 Extraction process of the liquid–liquid–solid microextraction technique. **a** Before extraction, **b** during extraction and **c** after extraction. Reprinted with the permission from Ref. [125]. Copyright 2009 Elsevier

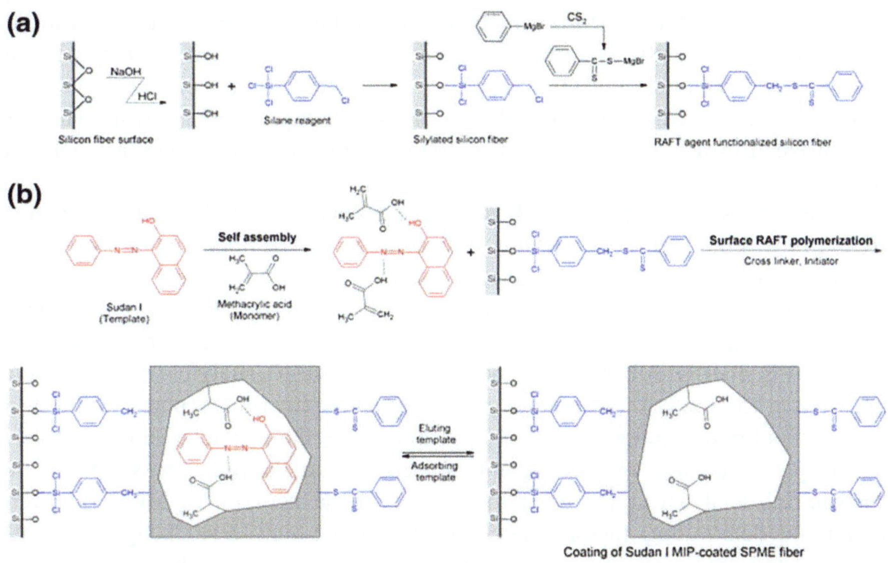

Fig. 2.8 Preparation of MIP coating through surface reversible addition-fragmentation chain transfer polymerization. Reprinted with the permission from Ref. [122]. Copyright 2012 Elsevier

The main drawbacks those limit the applications of MIPs are the incomplete removal of templates and incompletely accessible cavities. These limitations are mainly attributed to the high cross-linking nature of MIPs [122]. Preparing thin MIP coatings provides a solution for these problems. Recently, surface reversible addition-fragmentation chain transfer polymerization was applied to coat MIP on the surface of fused-silica fibers (Fig. 2.8) [122]. The coating prepared by this method was reported to be about only 0.55 μm thick, which resulted in effective removal of template and fast adsorption/desorption kinetics. Similarly, MIP coatings on microparticles with small thicknesses also exhibited fast adsorption/desorption kinetics [143, 144].

2.6 MOFs

MOFs are inorganic–organic hybrid microporous crystalline materials, in which metal ions and organic bridging ligands assemble autonomously to form orderly packed three-dimensional networks. With diverse metal ions and organic ligands available for preparing MOFs, MOFs exhibit extensive structures, topologies, and porosities. Recently, due to the unique properties of MOFs, such as high specific surface areas, uniform structured cavities, and good thermal stabilities, they are also extensively studied for the potential in sample preparation and chromatographic separation, including the use as SPME fiber coatings [147]. Generally, MOFs used as SPME fiber coatings for selectively extracting target analytes via: (a) van der Waals interactions between organic ligands and analytes, such as π–π stacking effect and hydrophobic interaction; (b) Lewis acid–base interactions between metal ions and analytes, when open metal sites are available in MOFs and the analytes contain electron-rich structures; and (c) size selectivity of the nanoscale pores in MOFs.

Table 2.5 lists the recent studies in using MOFs as SPME fiber coatings. MOF-199 was the first reported MOF utilized as an SPME fiber coating material. It was immobilized on etched stainless wire via an in situ hydrothermal method [148]. This homemade fiber exhibited wide linear ranges, low detection limits, high enrichment factors, and higher extraction efficiencies than the commercial SPME/DVB fiber with benzene homologues as the target analytes. The large surface area (1458 cm^2/g) and unique porous structure (two large pores of 12 Å in diameter and pore windows about 8 Å in diameter), the π–π interactions between the analyte molecules and the ligand molecules, and the π-complexation of the electron-rich analytes to the open metal sites in the network of MOF-199 were thought the determining factors that contributed to the high EEs. Chang et al. [149] demonstrated the importance of pore sizes in determining the selectivity of MOF-based coatings with a comparison study between ZIF-7-, ZIF-8-, and MOF-199-coated SPME fibers. In their study, ZIF-7-coated SPME fiber could not extract linear and branched alkanes since the pores of ZIF-7 were too small, whereas the ZIF-8 with the median-sized pores exhibited great selectivity to *n*-alkanes over branched alkanes, and MOF-199 exhibited no selectivity since its pores were too large. In other works, hydrophobic interactions [150–153], π–π stacking effects [150–152, 154], Lewis acid–base interactions [151, 154], size selectivity [150, 153], as well as dipole–dipole interactions [150], were also reported to play key roles in selectively and effectively extracting the analytes of interest.

He et al. [19] observed a profound selectivity to benzene homologues over phenols by using MAF-X8 as the coating material. Experimental and simulation results revealed that the selectivity originated from the differences in adsorption kinetics between benzene homologues and phenols. As shown in Fig. 2.9a, the adsorption of toluene can rapidly reach saturation within 8 min; however, the adsorption of 2-chlorophenol was almost neglected in the first 15 min. And as shown in Fig. 2.9b, the simulated diffusion resistance of 2-chlorophenol in MAF-X8 was much larger than that of toluene.

Owing to the long-range ordered structures of MOFs, the molecular mechanics simulation could be a good approach to explain the selectivity of other MOFs. Liu et al. [155] also observed good compatibility between the simulated results and experimental results.

The endurances of MOFs to humidity are very important for their applications as fiber coating materials. Cui et al. [148] observed that humidity decreased the extraction efficiencies, since the coordination structure would be unstable under certain humidity. Some MOFs are stable to humidity and water, such as MIL-53 [156] and $[Cd(L)_2(ClO_4)_2]\cdot H_2O$ (L = 4-amino-3,5-bis(4-pyridyl-3-phenyl)-

Table 2.5 The applications of metal-organic frameworks in the preparation of fiber coatings

References	MOF	Supporting substrate	Preparation method	Analytes	Sample matrix	SPME mode
[148]	MOF-199	SSW	In situ hydrothermal growth	Benzene homologues	Indoor air samples	HS
[149]	MOF-199, ZIF-7, ZIF-8	SSW	Layer by layer deposition	*n*-Alkanes	Diluted petroleum-based fuel and human serum	HS
[150]	MIL-88B	SSW	In situ hydrothermal growth	Polychlorinated biphenyls	Lake water, river water samples and soil	HS
[151]	UiO-66	SSW	Gluing	Phenols	River water and lake water	HS
[152]	$[Yb_4(AIP)_6(2,2'\text{-bipy})]_n^b$	SSW	In situ hydrothermal growth	Polycyclic aromatic hydrocarbons	Spring water, wastewater and soil	HS
[153]	$[Cu_3(\mu_3\text{-}O)(\mu\text{-}OH)(\text{triazolate})2]^+$	SSW	Gluing	Polychlorinated biphenyls	Tap water, river water, underground water and pond water	DI
[154]	MIL-101(Cr)	SSW	Gluing	Benzene homologues and polycyclic aromatic hydrocarbons	River water	HS
[19]	MAF-X8	SSW	In situ hydrothermal growth	Benzene homologues	Water	HS
[155]	bio-MOF-100, bio-MOF-101, bio-MOF-102	SSW	Gluing	Organochlorine pesticides and polycyclic aromatic hydrocarbons	Pond water and river water	DI
[156]	MIL-53	SSW	Gluing	Polycyclic aromatic hydrocarbons	Wastewater	DI
[157]	$[Cd(L)_2(ClO_4)_2]\cdot H_2O^a$	Fused-silica fiber	In situ growth under room temperature in aqueous solution	Benzene homologues	Seawater	DI

(continued)

Table 2.5 (continued)

References	MOF	Supporting substrate	Preparation method	Analytes	Sample matrix	SPME mode
[158]	IRMOF-3	SSW	In situ hydrothermal growth	Polycyclic aromatic hydrocarbons	Rainwater	HS
[159]	MOF-199/GO composite	Fused-silica fiber	Chemical bonding	Organochlorine pesticides	River water, soil, water convolvulus and longan	HS
[160]	MOF-199/CNT composite	Fused-silica fiber	Chemical bonding	Ethylene, methanol and ethanol	Durian husk, wampee, blueberry, and grape	HS
[161]	MOF-5	Porous copper foam	In situ solvothermal growth	Plant volatile sulfides	Chinese chive and garlic sprout	HS
[162]	MOF-177	SSW	Gluing	Polychlorinated biphenyls and polycyclic aromatic hydrocarbons	River water, lake water and wastewater	HS
[163]	MIL-101, MOF-199	SSW	Sol–gel	Polycyclic aromatic hydrocarbons	River water and lake water	HS
[164]	MIL-101(Cr)	SSW	Sol–gel	Volatile aromatic compounds	Tap water and lake water	HS

[a]L = 4-amino-3,5-bis(4-pyridyl-3-phenyl)-1,2,4-triazole; [b]AIP = 5-aminoisophthalic acid radical, 2,2′-bipy = 2,2′-bipyridine

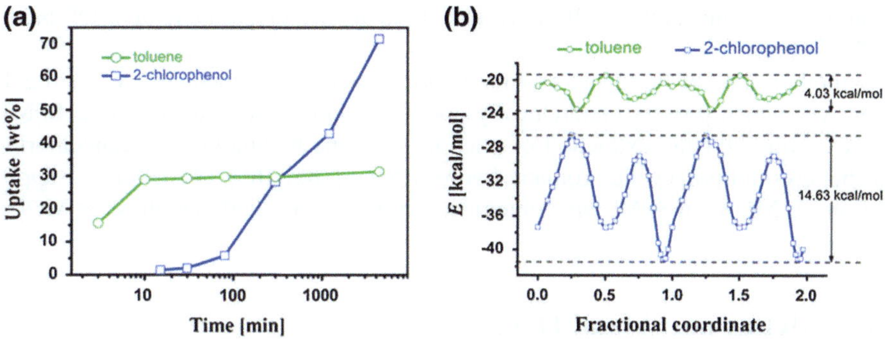

Fig. 2.9 **a** Adsorption kinetics of saturated toluene and 2-chlorophenol vapors on microcrystalline MAF-X8; **b** potential profiles and diffusion barriers of guest molecules passing through the 1D channels of MAF-X8. Reprinted with the permission from Ref. [19]. Copyright 2012 Royal Society of Chemistry

1,2,4-triazole) [157], which were used for DI SPME in aqueous solutions. On the other hand, MOFs, which are sensitive to humidity, can be protected by other humidity-resistant materials for guaranteeing their applications. For example, after covering bio-MOFs with silicone glue, the fibers were used for DI SPME in aqueous samples [155]. In another study [158], IRMOF-3, which is very sensitive to humidity, was used in the headspace of aqueous samples after being protected with ionic liquid and polydimethylsiloxane (PDMS).

The composites of MOFs and other materials have also been used as SPME fiber coating materials, which showed some intriguing properties. The MOF-199/graphene oxide (GO) composite exhibited higher extraction efficiencies than MOF-199 and graphene oxide, since new small pore spaces were created at the interface between graphite layers and MOF units after incorporating MOF-199 in GO [162]. In the MOF-199/carbon nanotubes (CNTs) composite coating [160], CNTs acted as hydrophobic shield to protect MOF-199, and the joint effects of the open metal sites of MOF-199, the oxide atoms surrounding the copper ions and the pores in MOF-199, and the conjugated aromatic rings of CNTs, together contributed to the high extraction efficiencies toward ethylene.

It is notable that due to the highly ordered structures of MOFs, the mass transfer of the target analytes in the MOF coatings might be very fast. He et al. [19] observed that the time needed to reach the SPME equilibrium was less than 10 min. And Li et al. [157] observed the equilibrium time was about 15 min, and 15 s was enough for effective thermal desorption with their MOF-based fiber.

In situ growth is popular for preparing MOF-based SPME fibers [19, 148, 150, 152, 157, 158, 161]. Generally, the substrates needed to be etched to provide rough surfaces to enhance the adhesive strength of the coatings. The idea of using porous copper foam to entrap the as-grown nanosized MOF particles in its pores would promote the mechanical stability of the fiber [161]. Other immobilization methods, such as physical deposition [149], gluing [153–156, 162], sol–gel [163, 164], chemical bonding [159, 160], were also used in the preparation of MOF-based SPME fibers.

Another class of inorganic–organic hybrid material, i.e., metal-organic aerogels (MOAs), which possess hierarchically porous structures, were also used as SPME fiber coatings for the extraction of phenols, benzene homologues, organochlorine pesticides, and polycyclic aromatic hydrocarbons [165, 166]. However, the applications of MOAs in SPME are contemporarily far less studied than those of MOFs.

2.7 Metals and Metal Oxides

Metals and metal oxides are also popular extraction materials utilized in SPME fibers (Table 2.6). The relatively facile strategies for preparation of diverse nanostructures and morphologies, and the excellent thermal, mechanical, and chemical stabilities facilitate the applications of metals and metal oxides in SPME fiber coatings. In addition, the unique selectivity toward analytes could be easily understood and possibly be anticipated due to the well-understood electronic structures of metals and metal oxides [167].

Noble metals could be reduced from their oxidation states to prepare SPME fiber coatings. For example, a porous silver coating was prepared on a SSW through the silver mirror reaction for the extraction of phthalate esters and polycyclic aromatic hydrocarbons [167]. And Au nanoparticles were prepared from $HAuCl_4$ and subsequently immobilized on SSWs via chemical bonding [23] or in situ deposition [168] for the preparation of SPME fibers. Black Pd coating was prepared on SSWs by reducing the Pd(II) cations in the aqueous solution with the SSW substrate [169]. The satisfactory EEs of the noble metal coatings to phthalate esters and polycyclic aromatic hydrocarbons might be owing to the hydrophobic interactions and the electron-transfer effect between the π–donor system and the empty valency shells of the noble metals [23, 167–169].

Due to the outstanding malleability of metals, metal wires at extremely small dimensions could be prepared. Gong et al. [170] used tungsten wires with the tip diameters of about 1 μm to sample metabolites in single *Allium cepa* cells. Subcellular analysis was realized in nucleus and cytoplasm.

SSW, which is widely used as supporting substrates of SPME fibers nowadays, was simply etched with hydrofluoric acid to be used as a SPME fiber [171]. The fiber possessed a porous and flower-like structure with Fe_2O_3, FeF_3, Cr_2O_3, and CrF_2 on the surface (Fig. 2.10), which exhibited high EEs to polycyclic aromatic hydrocarbons due to cation–π interaction.

Au and Ag can form valent bonds with sulfhydryl groups; this property was used to modify Au and Ag nanoparticles with organic functional groups to promote their EEs to the target analytes [172–174]. Tungsten wire was also functionalized with *n*-octadecyldimethyl[3-(trimethoxysilyl)propyl]ammonium chloride after being oxidized, the modified wire exhibited excellent EEs to perfluorinated compounds [175].

Table 2.6 The applications of metals and metal oxides in the preparation of SPME fibers

References	Coating	Substrate	Preparation procedure	Analyte	Sample matrix	SPME mode
[167]	Ag	SSW	Silver mirror reaction	Phthalate esters and polycyclic aromatic hydrocarbons	Aqueous extracts of disposable paper cup and instant noodle barrel	DI
[23]	Au nanoparticles	Au coated SSW	Chemical bonding	Polycyclic aromatic hydrocarbons	Rainwater and soil extract	DI
[168]	Au nanoparticles	SSW	Chemical deposition	Polycyclic aromatic hydrocarbons	River water and wastewater	DI
[169]	Black Pd	SSW	In situ oxidation–reduction	Polycyclic aromatic hydrocarbons and phthalate esters	Tap water, snow water, bottled water, and hot water in disposable paper cup	DI
[170]	Tungsten	Tungsten		Metabolites	Single *Allium cepa* cells	DI
[171]	Etched SSW	SSW	Hydrofluoric acid etching	Polycyclic aromatic hydrocarbons	River water	HS and DI
[172]	Mercaptoundecanol-functionalized Au nanoparticles	SSW	Chemical deposition followed by mercaptoundecanol-functionalization	Polycyclic aromatic hydrocarbons	River water, snow water, and influent wastewater	DI
[173]	Octanedithiol-functionalized Au nanoparticles	SSW	Electrochemical deposition followed by octanedithiol-functionalization	UV filters	River water, influent and effluent wastewater, and rainwater	DI

(continued)

Table 2.6 (continued)

References	Coating	Substrate	Preparation procedure	Analyte	Sample matrix	SPME mode
[174]	Propanethiol-functionalized Ag nanoparticles	Ag–Cu fiber	Galvanic displacement reaction followed by propanethiol-functionalization	Polycyclic aromatic hydrocarbons	Spring water	DI
[175]	*n*-Octadecyldimethyl[3-(trimethoxysilyl)propyl]ammonium chloride	Tungsten microdissecting needles	Chemical oxidation followed by silanization	Perfluorinated compounds	*Daphnia magna*	DI
[176]	Anodized aluminum	Aluminum wire	Anodization	Alcohols, benzene homologues and alkanes	Gaseous samples	HS
[177]	Ordered nanoporous anodic alumina	Aluminum wire	Anodization	Volatile organic compounds	Human exhaled breath	HS
[178]	Anodized aluminum	Aluminum wire	Anodization	Volatile compounds	Medicinal plant	HS
[179]	Alumina nanowires	Aluminum wire	Anodic oxidation followed by chemical etching	Volatile esters and alcohols	Banana	HS
[180]	Anodized zinc	Zinc wire	Anodization	Organic compounds	Gaseous samples and aqueous solutions	HS
[181]	TiO_2 nanosheets	Titanium wire	Anodization	Ultraviolet filters	River water and wastewater	DI
[182]	Phenyl-functionalized TiO_2 nanosheets	Ti wire	Anodization followed by phenyl-functionalization	Polycyclic aromatic hydrocarbons	River water and wastewater	DI
[183]	Nanostructured titania	Titanium wire	In situ oxidation with H_2O_2	Dichlorodiphenyltrichloroethane and its degradation products	Aqueous solutions	HS
[184]	Zn–ZnO	SSW	Electrochemical deposition followed by spontaneous oxidation	UV filters	River water, influent and effluent wastewater	DI

(continued)

Table 2.6 (continued)

References	Coating	Substrate	Preparation procedure	Analyte	Sample matrix	SPME mode
[29]	ZrO_2	NiTi alloy wire	Electrochemical deposition	Halophenols	Wastewater	HS
[185]	PbO_2 nanoparticles	Pt wire	Electrochemical deposition	Volatile organoselenium	Milk, apple juice, grape juice, urine, and plasma	HS
[186]	Nanostructured Co_3O_4	Platinum wire	Chemical bath deposition	Benzene homologues	River water, sewage water, and orange juice	HS
[187]	ZnO nanorod	Fused-silica fiber	Chemical deposition	Chlorobenzenes	Seawater, wastewater, soil	HS
[188]	TiO_2	SSW	Coating with titania sol, then heated in the resistance furnace at 400 °C for 10 min	Phosphopeptides	Protein digestion samples	DI
[189]	TiO_2 nanoparticles	SSW	Electrophoretic deposition	Phthalate esters	Bottled mineral water	DI
[190]	Mesoporous TiO_2 nanoparticles	SSW	Gluing	Organochlorine pesticides	Rainwater and lake water	DI

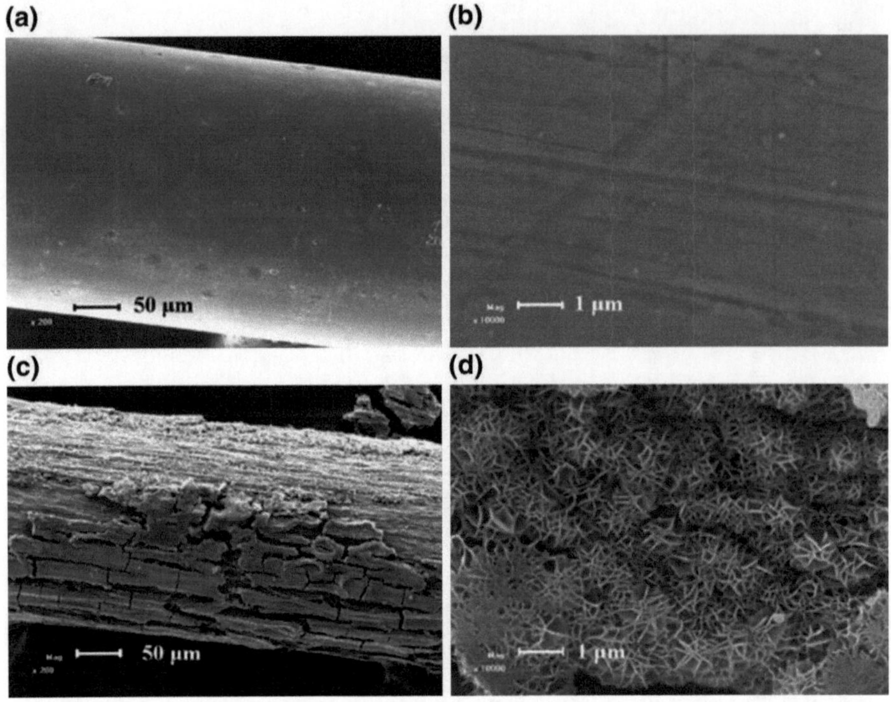

Fig. 2.10 Scanning electron micrographs of the surface of the SSW before (**a**, **b**) and after (**c**, **d**) etching. The images **a** and **c** are at a magnification of 200; images **b** and **d** are at a magnification of 10,000. Reprinted with the permission from Ref. [171]. Copyright 2009 American Chemical Society

Nanostructured metal oxide coatings could be readily prepared by anodizing metal wires. Aluminum oxide was one of the most studied coating materials; under different anodizing conditions, different morphologies were obtained [176–179]. Zhang et al. [179] even coupled chemical etching to anodization to prepare a nanowire coating. Similarly, zinc wire and titanium wire were also anodized to prepare ZnO and TiO_2 coatings [180, 181]. After TiO_2 coating formed by anodizing, phenyl groups could be subsequently functionalized on TiO_2 for extracting PAHs through π–π stacking effect [182]. In another work, TiO_2 coating was also prepared by oxidizing the surface of titanium wire with H_2O_2 [187]. Electrochemical deposition and chemical deposition were also widely used for preparing metal oxide coatings in situ [29, 184–187]. Heating titania sol coated on a SSW for the preparation of TiO_2 coating was also reported in the literature [188]. In other works, TiO_2 nanoparticles were firstly prepared and then immobilized on substrates by electrophoretic deposition [189] or gluing [190]. The metal oxide coatings could be used to extract analytes with hydroxyl groups and halogen atoms probably due to the Lewis acidic sites offered by the metal cations [184].

2.8 Conductive Polymers and Modified Silica

Among various conducting polymers, polypyrrole (PPY), polythiophene (PTH), polyaniline (PANI), and their derivatives have been utilized as SPME fiber coatings [191–197]. The introduction of these conducting polymers to SPME may be attributed to the ease and the controllability of their preparation. Typically, metal wires were used as working electrodes to perform electropolymerization. Based on the preparation protocol, the fibers were reported with thicknesses ranging from dozens to hundreds micrometers and often possessed rather rough and granulated morphologies. To date, various analytes including organochlorine pesticides [191], phenols [192], phthalate esters [193], antibiotic drugs [194–196], and adrenolytic drugs [197], have been analyzed in different sample matrices, even in human plasma [195–197] and whole blood [194]. It is notable that Olszowy et al. [196] have used several different poly(3-alkylthiophene)-coated SPME fibers to extract linezolid, which showed that the functional groups on the monomeric units could affect the EEs. In addition, PANI and PPY were often used as binder to immobilize other extractive materials on metal wires for the preparation of SPME fibers, as discussed in the sections above [51, 52, 80, 81].

Even though the application of silica fibers as the supporting substrates of SPME fibers is declined due to the fragility, the facile functionalization property of silica, however, makes it a versatile and promising material for preparing SPME fiber coatings. For example, nanoporous silica (SBA-15) was functionalized with 3-[bis (2-hydroxyethyl)amino] propyl-triethoxysilane and subsequently glued to copper wire for the preparation of a novel SPME fiber. This fiber exhibited much higher EEs to phenols than the commercial PDMS fiber [198]. In other works, phenyl-functionalized SBA-15 was used for the extraction of PAHs [199], and mercaptopropyl-functionalized nanoporous silica was used for the extraction of phenols [200] Query ID="Q3' Text="Please check and confirm the inserted citation of Table 2.6 is correct. If not, please suggest an alternative citation. Please note that Tables should be cited in sequential order in the text.'.

2.9 Fibers for DI SPME in Biological Matrices

Solid-phase microextraction of endogenous and exogenous bioactive compounds from various complex biological sample matrices, such as semisolid tissues of animals and plants, whole blood, and so on, has been a very popular method in recent years, which showed great potential in pharmacokinetic studies [3], metabolome detection [1], field sampling of pollutants in animals and plants [8, 201], food quality assessment [5], and so on. For the implementation of DI SPME in biological matrices, it is critical to select proper SPME fibers which, besides possessing high EEs to target analytes, are also required to be biofouling resistant, i.e., the fibers should avoid binding of biomacromolecules, such as proteins and lipids on their surfaces. In addition, for in vivo SPME in tissues of animals or plants, the fibers are also required to be biocompatible to avoid causing rejection reactions when the fibers are deployed in living tissues.

PDMS is one of the most widely used fiber coatings for DI SPME in biological matrices. A study showed that by overcoating the commercial PDMS/DVB fiber with a PDMS layer, the lifetime of the commercial fiber was greatly prolonged for DI SPME in grape pulp. This study demonstrated that PDMS was a material that was resistant to biofouling [202]. In addition, even though PDMS is thought to be low biocompatible due to its possible surface instability, PDMS was still widely used as a coating for in vivo SPME in fish bodies [4, 201], probably because PDMS was stable enough during the relatively short extraction processes. And since PDMS is low cost, disposable PDMS fibers were also prepared in multiple dimensions (dimeter of 640 or 300 μm) [4, 201].

Another class of fiber coatings, those are prepared by incorporating sorbents of high extraction capacities in biocompatible polymers such as polyacrylonitrile (PAN) and poly(ethylene glycol) (PEG), have also been widely used in SPME in complex biological sample matrices. Up to now, sorbents such as C18, HLB (hydrophilic–lipophilic balanced sorbent), and a mixed-mode sorbent (C18/benzenesulfonic acid) have been incorporated into biocompatible polymers. These fibers have been widely used in detecting pharmaceuticals and their metabolites in blood in vivo and in vitro, as well as metabolomes in human plasma and living rat brains [1–3].

Restricted access materials (RAMs) are silica particles with pore radius of about 3 nm, and the outer surfaces of RAMs are functionalized with hydrophilic diol groups, and the pore are modified with hydrophobic alkyl chains and/or ion exchange groups. RAMs were also used as SPME fiber coatings for extracting pharmaceuticals in whole blood [203]. In addition, as described in the last section, PPY- and PTH-coated fibers were also used to detect drugs in human plasma and whole blood [194–197].

To promote the EEs of in vivo SPME in animal tissues, a polystyrene@polydopamine (PS@PDA) core–sheath nanofibrous coating was reported [18]. PDA sheath was spontaneously covered on electrospun PS nanofibers under mild alkaline pH. PDA is of good biocompatibility. And the PDA sheath also made the coating hydrophilic that the body fluids could enter the inner pores of the fibrous structure under capillary effect, which made the inner surface of the coating accessible to analytes. In addition, the water layers surrounding the nanofibers could avoid the binding of biomacromolecules on the coating. In another work [204], polyelectrolyte was dispersed in silicone rubber to prepare a new fiber coating for the extraction of pharmaceuticals in living fish brains and muscle. Due to the incorporation of polyelectrolyte in silicone rubber, the novel coating could simultaneously extract neutral and protonated analytes, which also promoted the extraction efficiencies of analytes partially protonated under physiological pH.

Compared with analysis of several specific analytes or specific groups of analytes, metabolomics studies require extraction of metabolomes at wide polarity ranges. Vuckovic et al. [205] revealed the mixed-mode sorbents could be good candidates for extracting metabolomes. Subsequent studies successfully used the coating for extracting metabolomes from rat blood and rat brains. And the coating showed a great potential to extract metabolites missed by microdialysis [2].

2.10 Conclusions

Development of new fibers is one of the hottest research fields of SPME. A lot of extraction materials have been developed into fiber coatings and monolithic fibers. Among these materials, the aforementioned ones are the most studied in recent years. This chapter presents a brief review of the applications of aforementioned extraction materials in developing SPME fibers with an emphasis on: (1) the strategies for preparation of fibers with the aforementioned extraction materials, (2) the target analytes that each extraction material could effectively and selectively extract and the corresponding thermodynamic or kinetic mechanisms, and (3) the potential of promoting or modifying the affinities of the extraction materials to analytes by functionalizing different functional groups or compositing with other materials. With these key points reviewed in this chapter, we hope this chapter could be a guide to the future explorations of developing other extraction materials into SPME fibers.

In addition, recent studies of DI SPME in complex biological matrices put forth more restrictive requirements for the development of new fibers. As the fibers for DI SPME in complex biological matrices were not as diverse as those for conventional uses, further studies are needed.

Acknowledgement We acknowledge financial support from the projects of NNSFC (Grants 21225731, 21377172, and 21477166) and the NSF of Guangdong Province (Grant S2013030013474).

References

1. Vuckovic D, de Lannoy I, Gien B et al (2011) In vivo solid-phase microextraction: capturing the elusive portion of metabolome. Angew Chem Int Ed 50:5344–5348. doi:10.1002/anie.201006715
2. Cudjoe E, Bojko B, de Lannoy I et al (2013) Solid-phase microextraction: a complementary in vivo sampling method to microdialysis. Angew Chem Int Ed 52:12124–12126. doi:10.1002/anie.201304538
3. Musteata MF, Musteata ML, Pawliszyn J (2008) Fast in vivo microextraction: a new tool for clinical analysis. Clin Chem 52:708–715. doi:10.1373/clinchem.2005.064758
4. Xu J, Luo J, Ruan J et al (2014) In vivo tracing uptake and elimination of organic pesticides in fish muscle. Environ Sci Technol 48:8012–8020. doi:10.1021/es5009032
5. Souza-Silva ÉA, Gionfriddo E, Pawliszyn J (2015) A critical review of the state of the art of solid-phase microextraction of complex matrices II. Food analysis. Trends Anal Chem 71:236–248. doi:10.1016/j.trac.2015.04.018

6. Ouyang G, Vuckovic D, Pawliszyn J (2011) Nondestructive sampling of living systems using in vivo solid-phase microextraction. Chem Rev 111:2784–2814. doi:10.1021/cr100203t
7. Xu J, Zheng J, Tian J (2013) New materials in solid-phase microextraction. Trends Anal Chem 47:68–83. doi:10.1016/j.trac.2013.02.012
8. Spietelun A, Pilarczyk M, Kloskowski A et al (2010) Current trends in solid-phase microextraction (SPME) fibre coatings. Chem Soc Rev 39:4524–4537. doi:10.1039/C003335A
9. Gómez-Ríos GA, Pawliszyn J (2014) Development of coated blade spray ionization mass spectrometry for the quantitation of target analytes present in complex matrices. Angew Chem Int Ed 53:14503–14507. doi:10.1002/anie.201407057
10. Gómez-Ríos GA, Pawliszyn J (2014) Solid phase microextraction (SPME)-transmission mode (TM) pushes down detection limits in direct analysis in real time (DART). Chem Commun 50:12937–12940. doi:10.1039/C4CC05301J
11. Vuckovic D (2013) High-throughput solid-phase microextraction in multi-well-plate format. Trends Anal Chem 45:136–153. doi:10.1016/j.trac.2013.01.004
12. Jiang R, Pawliszyn J (2012) Thin-film microextraction offers another geometry for solid-phase microextraction. Trends Anal Chem 39:245–253. doi:10.1016/j.trac.2012.07.005
13. Wang X, Li X, Li Z et al (2014) Online coupling of in-tube solid-phase microextraction with direct analysis in real time mass spectrometry for rapid determination of triazine herbicides in water using carbon-nanotubes-incorporated polymer monolith. Anal Chem 86:4739–4747. doi:10.1021/ac500382x
14. Kataoka H, Inoue T, Saito K et al (2013) Analysis of heterocyclic amines in hair by on-line in-tube solid-phase microextraction coupled with liquid chromatography–tandem mass spectrometry. Anal Chim Acta 786:54–60. doi:10.1016/j.aca.2013.05.007
15. Chen Y, Sidisky LM (2014) A new interface for coupling solid phase microextraction with liquid chromatography. Anal Chim Acta 817:23–27. doi:10.1016/j.aca.2014.01.056
16. Deng J, Yang Y, Wang X et al (2014) Strategies for coupling solid-phase microextraction with mass spectrometry. Trends Anal Chem 55:55–67. doi:10.1016/j.trac.2013.12.004
17. Li L, Xia Z, Pawliszyn J (2015) Selective extraction and enrichment of glycoproteins based on boronate affinity SPME and determination by CIEF-WCID. Anal Chim Acta 886:83–90
18. Xu J, Huang S, Wu R et al (2015) Bioinspired polydopamine sheathed nanofibers for high-efficient in vivo solid-phase microextraction of pharmaceuticals in fish muscle. Anal Chem 87:3453–3459. doi:10.1021/ac5048357
19. He CT, Tian JY, Liu SY et al (2013) A porous coordination framework for highly sensitive and selective solid-phase microextraction of non-polar volatile organic compounds. Chem Sci 4:351–356. doi:10.1039/C2SC21181E
20. Aziz-Zanjani MO, Mehdinia A (2013) Electrochemically prepared solid-phase microextraction coatings—a review. Anal Chim Acta 781:1–13. doi:10.1016/j.aca.2013.03.012
21. Pang L, Liu J (2012) Development of a solid-phase microextraction fiber by chemical binding of polymeric ionic liquid on a silica coated stainless steel wire. J Chromatogr A 1230:8–14. doi:10.1016/j.chroma.2012.01.052
22. Feng J, Sun M, Xu L et al (2011) Preparation of a polymeric ionic liquid-coated solid-phase microextraction fiber by surface radical chain-transfer polymerization with stainless steel wire as support. J Chromatogr A 1218:7758–7764. doi:10.1016/j.chroma.2011.08.076
23. Feng J, Sun M, Liu H et al (2010) Au nanoparticles as a novel coating for solid-phase microextraction. J Chromatogr A 1217:8079–8086. doi:10.1016/j.chroma.2010.10.089
24. Liu H, Ji L, Li J et al (2011) Magnetron sputtering Si interlayer: a protocol to prepare solid phase microextraction coatings on metal-based fiber. J Chromatogr A 1218:2835–2840. doi:10.1016/j.chroma.2011.03.047
25. Zhao T, Guan X, Tang W et al (2015) Preparation of temperature sensitive molecularly imprinted polymer for solid-phase microextraction coatings on stainless steel fiber to measure ofloxacin. Anal Chim Acta 853:668–675. doi:10.1016/j.aca.2014.10.019

26. Zhang S, Du Z, Li G (2011) Layer-by-layer fabrication of chemical-bonded graphene coating for solid-phase microextraction. Anal Chem 83:7531–7541. doi:10.1021/ac201864f
27. Minet I, Hevesi L, Azenha M et al (2010) Preparation of a polyacrylonitrile/multi-walled carbon nanotubes composite by surface-initiated atom transfer radical polymerization on a stainless steel wire for solid-phase microextraction. J Chromatogr A 1217:2758–2767. doi:10.1016/j.chroma.2010.02.030
28. Bagheri H, Ayazi Z, Sistani H (2011) Chemically bonded carbon nanotubes on modified gold substrate as novel unbreakable solid phase microextraction fiber. Microchim Acta 174:295–301. doi:10.1007/s00604-011-0621-4
29. Budziak D, Martendal E, Carasek E et al (2007) Application of NiTi alloy coated with ZrO_2 as a new fiber for solid-phase microextraction for determination of halophenols in water samples. Anal Chim Acta 598:254–260. doi:10.1016/j.aca.2007.07.061
30. Mei M, Huang X, Yuan D (2014) Multiple monolithic fiber solid-phase microextraction: a new extraction approach for aqueous samples. J Chromatogr A 1345:29–36. doi:10.1016/j.chroma.2014.04.029
31. Liu Q, Cheng M, Long Y et al (2014) Graphenized pencil lead fiber: facile preparation and application in solid-phase microextraction. J Chromatogr A 1325:1–7. doi:10.1016/j.chroma.2013.11.051
32. Souza-Silva ÉA, Pawliszyn J (2012) Optimization of fiber coating structure enables direct immersion solid phase microextraction and high-throughput determination of complex samples. Anal Chem 84:6933–6938. doi:10.1021/ac301305u
33. Armand M, Endres F, MacFarlane DR et al (2009) Ionic-liquid materials for the electrochemical challenges of the future. Nat Mater 8:621–629. doi:10.1038/nmat2448
34. Galiński M, Lewandowski A, Stępniak I (2006) Ionic liquids as electrolytes. Electrochim Acta 51:5567–5580. doi:10.1016/j.electacta.2006.03.016
35. Sun P, Armstrong DW (2010) Ionic liquids in analytical chemistry. Anal Chim Acta 661:1–16. doi:10.1016/j.aca.2009.12.007
36. Liu J, Li N, Jiang G et al (2005) Disposable ionic liquid coating for headspace solid-phase microextraction of benzene, toluene, ethylbenzene, and xylenes in paints followed by gas chromatography–flame ionization detection. J Chromatogr A 1066:27–32. doi:10.1016/j.chroma.2005.01.024
37. Hsieh YN, Huang PC, Sun IW et al (2006) Nafion membrane-supported ionic liquid–solid phase microextraction for analyzing ultra trace PAHs in water samples. Anal Chim Acta 557:321–328. doi:10.1016/j.aca.2005.10.019
38. Huang KP, Wang GR, Huang BY et al (2009) Preparation and application of ionic liquid-coated fused-silica capillary fibers for solid-phase microextraction. Anal Chim Acta 645:42–47. doi:10.1016/j.aca.2009.04.037
39. Cui M, Qiu J, Li Z et al (2015) An etched stainless steel wire/ionic liquid–solid phase microextraction technique for the determination of alkylphenols in river water. Talanta 132:564–571. doi:10.1016/j.talanta.2014.09.012
40. Ho TT, Chen CY, Li ZG et al (2012) Determination of chlorophenols in landfill leachate using headspace sampling with ionic liquid-coated solid-phase microextraction fibers combined with gas chromatography–mass spectrometry. Anal Chim Acta 712:72–77. doi:10.1016/j.aca.2011.11.025
41. He Y, Pohl J, Engel R et al (2009) Preparation of ionic liquid based solid-phase microextraction fiber and its application to forensic determination of methamphetamine and amphetamine in human urine. J Chromatogr A 1216:4824–4830. doi:10.1016/j.chroma.2009.04.028
42. Carda-Broch S, Ruiz-Angel MJ, Armstrong DW et al (2010) Ionic liquid based headspace solid-phase microextraction–gas chromatography for the determination of volatile polar organic compounds. Sep Sci Technol 45:2322–2328. doi:10.1080/01496395.2010.497526
43. Wanigasekara E, Perera S, Crank JA et al (2010) Bonded ionic liquid polymeric material for solid-phase microextraction GC analysis. Anal Bioanal Chem 396:511–524. doi:10.1007/s00216-009-3254-2

44. Amini R, Rouhollahi A, Adibi M et al (2011) A novel reusable ionic liquid chemically bonded fused-silica fiber for headspace solid-phase microextraction/gas chromatography-flame ionization detection of methyl *tert*-butyl ether in a gasoline sample. J Chromatogr A 1218:130–136. doi:10.1016/j.chroma.2010.10.114
45. Liu M, Zhou X, Chen Y et al (2010) Innovative chemically bonded ionic liquids-based sol–gel coatings as highly porous, stable and selective stationary phases for solid phase microextraction. Anal Chim Acta 683:96–106. doi:10.1016/j.aca.2010.10.004
46. Zhou X, Xie P, Wang J et al (2011) Preparation and characterization of novel crown ether functionalized ionic liquid-based solid-phase microextraction coatings by sol–gel technology. J Chromatogr A 1218:3571–3580. doi:10.1016/j.chroma.2011.03.048
47. Zhou X, Shao X, Shu J et al (2012) Thermally stable ionic liquid-based sol–gel coating for ultrasonic extraction–solid-phase microextraction–gas chromatography determination of phthalate esters in agricultural plastic films. Talanta 89:129–135. doi:10.1016/j.talanta.2011.12.001
48. Gao Z, Deng Y, Hu X et al (2013) Determination of organophosphate esters in water samples using an ionic liquid-based sol–gel fiber for headspace solid-phase microextraction coupled to gas chromatography-flame photometric detector. J Chromatogr A 1300:141–150. doi:10.1016/j.chroma.2013.02.089
49. Shu J, Xie P, Lin D et al (2014) Two highly stable and selective solid phase microextraction fibers coated with crown ether functionalized ionic liquids by different sol–gel reaction approaches. Anal Chim Acta 806:152–164. doi:10.1016/j.aca.2013.11.006
50. Shu J, Li C, Liu M et al (2012) Role of counteranions in sol–gel-derived alkoxyl-functionalized ionic-liquid-based organic-inorganic hybrid coatings for SPME. Chromatographia 75:1421–1433. doi:10.1007/s10337-012-2323-4
51. Gao Z, Li W, Liu B et al (2011) Nano-structured polyaniline-ionic liquid composite film coated steel wire for headspace solid-phase microextraction of organochlorine pesticides in water. J Chromatogr A 1218:6285–6291. doi:10.1016/j.chroma.2011.07.041
52. Zhao F, Wang M, Ma Y et al (2011) Electrochemical preparation of polyaniline–ionic liquid based solid phase microextraction fiber and its application in the determination of benzene derivatives. J Chromatogr A 1218:387–391. doi:10.1016/j.chroma.2010.12.017
53. Abolghasemi MM, Karimi B, Yousefi V (2013) Periodic mesoporous organosilica with ionic liquid framework as a novel fiber coating for headspace solid-phase microextraction of polycyclic aromatic hydrocarbons. Anal Chim Acta 804:280–286. doi:10.1016/j.aca.2013.10.022
54. Ebrahimi M, Es'haghi Z, Samadi F et al (2011) Ionic liquid mediated sol–gel sorbents for hollow fiber solid-phase microextraction of pesticide residues in water and hair samples. J Chromatogr A 1218:8313–8321. doi:10.1016/j.chroma.2011.09.058
55. Sarafraz-Yazdi A, Vatani H (2013) A solid phase microextraction coating based on ionic liquid sol–gel technique for determination of benzene, toluene, ethylbenzene and *o*-xylene in water samples using gas chromatography flame ionization detector. J Chromatogr A 1300:104–111. doi:10.1016/j.chroma.2013.03.039
56. Pena-Pereira F, Marcinkowski Ł, Kloskowski A et al (2014) Silica-based ionogels: nanoconfined ionic liquid-rich fibers for headspace solid-phase microextraction coupled with gas chromatography-barrier discharge ionization detection. Anal Chem 86:11640–11648. doi:10.1021/ac502666f
57. Zhao F, Meng Y, Anderson JL (2008) Polymeric ionic liquids as selective coatings for the extraction of esters using solid-phase microextraction. J Chromatogr A 1208:1–9. doi:10.1016/j.chroma.2008.08.071
58. Zhao Q, Wajert JC, Anderson JL (2010) Polymeric ionic liquids as CO_2 selective sorbent coatings for solid-phase microextraction. Anal Chem 82:707–713. doi:10.1021/ac902438k
59. Zhao Q, Anderson JL (2010) Selective extraction of CO_2 from simulated flue gas using polymeric ionic liquid sorbent coatings in solid-phase microextraction gas chromatography. J Chromatogr A 1217:4517–4522. doi:10.1016/j.chroma.2010.04.080

60. López-Darias J, Pino V, Anderson JL et al (2010) Determination of water pollutants by direct-immersion solid-phase microextraction using polymeric ionic liquid coatings. J Chromatogr A 1217:1236–1243. doi:10.1016/j.chroma.2009.12.041
61. López-Darias J, Anderson JL, Pino V et al (2011) Developing qualitative extraction profiles of coffee aromas utilizing polymeric ionic liquid sorbent coatings in headspace solid-phase microextraction gas chromatography–mass spectrometry. Anal Bioanal Chem 401:2965–2976. doi:10.1007/s00216-011-5394-4
62. Meng Y, Anderson JL (2010) Tuning the selectivity of polymeric ionic liquid sorbent coatings for the extraction of polycyclic aromatic hydrocarbons using solid-phase microextraction. J Chromatogr A 1217:6143–6152. doi:10.1016/j.chroma.2010.08.007
63. Meng Y, Pino V, Anderson JL (2011) Role of counteranions in polymeric ionic liquid-based solid-phase microextraction coatings for the selective extraction of polar compounds. Anal Chim Acta 687:141–149. doi:10.1016/j.aca.2010.11.046
64. Ho TD, Yu H, Cole WTS et al (2012) Ultraviolet photoinitiated on-fiber copolymerization of ionic liquid sorbent coatings for headspace and direct immersion solid-phase microextraction. Anal Chem 84:9520–9528. doi:10.1021/ac302316c
65. Feng J, Sun M, Wang X et al (2012) Ionic liquids-based crosslinked copolymer sorbents for headspace solid-phase microextraction of polar alcohols. J Chromatogr A 1245:32–38. doi:10.1016/j.chroma.2012.05.026
66. Feng J, Sun M, Li L et al (2014) Multiwalled carbon nanotubes-doped polymeric ionic liquids coating for multiple headspace solid-phase microextraction. Talanta 123:18–24. doi:10.1016/j.talanta.2014.01.030
67. Feng J, Sun M, Xu L et al (2012) Novel double-confined polymeric ionic liquids as sorbents for solid-phase microextraction with enhanced stability and durability in high-ionic-strength solution. J Chromatogr A 1268:16–21. doi:10.1016/j.chroma.2012.10.037
68. Jia J, Liang X, Wang L et al (2013) Nanoporous array anodic titanium-supported co-polymeric ionic liquids as high performance solid-phase microextraction sorbents for hydrogen bonding compounds. J Chromatogr A 1320:1–9. doi:10.1016/j.chroma.2013.10.042
69. Chen C, Liang X, Wang J et al (2014) Development of a polymeric ionic liquid coating for direct-immersion solid-phase microextraction using polyhedral oligomeric silsesquioxane as cross-linker. J Chromatogr A 1348:80–86. doi:10.1016/j.chroma.2014.04.098
70. Ho TD, Toledo BR, Hantao LW et al (2014) Chemical immobilization of crosslinked polymeric ionic liquids on nitinol wires produces highly robust sorbent coatings for solid-phase microextraction. Anal Chim Acta 843:18–26. doi:10.1016/j.aca.2014.07.034
71. Toledo BR, Hantao LW, Ho TD et al (2014) A chemometric approach toward the detection and quantification of coffee adulteration by solid-phase microextraction using polymeric ionic liquid sorbent coatings. J Chromatogr A 1346:1–7. doi:10.1016/j.chroma.2014.04.035
72. Amini R, Rouhollahi A, Adibi M et al (2011) A new disposable ionic liquid based coating for headspace solid-phase microextraction of methyl tert-butyl ether in a gasoline sample followed by gas chromatography–flame ionization detection. Talanta 84:1–6. doi:10.1016/j.talanta.2010.10.043
73. López-Darias J, Pino V, Meng Y et al (2010) Utilization of a benzyl functionalized polymeric ionic liquid for the sensitive determination of polycyclic aromatic hydrocarbons; parabens and alkylphenols in waters using solid-phase microextraction coupled to gas chromatography–flame ionization detection. J Chromatogr A 1217:7189–7197. doi:10.1016/j.chroma.2010.09.016
74. Chen J, Zou J, Zeng J et al (2010) Preparation and evaluation of graphene-coated solid-phase microextraction fiber. Anal Chim Acta 678:44–49. doi:10.1016/j.aca.2010.08.008
75. Ponnusamy VK, Jen JF (2011) A novel graphene nanosheets coated stainless steel fiber for microwave assisted headspace solid phase microextraction of organochlorine pesticides in aqueous samples followed by gas chromatography with electron capture detection. J Chromatogr A 1218:6861–6868. doi:10.1016/j.chroma.2011.08.019
76. Wang Y, Wang X, Guo Z et al (2014) Ultrafast coating procedure for graphene on solid-phase microextraction fibers. Talanta 119:517–523. doi:10.1016/j.talanta.2013.11.047

77. Zhang H, Lee HK (2011) Plunger-in-needle solid-phase microextraction with graphene-based sol–gel coating as sorbent for determination of polybrominated diphenyl ethers. J Chromatogr A 1218:4509–4516. doi:10.1016/j.chroma.2011.05.016
78. Ke Y, Zhu F, Zeng F et al (2013) Preparation of graphene-coated solid-phase microextraction fiber and its application on organochlorine pesticides determination. J Chromatogr A 1300:187–192. doi:10.1016/j.chroma.2012.11.072
79. Zhang G, Li Z, Zang X et al (2014) Solid-phase microextraction with a graphene-composite-coated fiber coupled with GC for the determination of some halogenated aromatic hydrocarbons in water samples. J Sep Sci 37:440–446. doi:10.1002/jssc.201301183
80. Zou J, Song X, Ji J et al (2011) Polypyrrole/graphene composite-coated fiber for the solid-phase microextraction of phenols. J Sep Sci 34:2765–2772. doi:10.1002/jssc.201100303
81. Abedi H, Ebrahimzadeh H, Ghasemi JB (2015) Solid phase headspace microextraction of tricyclic antidepressants using a directly prepared nanocomposite consisting of graphene, CTAB and polyaniline. Microchim Acta 182:633–641. doi:10.1007/s00604-014-1367-6
82. Zhang B, Li H, Zheng X et al (2014) Preparation of durable graphene-bonded titanium fibers for efficient microextraction of phthalates from aqueous matrices and analysis with gas chromatography–mass spectrometry. J Chromatogr A 1370:9–16. doi:10.1016/j.chroma.2014.10.034
83. Sun M, Feng J, Bu Y et al (2015) Graphene coating bonded onto stainless steel wire as a solid-phase microextraction fiber. Talanta 134:200–205. doi:10.1016/j.talanta.2014.11.005
84. Banitaba MH, Davarani SSH, Ahmar H et al (2014) Application of a new fiber coating based on electrochemically reduced graphene oxide for the cold-fiber headspace solid-phase microextraction of tricyclic antidepressants. J Sep Sci 37:1162–1169. doi:10.1002/jssc.201301369
85. Luo Y, Yuan B, Yu Q et al (2012) Substrateless graphene fiber: a sorbent for solid-phase microextraction. J Chromatogr A 1268:9–15. doi:10.1016/j.chroma.2012.10.035
86. Zhang S, Du Z, Li G (2012) Graphene-supported zinc oxide solid-phase microextraction coating with enhanced selectivity and sensitivity for the determination of sulfur volatiles in *Allium* species. J Chromatogr A 1260:1–8. doi:10.1016/j.chroma.2012.08.045
87. Li Z, Ma R, Bai S et al (2014) A solid phase microextraction fiber coated with graphene–poly(ethylene glycol) composite for the extraction of volatile aromatic compounds from water samples. Talanta 119:498–504. doi:10.1016/j.talanta.2013.11.068
88. Banitaba MH, Davarani SSH, Movahed SK (2014) Comparison of direct, headspace and headspace cold fiber modes in solid phase microextraction of polycyclic aromatic hydrocarbons by a new coating based on poly(3,4-ethylenedioxythiophene)/graphene oxide composite. J Chromatogr A 1325:23–30. doi:10.1016/j.chroma.2013.11.056
89. Zhang C, Zhang Z, Li G (2014) Preparation of sulfonated graphene/polypyrrole solid-phase microextraction coating by in situ electrochemical polymerization for analysis of trace terpenes. J Chromatogr A 1346:8–15. doi:10.1016/j.chroma.2014.04.043
90. Jafari M, Ebrahimzadeh H, Banitaba MH et al (2014) Solid-phase microextraction of phthalate esters by a new coating based on a thermally stable polypyrrole/graphene oxide composite. J Sep Sci 37:3142–3149. doi:10.1002/jssc.201400664
91. Wang JX, Jiang DQ, Gu ZY et al (2006) Multiwalled carbon nanotubes coated fibers for solid-phase microextraction of polybrominated diphenyl ethers in water and milk samples before gas chromatography with electron-capture detection. J Chromatogr A 1137:8–14. doi:10.1016/j.chroma.2006.10.003
92. Liu X, Ji Y, Zhang Y et al (2007) Oxidized multiwalled carbon nanotubes as a novel solid-phase microextraction fiber for determination of phenols in aqueous samples. J Chromatogr A 1165:10–17. doi:10.1016/j.chroma.2007.07.057
93. Lü J, Liu J, Wei Y et al (2007) Preparation of single-walled carbon nanotube fiber coating for solid-phase microextraction of organochlorine pesticides in lake water and wastewater. J Sep Sci 30:2136–2143. doi:10.1002/jssc.200700083

94. Rastkari N, Ahmadkhaniha R, Samadi N et al (2010) Single-walled carbon nanotubes as solid-phase microextraction adsorbent for the determination of low-level concentrations of butyltin compounds in seawater. Anal Chim Acta 662:90–96. doi:10.1016/j.aca.2009.12.035
95. Wu F, Lu W, Chen J et al (2010) Oxidized multiwalled carbon nanotubes as a novel solid-phase microextraction fiber for determination of phenols in aqueous samples. Talanta 82:1038–1043. doi:10.1016/j.chroma.2007.07.057
96. Chen W, Zeng J, Chen J et al (2009) High extraction efficiency for polar aromatic compounds in natural water samples using multiwalled carbon nanotubes/Nafion solid-phase microextraction coating. J Chromatogr A 1216:9143–9148. doi:10.1016/j.chroma.2009.07.025
97. Zeng J, Chen J, Song X et al (2010) An electrochemically enhanced solid-phase microextraction approach based on a multi-walled carbon nanotubes/Nafion composite coating. J Chromatogr A 1217:1735–1741. doi:10.1016/j.chroma.2010.01.034
98. Kueseng P, Pawliszyn J (2013) Carboxylated multiwalled carbon nanotubes/polydimethylsiloxane, a new coating for 96-blade solid-phase microextraction for determination of phenolic compounds in water. J Chromatogr A 1317:199–202. doi:10.1016/j.chroma.2013.08.038
99. Matin AA, Biparva P, Gheshlaghi M (2013) Environmental monitoring of complex hydrocarbon mixtures in water and soil samples after solid phase microextraction using PVC/MWCNTs nanocomposite fiber. Chemosphere 93:1920–1926. doi:10.1016/j.chemosphere.2013.06.072
100. Zhang W, Sun Y, Wu C et al (2009) Polymer-functionalized single-walled carbon nanotubes as a novel sol–gel solid-phase micro-extraction coated fiber for determination of poly-brominated diphenyl ethers in water samples with gas chromatography–electron capture detection. Anal Chem 81:2912–2920. doi:10.1021/ac802123s
101. Jiang R, Zhu F, Luan T et al (2009) Carbon nanotube-coated solid-phase microextraction metal fiber based on sol–gel technique. J Chromatogr A 1216:4641–4647. doi:10.1016/j.chroma.2009.03.076
102. Sarafraz-Yazdi A, Abbasian M, Amiri A (2012) Determination of furan in food samples using two solid phase microextraction fibers based on sol–gel technique with gas chromatography–flame ionisation detector. Food Chem 131:698–704. doi:10.1016/j.foodchem.2011.09.019
103. Sarafraz-Yazdi A, Amiri A, Rounaghi G et al (2011) A novel solid-phase microextraction using coated fiber based sol–gel technique using poly(ethylene glycol) grafted multi-walled carbon nanotubes for determination of benzene, toluene, ethylbenzene and *o*-xylene in water samples with gas chromatography-flam ionization detector. J Chromatogr A 1218:5757–5764. doi:10.1016/j.chroma.2011.06.099
104. Li Q, Wang X, Yuan D (2009) Preparation of solid-phase microextraction fiber coated with single-walled carbon nanotubes by electrophoretic deposition and its application in extracting phenols from aqueous samples. J Chromatogr A 1216:1305–1311. doi:10.1016/j.chroma.2008.12.082
105. Li Q, Ding Y, Yuan D (2011) Electrosorption-enhanced solid-phase microextraction of trace anions using a platinum plate coated with single-walled carbon nanotubes. Talanta 85:1148–1153. doi:10.1016/j.talanta.2011.05.042
106. Du W, Zhao F, Zeng B (2009) Novel multiwalled carbon nanotubes–polyaniline composite film coated platinum wire for headspace solid-phase microextraction and gas chromatographic determination of phenolic compounds. J Chromatogr A 1216:3751–3757. doi:10.1016/j.chroma.2009.03.013
107. Asadollahzadeh H, Noroozian E, Maghsoudi S (2010) Solid-phase microextraction of phthalate esters from aqueous media by electrochemically deposited carbon nanotube/polypyrrole composite on a stainless steel fiber. Anal Chim Acta 669:32–38. doi:10.1016/j.aca.2010.04.029
108. Chen L, Chen W, Ma C et al (2011) Electropolymerized multiwalled carbon nanotubes/polypyrrole fiber for solid-phase microextraction and its applications in the determination of pyrethroids. Talanta 84:104–108. doi:10.1016/j.talanta.2010.12.027

109. Behzadi M, Noroozian E, Mirzaei M (2013) A novel coating based on carbon nanotubes/poly-ortho-phenylenediamine composite for headspace solid-phase microextraction of polycyclic aromatic hydrocarbons. Talanta 108:66–73. doi:10.1016/j.talanta.2013.02.040
110. Liu H, Li J, Liu X et al (2009) A novel multiwalled carbon nanotubes bonded fused-silica fiber for solid phase microextraction–gas chromatographic analysis of phenols in water samples. Talanta 78:929–935. doi:10.1016/j.talanta.2008.12.061
111. Feng J, Sun M, Xu L et al (2011) Preparation of metal wire supported solid-phase microextraction fiber coated with multi-walled carbon nanotubes. J Sep Sci 34:2482–2488. doi:10.1002/jssc.201100375
112. Song XY, Shi YP, Chen J (2013) Carbon nanotubes reinforced hollow fiber solid phase microextraction for the determination of strychnine and brucine in urine. Talanta 116:188–194. doi:10.1016/j.talanta.2013.05.025
113. Song XY, Ha W, Chen J et al (2014) Application of β-cyclodextrin-modified, carbon nanotube-reinforced hollow fiber to solid-phase microextraction of plant hormones. J Chromatogr A 1374:23–30. doi:10.1016/j.chroma.2014.11.029
114. Zhao RS, Liu YL, Zhou JB et al (2013) Bamboo charcoal as a novel solid-phase microextraction coating material for enrichment and determination of eleven phthalate esters in environmental water samples. Anal Bioanal Chem 405:4993–4996. doi:10.1007/s00216-013-6865-6
115. Sun M, Feng J, Qiu H et al (2013) A solid-phase microextraction fiber with carbon nanoparticles as sorbent material prepared by a simple flame-based preparation process. J Chromatogr A 1300:173–179. doi:10.1016/j.chroma.2013.04.061
116. Lin Y, Wu L, Xu K et al (2015) In situ synthesis of porous carbons by using room-temperature, atmospheric-pressure dielectric barrier discharge plasma as high-performance adsorbents for solid-phase microextraction. Chem Eur J 21:13618–13624. doi:10.1002/chem.201500814
117. Huang Z, Chua PE, Lee HK (2015) Carbonized polydopamine as coating for solid-phase microextraction of organochlorine pesticides. J Chromatogr A 1399:8–17. doi:10.1016/j.chroma.2015.04.028
118. Whitcombe MJ, Rodriguez ME, Villar P et al (1995) A new method for the introduction of recognition site functionality into polymers prepared by molecular imprinting: synthesis and characterization of polymeric receptors for cholesterol. J Am Chem Soc 117:7105–7111. doi:10.1021/ja00132a010
119. Li MKY, Lei NY, Gong C et al (2009) An organically modified silicate molecularly imprinted solid-phase microextraction device for the determination of polybrominated diphenyl ethers. Anal Chim Acta 633:197–203. doi:10.1016/j.aca.2008.11.060
120. Li JW, Wang YL, Yan S et al (2016) Molecularly imprinted calixarene fiber for solid-phase microextraction of four organophosphorous pesticides in fruits. Food Chem 192:260–267. doi:10.1016/j.foodchem.2015.07.018
121. Huang J, Hu Y, Hu Y et al (2011) Development of metal complex imprinted solid-phase microextraction fiber for 2,2′-dipyridine recognition in aqueous medium. Talanta 83:1721–1729. doi:10.1016/j.talanta.2010.12.001
122. Hu X, Fan Y, Zhang Y et al (2012) Molecularly imprinted polymer coated solid-phase microextraction fiber prepared by surface reversible addition–fragmentation chain transfer polymerization for monitoring of Sudan dyes in chilli tomato sauce and chilli pepper samples. Anal Chim Acta 731:40–48. doi:10.1016/j.aca.2012.04.013
123. Qiu L, Liu W, Huang M et al (2010) Preparation and application of solid-phase microextraction fiber based on molecularly imprinted polymer for determination of anabolic steroids in complicated samples. J Chromatogr A 1217:7461–7470. doi:10.1016/j.chroma.2010.08.056
124. Hu Y, Wang Y, Chen X et al (2010) A novel molecularly imprinted solid-phase microextraction fiber coupled with high performance liquid chromatography for analysis of trace estrogens in fishery samples. Talanta 80:2099–2105. doi:10.1016/j.talanta.2009.11.015

125. Hu Y, Wang Y, Hu Y et al (2009) Liquid–liquid–solid microextraction based on membrane-protected molecularly imprinted polymer fiber for trace analysis of triazines in complex aqueous samples. J Chromatogr A 1216:8304–8311. doi:10.1016/j.chroma.2009.09.063
126. Tan F, Zhao H, Li X et al (2009) Preparation and evaluation of molecularly imprinted solid-phase microextraction fibers for selective extraction of bisphenol A in complex samples. J Chromatogr A 1216:5647–5654. doi:10.1016/j.chroma.2009.06.007
127. Hu X, Pan J, Hu Y et al (2009) Preparation and evaluation of propranolol molecularly imprinted solid-phase microextraction fiber for trace analysis of β-blockers in urine and plasma samples. J Chromatogr A 1216:190–197. doi:10.1016/j.chroma.2008.11.064
128. Hu X, Pan J, Hu Y et al (2008) Preparation and evaluation of solid-phase microextraction fiber based on molecularly imprinted polymers for trace analysis of tetracyclines in complicated samples. J Chromatogr A 1188:97–107. doi:10.1016/j.chroma.2008.02.062
129. Hu X, Hu Y, Li G (2007) Development of novel molecularly imprinted solid-phase microextraction fiber and its application for the determination of triazines in complicated samples coupled with high-performance liquid chromatography. J Chromatogr A 1147:1–9. doi:10.1016/j.chroma.2007.02.037
130. Hu Y, Song C, Li G (2012) Fiber-in-tube solid-phase microextraction with molecularly imprinted coating for sensitive analysis of antibiotic drugs by high performance liquid chromatography. J Chromatogr A 1263:21–27
131. Prasad BB, Tiwari K, Singh M et al (2008) Molecularly imprinted polymer-based solid-phase microextraction fiber coupled with molecularly imprinted polymer-based sensor for ultratrace analysis of ascorbic acid. J Chromatogr A 1198–1199:59–66. doi:10.1016/j.chroma.2008.05.059
132. Wang YL, Gao YL, Wang PP et al (2013) Sol–gel molecularly imprinted polymer for selective solid phase microextraction of organophosphorous pesticides. Talanta 115:920–927. doi:10.1016/j.talanta.2013.06.056
133. Turiel E, Tadeo JL, Martin-Esteban A (2007) Molecularly imprinted polymeric fibers for solid-phase microextraction. Anal Chem 79:3099–3104. doi:10.1021/ac062387f
134. Xu S, Zhang X, Sun Y et al (2013) Microwave-assisted preparation of monolithic molecularly imprinted polymeric fibers for solid phase microextraction. Analyst 138:2982–2987. doi:10.1039/C3AN00003F
135. Zhang X, Xu S, Lim JM et al (2012) Molecularly imprinted solid phase microextraction fiber for trace analysis of catecholamines in urine and serum samples by capillary electrophoresis. Talanta 99:270–276. doi:10.1016/j.talanta.2012.05.050
136. Deng DL, Zhang JY, Chen C et al (2012) Monolithic molecular imprinted polymer fiber for recognition and solid phase microextraction of ephedrine and pseudoephedrine in biological samples prior to capillary electrophoresis analysis. J Chromatogr A 1219:195–200. doi:10.1016/j.chroma.2011.11.016
137. He J, Lv R, Zhan H et al (2010) Preparation and evaluation of molecularly imprinted solid-phase micro-extraction fibers for selective extraction of phthalates in an aqueous sample. Anal Chim Acta 674:53–58. doi:10.1016/j.aca.2010.06.018
138. Djozan D, Ebrahimi B (2008) Preparation of new solid phase micro extraction fiber on the basis of atrazine-molecular imprinted polymer: application for GC and GC/MS screening of triazine herbicides in water, rice and onion. Anal Chim Acta 616:152–159. doi:10.1016/j.aca.2008.04.037
139. Djozan D, Mahkam M, Ebrahimi B (2009) Preparation and binding study of solid-phase microextraction fiber on the basis of ametryn-imprinted polymer: application to the selective extraction of persistent triazine herbicides in tap water, rice, maize and onion. J Chromatogr A 1216:2211–2219. doi:10.1016/j.chroma.2008.12.101
140. Barahona F, Turiel E, Martín-Esteban A (2011) Supported liquid membrane-protected molecularly imprinted fibre for solid-phase microextraction of thiabendazole. Anal Chim Acta 694:83–89. doi:10.1016/j.aca.2011.03.052

141. Hu X, Cai Q, Fan Y et al (2012) Molecularly imprinted polymer coated solid-phase microextraction fibers for determination of Sudan I–IV dyes in hot chili powder and poultry feed samples. J Chromatogr A 1219:39–46. doi:10.1016/j.chroma.2011.10.089
142. Szultka M, Szeliga J, Jackowski M et al (2012) Development of novel molecularly imprinted solid-phase microextraction fibers and their application for the determination of antibiotic drugs in biological samples by SPME–LC/MS^n. Anal Bioanal Chem 403:785–796. doi:10.1007/s00216-012-5901-2
143. Lan H, Gan N, Pan D et al (2014) An automated solid-phase microextraction method based on magnetic molecularly imprinted polymer as fiber coating for detection of trace estrogens in milk powder. J Chromatogr A 1331:10–18. doi:10.1016/j.chroma.2014.01.016
144. Lan H, Gan N, Pan D et al (2014) Development of a novel magnetic molecularly imprinted polymer coating using porous zeolite imidazolate framework-8 coated magnetic iron oxide as carrier for automated solid phase microextraction of estrogens in fish and pork samples. J Chromatogr A 1365:35–44. doi:10.1016/j.chroma.2014.08.096
145. Ye N, Gao T, Li J (2014) Hollow fiber-supported graphene oxide molecularly imprinted polymers for the determination of dopamine using HPLC-PDA. Anal Methods 6:7518–7524. doi:10.1039/C4AY01017E
146. Golsefidia MA, Es'haghia Z, Sarafraz-Yazdic A (2012) Design, synthesis and evaluation of a molecularly imprinted polymer for hollow fiber–solid phase microextraction of chlorogenic acid in medicinal plants. J Chromatogr A 1229:24–29. doi:10.1016/j.chroma.2012.01.019
147. Gu ZY, Yang CX, Chang N et al (2012) Metal-organic frameworks for analytical chemistry: from sample collection to chromatographic separation. Acc Chem Res 45:734–745. doi:10.1021/ar2002599
148. Cui XY, Gu ZY, Jiang DQ et al (2009) In situ hydrothermal growth of metal-organic framework 199 films on stainless steel fibers for solid-phase microextraction of gaseous benzene homologues. Anal Chem 81:9771–9777. doi:10.1021/ac901663x
149. Chang N, Gu ZY, Wang HF et al (2011) Metal-organic-framework-based tandem molecular sieves as a dual platform for selective microextraction and high-resolution gas chromatographic separation of *n*-alkanes in complex matrixes. Anal Chem 83:7094–7101. doi:10.1021/ac2014004
150. Wu YY, Yang CX, Yan XP (2014) Fabrication of metal-organic framework MIL-88B films on stainless steel fibers for solid-phase microextraction of polychlorinated biphenyls. J Chromatogr A 1334:1–8. doi:10.1016/j.chroma.2014.01.079
151. Shang HB, Yang CX, Yan XP (2014) Metal-organic framework UiO-66 coated stainless steel fiber for solid-phase microextraction of phenols in water samples. J Chromatogr A 1357:165–171. doi:10.1016/j.chroma.2014.05.027
152. Li QL, Wang X, Chen XF et al (2015) In situ hydrothermal growth of ytterbium-based metal-organic framework on stainless steel wire for solid-phase microextraction of polycyclic aromatic hydrocarbons from environmental samples. J Chromatogr A 1415:11–19. doi:10.1016/j.chroma.2015.08.036
153. Li QL, Wang X, Liu YL et al (2014) Feasibility of metal-organic nanotubes $[Cu_3(\mu_3\text{-}O)(\mu\text{-}OH)(\text{triazolate})_2]^+$-coated fibers for solid-phase microextraction of polychlorinated biphenyls in water samples. J Chromatogr A 1374:58–65. doi:10.1016/j.chroma.2014.11.058
154. Xie L, Liu S, Han Z et al (2015) Preparation and characterization of metal-organic framework MIL-101(Cr)-coated solid-phase microextraction fiber. Anal Chim Acta 853:303–310. doi:10.1016/j.aca.2014.09.048
155. Liu S, Zhou Y, Zheng J et al (2015) Isoreticular bio-MOF 100–102 coated solid-phase microextraction fibers for fast and sensitive determination of organic pollutants by the pore structure dominated mechanism. Analyst 140:4384–4387. doi:10.1039/C5AN00775E
156. Chen XF, Zang H, Wang X et al (2012) Metal-organic framework MIL-53(Al) as a solid-phase microextraction adsorbent for the determination of 16 polycyclic aromatic hydrocarbons in water samples by gas chromatography–tandem mass spectrometry. Analyst 137:5411–5419. doi:10.1039/C2AN35806A

157. Li YA, Yang F, Liu ZC et al (2014) A porous Cd(II)-MOF-coated quartz fiber for solid-phase microextraction of BTEX. J Mater Chem A 2:13868–13872. doi:10.1039/C4TA01940G
158. Zheng J, Li S, Wang Y et al (2014) *In situ* growth of IRMOF-3 combined with ionic liquids to prepare solid-phase microextraction fibers. Anal Chim Acta 829:22–27. doi:10.1016/j.aca.2014.04.039
159. Zhang S, Du Z, Li G (2013) Metal-organic framework-199/graphite oxide hybrid composites coated solid-phase microextraction fibers coupled with gas chromatography for determination of organochlorine pesticides from complicated samples. Talanta 115:32–39. doi:10.1016/j.talanta.2013.04.029
160. Zhang Z, Huang Y, Ding W et al (2014) Multilayer interparticle linking hybrid MOF-199 for noninvasive enrichment and analysis of plant hormone ethylene. Anal Chem 86:3533–3540. doi:10.1021/ac404240n
161. Hu Y, Lian H, Zhou L et al (2015) In situ solvothermal growth of metal-organic framework-5 supported on porous copper foam for noninvasive sampling of plant volatile sulfides. Anal Chem 87:406–412. doi:10.1021/ac502146c
162. Wang G, Lei Y, Song H (2015) Exploration of metal-organic framework MOF-177 coated fibers for headspace solid-phase microextraction of polychlorinated biphenyls and polycyclic aromatic hydrocarbons. Talanta 144:369–374. doi:10.1016/j.talanta.2015.06.058
163. Zhang G, Zang X, Li Z et al (2014) Polydimethylsiloxane/metal-organic frameworks coated fiber for solid-phase microextraction of polycyclic aromatic hydrocarbons in river and lake water samples. Talanta 129:600–605. doi:10.1016/j.talanta.2014.06.013
164. Zang X, Zhang G, Chang Q et al (2015) Metal organic framework MIL-101 coated fiber for headspace solid phase microextraction of volatile aromatic compounds. Anal Methods 7:918–923. doi:10.1039/C4AY02540G
165. Li L, Xiang S, Cao S et al (2013) A synthetic route to ultralight hierarchically micro/mesoporous Al(III)-carboxylate metal-organic aerogels. Nat Commun 4:1774. doi:10.1038/ncomms2757
166. Zhang J, Liu L, Liu H et al (2015) Highly porous aerogels based on imine chemistry: syntheses and sorption properties. J Mater Chem A 3:10990–10998. doi:10.1039/C5TA00557D
167. Feng J, Sun M, Li J et al (2011) A novel silver-coated solid-phase microextraction metal fiber based on electroless plating technique. Anal Chim Acta 701:174–180. doi:10.1016/j.aca.2011.05.040
168. Zhang Y, Yang Y, Li Y et al (2015) Growth of cedar-like Au nanoparticles coating on an etched stainless steel wire and its application for selective solid-phase microextraction. Anal Chim Acta 876:55–62. doi:10.1016/j.aca.2015.03.044
169. Sun M, Feng J, Bu Y et al (2015) Palladium-coated stainless-steel wire as a solid-phase microextraction fiber. J Sep Sci 38:1584–1590. doi:10.1002/jssc.201401283
170. Gong X, Zhao Y, Cai S et al (2014) Single cell analysis with probe ESI–mass spectrometry: detection of metabolites at cellular and subcellular levels. Anal Chem 86:3809–3816. doi:10.1021/ac500882e
171. Xu HL, Li Y, Jiang DQ et al (2009) Hydrofluoric acid etched stainless steel wire for solid-phase microextraction. Anal Chem 81:4971–4977. doi:10.1021/ac900743s
172. Zhang Y, Song W, Yang Y et al (2015) Self-assembly of mercaptoundecanol on cedar-like Au nanoparticle coated stainless steel fiber for selective solid-phase microextraction. Anal Methods 7:7680–7689. doi:10.1039/C5AY01184A
173. Yang Y, Li Y, Liu H et al (2014) Electrodeposition of gold nanoparticles onto an etched stainless steel wire followed by a self-assembled monolayer of octanedithiol as a fiber coating for selective solid-phase microextraction. J. Chromatogr. A 1372:25–33. doi:10.1016/j.chroma.2014.10.095
174. Liu C, Zhang X, Li L et al (2015) Silver nanoparticle aggregates on metal fibers for solid phase microextraction–surface enhanced Raman spectroscopy detection of polycyclic aromatic hydrocarbons. Analyst 140:4668–4675. doi:10.1039/C5AN00590F

175. Deng J, Yang Y, Xu M et al (2015) Surface-coated probe nanoelectrospray ionization mass spectrometry for analysis of target compounds in individual small organisms. Anal Chem 87:9923–9930. doi:10.1021/acs.analchem.5b03110
176. Djozan D, Assadi Y, Haddadi SH (2001) Anodized aluminum wire as a solid-phase microextraction fiber. Anal Chem 73:4054–4058. doi:10.1021/ac0100188
177. Zhang G, Zou L, Xu H (2015) Anodic alumina coating for extraction of volatile organic compounds in human exhaled breath vapor. Talanta 132:528–534. doi:10.1016/j.talanta.2014.09.035
178. Gholivand MB, Piryaei M, Abolghasemi MM (2011) Anodized aluminum wire as a solid-phase microextraction fiber for rapid determination of volatile constituents in medicinal plant. Anal Chim Acta 701:1–5. doi:10.1016/j.aca.2011.05.046
179. Zhang Z, Ma Y, Wang Q et al (2013) Preparation of novel alumina nanowire solid-phase microextraction fiber coating for ultra-selective determination of volatile esters and alcohols from complicated food samples. J Chromatogr A 1290:27–35. doi:10.1016/j.chroma.2013.03.052
180. Djozan D, Abdollahi L (2003) Anodized zinc wire as a solid-phase microextraction fiber. Chromatographia 57:799–804. doi:10.1007/BF02491768
181. Li Y, Zhang M, Yang Y et al (2014) Electrochemical in situ fabrication of titanium dioxide-nanosheets on a titanium wire as a novel coating for selective solid-phase microextraction. J Chromatogr A 1358:60–67. doi:10.1016/j.chroma.2014.06.094
182. Guo M, Song W, Wang T et al (2015) Phenyl-functionalization of titanium dioxide-nanosheets coating fabricated on a titanium wire for selective solid-phase microextraction of polycyclic aromatic hydrocarbons from environment water samples. Talanta 144:998–1006. doi:10.1016/j.talanta.2015.07.064
183. Cao D, Lü J, Liu J et al (2008) In situ fabrication of nanostructured titania coating on the surface of titanium wire: a new approach for preparation of solid-phase microextraction fiber. Anal Chim Acta 611:56–61. doi:10.1016/j.aca.2008.01.067
184. Song W, Zhang Y, Guo M et al (2015) Rapid electrochemical preparation of porous sponge-like zinc–zinc oxide coating deposited on an etched stainless steel fiber for selective determination of UV filters in environmental water samples. Anal Methods 7:6619–6628. doi:10.1039/C5AY01248A
185. Ghasemia E, Farahani H (2012) Head space solid phase microextraction based on nano-structured lead dioxide: application to the speciation of volatile organoselenium in environmental and biological samples. J Chromatogr A 1258:16–20. doi:10.1016/j.chroma.2012.08.027
186. Gholivand MB, Shamsipur M, Shamizadeh M et al (2014) Cobalt oxide nanoparticles as a novel high-efficiency fiber coating for solid phase microextraction of benzene, toluene, ethylbenzene and xylene from aqueous solutions. Anal Chim Acta 822:30–36. doi:10.1016/j.aca.2014.02.032
187. Ghasemia E, Sillanpää M (2014) Optimization of headspace solid phase microextraction based on nano-structured ZnO combined with gas chromatography–mass spectrometry for preconcentration and determination of ultra-traces of chlorobenzenes in environmental samples. Talanta 130:322–327. doi:10.1016/j.talanta.2014.06.030
188. Zhao Y, Gong X, Si X et al (2015) Coupling a solid phase microextraction (SPME) probe with ambient MS for rapid enrichment and detection of phosphopeptides in biological samples. Analyst 140:2599–2602. doi:10.1039/C4AN02156H
189. Banitaba MH, Davarani SSH, Pourahadi H (2013) Solid-phase microextraction of phthalate esters from aqueous media by electrophoretically deposited TiO_2 nanoparticles on a stainless steel fiber. J Chromatogr A 1283:1–8. doi:10.1016/j.chroma.2013.01.092
190. Liu S, Xie L, Zheng J et al (2015) Mesoporous TiO_2 nanoparticles for highly sensitive solid-phase microextraction of organochlorine pesticides. Anal Chim Acta 878:109–117. doi:10.1016/j.aca.2015.03.054
191. Li X, Li C, Chen J et al (2008) Polythiophene as a novel fiber coating for solid-phase microextraction. J Chromatogr A 1198–1199:7–13. doi:10.1016/j.chroma.2008.05.051

192. Bagheri H, Mir A, Bahanezhad E (2005) An electropolymerized aniline-based fiber coating for solid phase microextraction of phenols from water. Anal Chim Acta 532:89–95. doi:10.1016/j.aca.2004.10.040
193. Mehdinia A, Bashour F, Roohi F et al (2012) Preparation and evaluation of thermally stable nano-structured self-doped polythiophene coating for analysis of phthalate ester trace levels. J Sep Sci 35:563–570. doi:10.1002/jssc.201100713
194. Szultka M, Kegler R, Fuchs P et al (2010) Polypyrrole solid phase microextraction: a new approach to rapid sample preparation for the monitoring of antibiotic drugs. Anal Chim Acta 667:77–82. doi:10.1016/j.aca.2010.04.011
195. Buszewski B, Szultka M, Olszowy P et al (2011) A novel approach to the rapid determination of amoxicillin in human plasma by solid phase microextraction and liquid chromatography. Analyst 136:2635–2642. doi:10.1039/C1AN00005E
196. Olszowy P, Szultka M, Buszewski B (2011) Poly(3-alkylthiophenes): new sorption materials for solid phase microextraction of drugs isolated from human plasma. Anal Bioanal Chem 401:1377–1384. doi:10.1007/s00216-011-5159-0
197. Olszowy P, Szultka M, Ligor T et al (2010) Fibers with polypyrrole and polythiophene phases for isolation and determination of adrenolytic drugs from human plasma by SPME–HPLC. J Chromatogr B 878:2226–2234. doi:10.1016/j.jchromb.2010.06.033
198. Hashemi P, Shamizadeh M, Badiei A et al (2009) Amino ethyl-functionalized nanoporous silica as a novel fiber coating for solid-phase microextraction. Anal Chim Acta 646:1–5. doi:10.1016/j.aca.2009.04.023
199. Shamsipur M, Gholivand MB, Shamizadeh M et al (2015) Preparation and evaluation of a novel solid-phase microextraction fiber based on functionalized nanoporous silica coating for extraction of polycyclic aromatic hydrocarbons from water samples followed by GC–MS detection. Chromatographia 78:795–803. doi:10.1007/s10337-015-2896-9
200. Shamizadeh M, Gholivand MB, Shamsipur M et al (2015) Mercaptopropyl-functionalized nanoporous silica as a novel coating for solid-phase microextraction fibers. Anal Methods 7:2505–2513. doi:10.1039/C4AY02826K
201. Ouyang G, Oakes KD, Bragg L et al (2011) Sampling-rate calibration for rapid and nonlethal monitoring of organic contaminants in fish muscle by solid-phase microextraction. Environ Sci Technol 45:7792–7798. doi:10.1021/es201709j
202. Silva ÉAS, Pawliszyn J (2012) Optimization of fiber coating structure enables direct immersion solid phase microextraction and high-throughput determination of complex samples. Anal Chem 84:6933–6938. doi:10.1021/ac301305u
203. Musteata FM, Walles M, Pawliszyn J (2005) Fast assay of angiotensin 1 from whole blood by cation-exchange restricted-access solid-phase microextraction. Anal Chim Acta 537:231–237. doi:10.1016/j.aca.2005.01.028
204. Xu J, Wu R, Huang S et al (2015) Polyelectrolyte microcapsules dispersed in silicone rubber for in vivo sampling in fish brains. Anal Chem 87:10593–10599. doi:10.1021/acs.analchem.5b03036
205. Vuckovic D, Pawliszyn J (2011) Systematic evaluation of solid-phase microextraction coatings for untargeted metabolomic profiling of biological fluids by liquid chromatography–mass spectrometry. Anal Chem 83:1944–1954. doi:10.1021/ac102614v

Chapter 3
Application of Solid-Phase Microextraction in Gas Sampling

Xiang Li

Abstract Over the last two decades, solid-phase microextraction (SPME) technology has been widely used in the analysis of complex environmental matrices due to its generally recognized advantages over other sampling methods. Here, we mainly focus on the application of SPME in gas sampling and offer the readers an introductory, concise discussion of the applications in four specific fields: urban air, vehicle exhaust gas, indoor air, and breath analysis. Reviewing the recent studies, we give particular attention to the typical SPME sampling devices applied to the four fields, with an emphasis on the recent developed SPME devices in the online breath analysis. A novel graphene/polyaniline electrodeposited coating developed for online in-tube SPME to measure aldehyde components in human exhaled breath condensate is also introduced in this chapter.

Keywords SPME · Urban air · Vehicle exhaust · Indoor air · Breath analysis

3.1 SPME in Urban Air

The solid-phase microextraction (SPME) method, as a potential sample preparation technique, has attracted more attention in recent years. Today, SPME has been widely used for trace gas compound analysis under solvent-free condition, and for aerosol samples, analytes can be extracted by SPME fiber, including direct exposure to aerosol or the use of the headspace method [1]. Thus, several equipments based on SPME have been made for air monitoring, including combination with gas chromatography with mass spectrometry (GC–MS), liquid chromatography with MS (LC–MS), gas chromatography coupled to triple quadrupole mass spectrometry (GC–QqQ-MS/MS), and nuclear magnetic resonance (NMR) spectroscopy [2].

X. Li (✉)
Shanghai Key Laboratory of Atmospheric Particle Pollution and Prevention (LAP3), Department of Environmental Science and Engineering, Fudan University, Shanghai 200032, China
e-mail: lixiang@fudan.edu.cn; xiangly2012@163.com

G. Ouyang and R. Jiang (eds.), *Solid Phase Microextraction*,
DOI 10.1007/978-3-662-53598-1_3

Several literatures have reported that many organic pollutants can cause genetic changes in living organisms and some are carcinogenic, mutagenic, and teratogenic, impacting on human health and the environment. So the analysis of the components of air pollutants are necessary [3]. Today, SPME has been widely used to collect air organic compounds (e.g., BTEX, monoterpene, and PAHs), and samples are directly analyzed by GC/MS [4–6]. Using this method, the analytes' limit of detection can reach part per trillion (ppt) to subpart per billion (ppb) concentration levels [4, 7].

Over the past few decades, trace gases and aerosols have received more attention from all areas due to the worse air quality [9]. Currently, the direct determination of gas and particle concentrations of atmospheric organic pollutants is still a challenging problem [8]. Many studies use polyurethane foam (PUF) and glass fiber filter (GFF) to measure the gas and particle concentrations [10, 11], which waste time and solvents, and more importantly, the problem of sample loss is serious. Thus, the new method of collecting the gas and particle concentrations of atmospheric organic pollutants was developed using needle trap device (NTD) and SPME [8, 9]. Figure 3.1 shows the sampling system using NTD and SPME. Now,

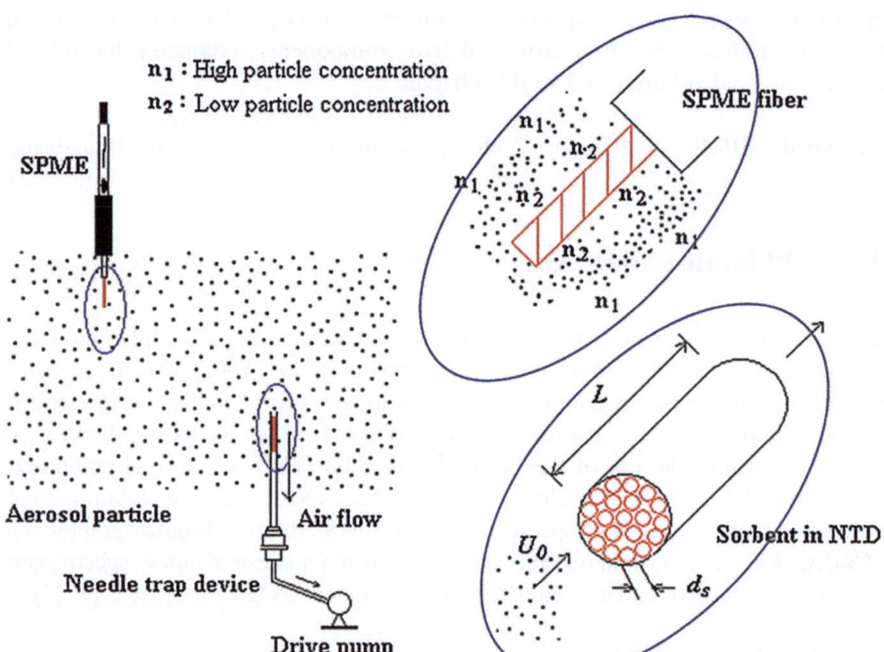

Fig. 3.1 Schematic diagram of the sampling system for aerosol particle experiments using NTD and SPME devices. Reprinted with the permission from Ref. [8]. Copyright 2010 American Chemical Society

these devices are well used to collect diesel exhaust, plant emission, and aerosol samples. Compared with conventional methods, NTD and SPME devices have the advantages of ease of use in field sampling, reusability, and saving time [6, 12].

3.2 SPME in Vehicle Exhaust Gas

Vehicle exhaust gas is considered as one of the most detrimental air pollution sources in many cities around the world [13]. Typical vehicle exhaust gas consists of priority air pollutants, unburned fossil fuel compositions, and their oxidation products. Their accurate separation and quantification plays a crucial role in assessment of air pollution and human health risk.

To date, a number of methods have been developed for determination of SVOC or VOC concentrations in exhaust gases. Conventional sampling methods include active sampling and passive sampling [15]. Active sampling involves absorption of SVOCs or VOCs into sorbent through drawing a certain volume of air. It is usually followed by thermal desorption and then measured by gas chromatograph (GC) coupled to various types of detectors such as flame ionization detector (FID) and mass spectrometer (MS). As an alternative to the active sampling, passive sampling using charcoal bed is also commonly applied to sampling [16]. However, both active sampling and passive sampling demand either an expensive sample desorption part or poisonous solvents in extraction process.

Solid-phase microextraction (SPME) is a comparatively fast, reliable, novel, and cost-effective method for measuring SVOCs or VOCs in ambient air under either static condition or dynamic condition [18]. A typical schematic vehicle exhaust gas sampling with SPME fibers is shown in Fig. 3.2. Vehicle exhaust gas sampling setup consisted of a galvanized steel pipe with five gas sampling ports equipped with Thermogreen septa, three of them for simultaneous SVOC/VOC SPME sampling with three fibers, one for direct gas sampling with syringe, and the last one for measuring temperature of exhaust gases by a digital HH501 DK thermocouple thermometer. A much simpler customer-made device for vehicle exhaust gas sampling with SPME is shown in Fig. 3.3. Simultaneous measurement of gas flow rate, temperature, and SVOCs/VOCs could be, respectively, achieved by using a flowmeter, a thermocouple, and gas analyzer denoted in Fig. 3.3. Most commonly, the SPME fibers are directly exposed to the undiluted exhaust, acting as a sampling device to trap the compounds in the gas/aerosol emitted. It has been first applied to the determination of PAHs in vehicle exhaust gas [7, 12]. Recently, SPME combined with methods such as thermal extraction [19] was successfully applied in the determination of PAHs in diesel particulate matter. In addition to PAHs, other compounds in vehicle exhaust such as BTEX, aldehydes, and *n*-alkanes could also be monitored using SPME method elsewhere [17, 20].

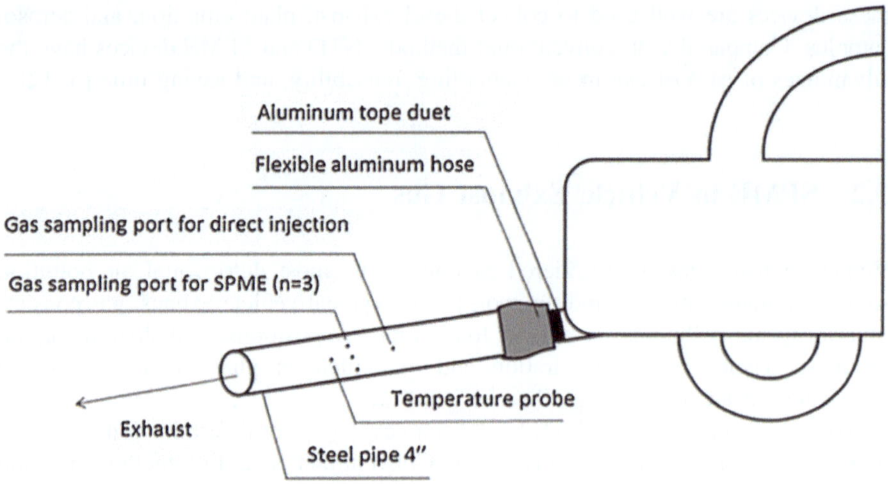

Fig. 3.2 A typical schematic vehicle exhaust gas sampling with retracted SPME fibers in triplicates. Reprinted from Ref. [14], Copyright 2015, with the permission from Elsevier

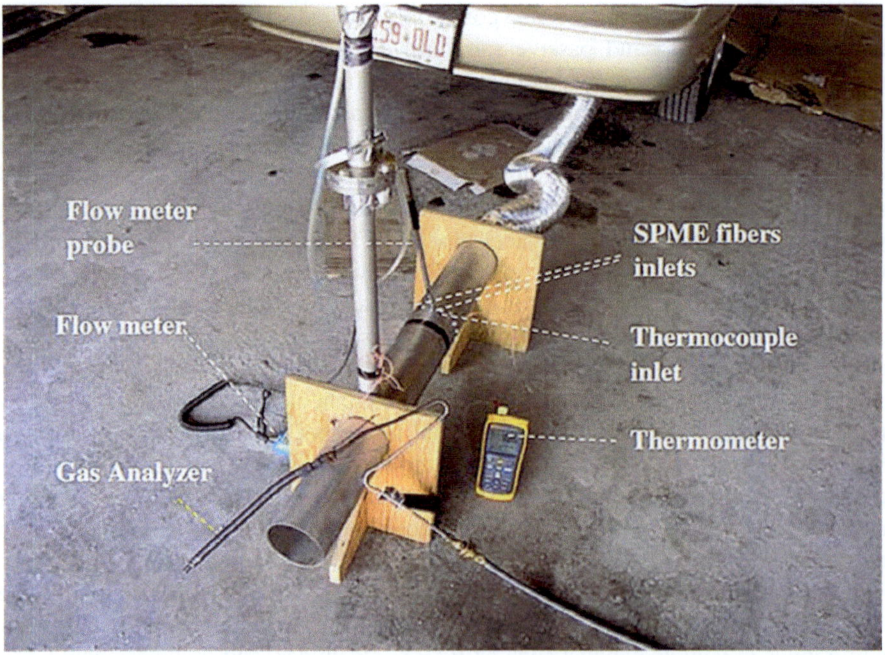

Fig. 3.3 A simple customer-made device for vehicle exhausts sampling with SPME. Reprinted from Ref. [17], with kind permission from Springer Science + Business Media

3.3 SPME in Indoor Air

Indoor air is a dominant exposure for humans. People in developed countries tend to spend 80–90% of their time staying indoors; thus, the indoor exposure has a large contribution to people's personal exposure [21, 22]. Indoor air quality (IAQ) is probably not one of the most concerned environmental issues today. But, there is increasing evidence that exposure to IAQ is the cause of excessive morbidity and mortality [23]. World Health Organization has realized the importance of indoor environment exposures and issued the guidelines for IAQ according to the health effects related to the indoor pollutant exposures.

Indoor air pollution varies greatly in concentration among different indoor environments. The pollution degree may also vary within a specific environment as a function of time and location. The degree of these variations depends on factors such as emission characteristics, sources, behavior of occupants and microclimate, and conditions of ventilation [22]. Thus, indoor air pollution is a highly dynamic process rather than a static phenomenon. The most concerned pollutants can be grouped as follows: aldehydes, particles, carbon dioxide (CO_2), carbon monoxide (CO), volatile organic compounds (VOCs), nicotine, pesticides, polycyclic aromatic hydrocarbons (PAHs), nitrogen dioxide (NO_2), radon, and brominated flame retardants (BFRs).

The detection and monitoring of chemical pollutants are necessary but can also be very challenging [24]. Thus, it is required that simple and cost-effective sampling tools are employed to meet the growing demand of on-site indoor air quality diagnostics [25]. SPME is one of the most commonly used techniques for indoor chemical pollutant detection. It is well functioned in indoor measurements for the stable indoor airflow. It presents many advantages over other analytical methods by combining sampling, isolation, and pre-concentration [26, 27]. SPME is particularly suitable for sensitive environmental analysis because it is noninvasive, noiseless, and easy to use. Apart from those mentioned above, SPME is also an environmentally friendly process widely used because it is solvent-free and reusable [28].

The application of SPME to air monitoring in the workplace is still an emerging field. Until now, two different sample collection techniques have been used [30]. One technique is grab sampling, in which the sampler is totally exposed to the air sample. It allows the determination at a given sampling site and time in which the analyte concentration could be considered as nearly constant in time. The second technique uses SPME for time-weighted average (TWA) sampling of airborne organic compounds (Fig. 3.4) [29]. SPME usually functions with a thermal desorption device coupled with a gas chromatography for high desorption efficiency and analytical sensitivity [31]. It is an alternative passive sampling method for environmental monitoring [32].

Apart from health concerns, indoor air is also studied with interests in other potential damage to our cultural heritage. SPME was widely used in relevant researches including cultural relics in the museum, books and archives in the library, and ancient architectural buildings [33]. More emphasis should be placed

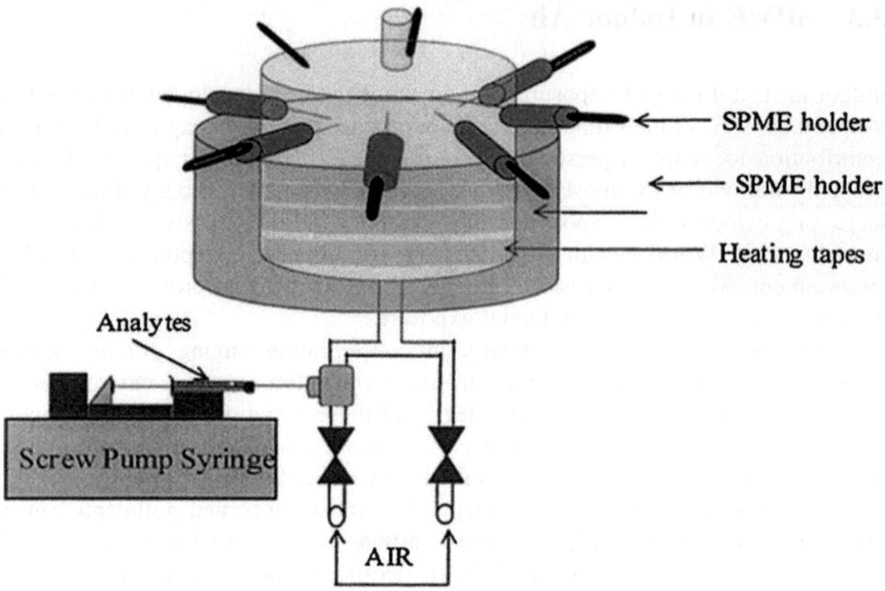

Fig. 3.4 Schematic representation of the standard gas generating device using SPME for time-weighted average (TWA) sampling. Reprinted from Ref. [29], Copyright 2000, with the permission from Elsevier

on IAQ. SPME can play a more important role in getting required data for IAQ management. Special treatments on IAQ are required for the protection of both human health and culture heritage based on the data collected.

3.4 SPME in Breath Analysis

3.4.1 Exhaled Breath Studies

The presence of trace compounds in human exhaled breath has been identified by gas chromatography and mass spectrometry in recent years [34, 35]. These compounds are produced by metabolic processes and partition from the blood flow via pulmonary alveoli into the exhaled breath air. This indicated that the concentrations measured in exhaled breath are closely related to the concentration in blood. Therefore, the analysis of exhaled breath can be used as an important tool, because the change in compound blood concentrations has been associated with diseases, biomarkers, and environmental exposure assessment [36].

3.4.2 Pre-concentration Procedure

Solid-phase microextraction has been used in pre-concentration procedure of exhaled breath analysis; the method involves some advantages: efficient, rapid, and inexpensive. It is noteworthy that several factors such as the type of fiber, extraction time, temperature, and humidity can influence the rate of SPME extraction. These factors should be taken into consideration and optimized before breath detection [37]. It is essential to note that thermal desorption unit with cryo-focusing has been deemed to be another important method in the exhaled breath analysis.

3.4.3 SPME in Human Breath Analysis

The use of SPME in exhaled breath detection was first introduced by Pawliszyn et al. [38]. With SPME technology, sampling of exhaled breath, either passive or active, is achievable. Briefly, in active sampling, the exhaled breath can be measured directly on the fiber; however, this sampling method has several uncontrollable factors. For passive sampling, exhaled breath sample was collected in the Tedlar or Teflon bags which were cleaned by flushing with nitrogen gas. In the pre-concentration procedure, the SPME needle was inserted into the gas sample and the fiber was exposed to the sample for approximately 10 min. Meanwhile, the pre-concentration procedure is determined to a great extent by fiber type and extraction time. Buszewski et al., Filipiak et al., Hyspler et al., and Ligor et al. used carboxen/polydimethylsiloxane (CAR/PDMS) (75 μm)-coated fiber to extract exhaled breath samples [34, 39–41], and 65 μm CAR/PDMS was also in use [42, 43]. Meanwhile, the work of Grote and Pawliszyn demonstrated that the most efficient fibers with short sampling time were PDMS, PDMS/DVB, and Carbowax/DVB, because they can reach equilibrium quickly [44], where SPME extraction temperature is approximately 37 °C in terms of human body temperature. And the extraction time of SPME fiber is always controlled in 10 or 15 min. Likewise, before the SPME procedure, the SPME fiber should also be treated in the injection port of a gas chromatography at 250 °C for 10 min or other conditions. After pre-concentration, the SPME fiber was removed and transferred into the port of a capillary gas chromatography, and exhaled breath compounds were quantified by mass spectrometry. Figure 3.5 shows the modified SPME device.

3.4.4 The New Views of SPME

In recent years, online in-tube SPME coupled with liquid chromatography has been applied to measure human exhaled breath condensate [45], as shown in Fig. 3.6. In parallel with the application of exhaled breath analysis, the analysis of organic

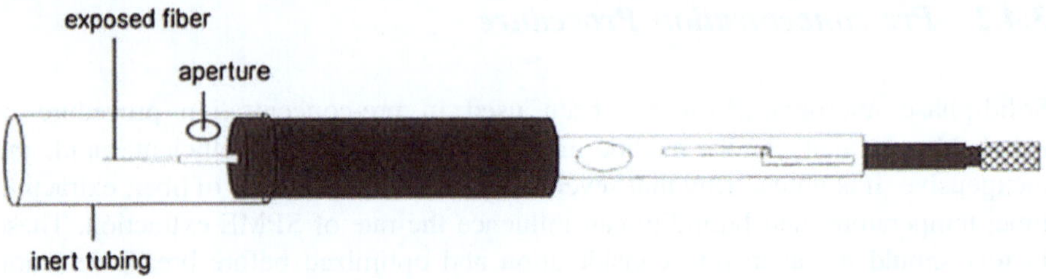

Fig. 3.5 SPME device modified for breath analysis. Reprinted with the permission from Ref. [44]. Copyright 1997 American Chemical Society

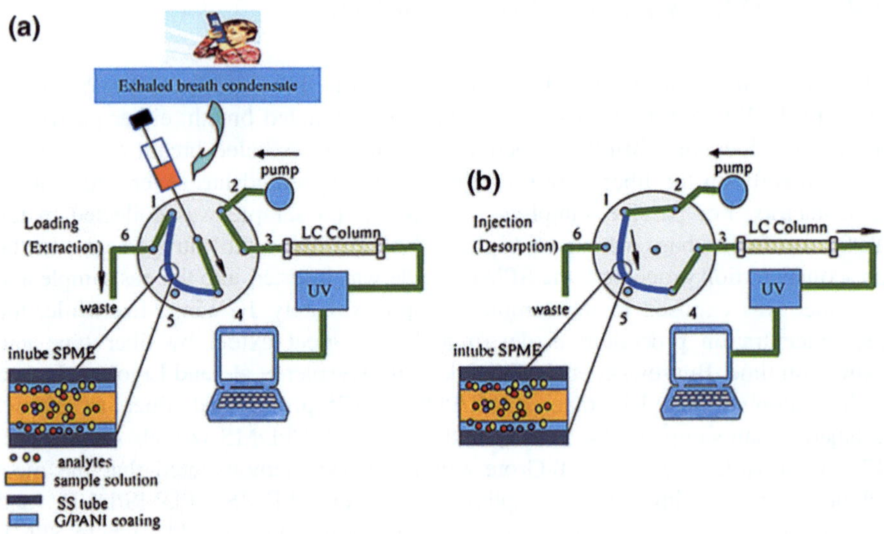

Fig. 3.6 Schematic diagram of online IT–SPME–HPLC. Injector valve position: **a** loading and **b** injection. Reprinted from Ref. [45], Copyright 2015, with the permission from Elsevier

compounds in human blood, tissues, and feces is also being developed with the development of SPME technology [46, 47].

References

1. Pandey SK, Kim KH (2010) A review of environmental tobacco smoke and its determination in air. Trac-Trends Anal Chem 29(8):804–819. doi:10.1016/j.trac
2. Popiel S, Sankowska M (2011) Determination of chemical warfare agents and related compounds in environmental samples by solid-phase microextraction with gas chromatography. J Chromatogr A 1218(47):8457–8479. doi:10.1016/j.chroma

3. Namiesnik J, Zygmunt B, Jastrzebska A (2000) Application of solid-phase microextraction for determination of organic vapours in gaseous matrices. J Chromatogr A 885(1–2):405–418. doi:10.1016/s0021-9673(99)01157-7
4. Namiesnik J, Gorlo D, Wolska L et al (1998) Analysis of selected gaseous organic micro-pollutants in air by means of solid phase microextraction gas chromatography mass spectrometry, calibration problems. Analusis 26(4):170–174. doi:10.1051/analusis
5. Odziemkowski M, Koziel JA, Irish DE et al (2001) Sampling and Raman confocal microspectroscopic analysis of airborne particulate matter using poly(dimethylsiloxane) solid phase microextraction fibers. Anal Chem 73(13):3131–3139. doi:10.1021/ac001141m
6. Reyes-Garces N, Gomez-Rios GA, Silva EAS et al (2013) Coupling needle trap devices with gas chromatography-ion mobility spectrometry detection as a simple approach for on-site quantitative analysis. J Chromatogr A 1300:193–198. doi:10.1016/j.chroma
7. Chai M, Pawliszyn J (1995) Analysis of environmental air samples by solid-phase microextraction and gas-chromatography ion-trap mass-spectrometry. Environ Sci Technol 29(3):693–701. doi:10.1021/es00003a017
8. Li XA, Ouyang GF, Lord H et al (2010) Theory and validation of solid-phase microextraction and needle trap devices for aerosol sample. Anal Chem 82(22):9521–9527. doi:10.1021/ac1022265
9. Niri VH, Eom IY, Kermani FR et al (2009) Sampling free and particle-bound chemicals using solid-phase microextraction and needle trap device simultaneously. J Sep Sci 32(7):1075–1080. doi:10.1002/jssc.200800603
10. Bi XH, Sheng GY, Peng P et al (2003) Distribution of particulate- and vapor-phase *n*-alkanes and polycyclic aromatic hydrocarbons in urban atmosphere of Guangzhou, China. Atmos Environ 37(2):289–298. doi:10.1016/s1352-2310(02)00832-4
11. Terzi E, Samara C (2004) Gas-particle partitioning of polycyclic aromatic hydrocarbons in urban, adjacent coastal, and continental background sites of western Greece. Environ Sci Technol 38(19):4973–4978. doi:10.1021/es040042d
12. Koziel JA, Odziemkowski M, Pawliszyn J (2001) Sampling and analysis of airborne particulate matter and aerosols using in-needle trap and SPME fiber devices. Anal Chem 73 (1):47–54. doi:10.1021/ac000835s
13. Fujita EM, Watson JG, Chow JC et al (1994) Validation of the chemical mass balance receptor model applied to hydrocarbon source apportionment in the Southern California air quality study. Environ Sci Technol 28(9):1633–1649. doi:10.1021/es00058a016
14. Baimatova N, Koziel JA, Kenessov B (2015) Quantification of benzene, toluene, ethylbenzene and o-xylene in internal combustion engine exhaust with time-weighted average solid phase microextraction and gas chromatography mass spectrometry. Anal Chim Acta 873:38–50. doi:10.1016/j.aca
15. Giardino NJ, Andelman JB, Borrazzo JE et al (1988) Sulfur hexafluoride as a surrogate for volatilization of organics from indoor water uses. JAPCA 38(3):278–280. doi:10.1080/08940630.1988.10466379
16. Scott RPW (1998) Introduction to analytical gas chromatography. Marcel Dekker, New York
17. Niri VH, Mathers JB, Musteata MF et al (2009) Monitoring BTEX and aldehydes in car exhaust from a gasoline engine during the use of different chemical cleaners by solid phase microextraction-gas chromatography. Water Air Soil Pollut 204(1–4):205–213. doi:10.1007/s11270-009-0038-4
18. Augusto F, Koziel J, Pawliszyn J (2001) Design and validation of portable SPME devices for rapid field air sampling and diffusion based calibration. Anal Chem 73(3):481–486. doi:10.1021/ac000629k
19. Ballesteros R, Hernandez JJ, Lyons LL (2009) Determination of PAHs in diesel particulate matter using thermal extraction and solid phase micro-extraction. Atmos Environ 43(3):655–662. doi:10.1016/j.atmosenv
20. Hussam A, Alauddin M, Khan AH et al (2007) Solid phase microextraction: measurement of volatile organic compounds (VOCs) in Dhaka city air pollution. J Environ Sci Health A 37 (7):1223–1239. doi:10.1081/ese-120005982

21. Colbeck I, Nasir ZA, Ali Z (2010) The state of indoor air quality in Pakistan—a review. Environ Sci Pollut Res 17(6):1187–1196. doi:10.1007/s11356-010-0293-3
22. de Gennaro G, Dambruoso PR, Loiotile AD et al (2014) Indoor air quality in schools. Environ Chem Lett 12(4):467–482. doi:10.1007/s10311-014-0470-6
23. Sundell J (2004) On the history of indoor air quality and health. Indoor Air 14:51–58. doi:10.1111/j.1600-0668.2004.00273.x
24. Souza-Silva EA, Jiang R, Rodriguez-Lafuente A et al (2015) A critical review of the state of the art of solid-phase microextraction of complex matrices I. Environmental analysis. Trac-Trends Anal Chem 71:224–235. doi:10.1016/j.trac
25. Gorlo D, Zygmunt B, Dudek M et al (1999) Application of solid-phase microextraction to monitoring indoor air quality. Fresen J Anal Chem 363(7):696–699. doi:10.1007/s002160051273
26. Ouyang G, Pawliszyn J (2006) Recent developments in SPME for on-site analysis and monitoring. Trac-Trends Anal Chem 25(7):692–703. doi:10.1016/j.trac
27. Eom I-Y, Risticevic S, Pawliszyn J (2012) Simultaneous sampling and analysis of indoor air infested with *Cimex lectularius* L. (Hemiptera: Cimicidae) by solid phase microextraction, thin film microextraction and needle trap device. Anal Chim Acta 716:2–10. doi:10.1016/j.aca
28. Spietelun A, Marcinkowski L, de la Guardia M et al (2013) Recent developments and future trends in solid phase microextraction techniques towards green analytical chemistry. J Chromatogr A 1321:1–13. doi:10.1016/j.chroma
29. Khaled A, Pawliszyn J (2000) Time-weighted average sampling of volatile and semi-volatile airborne organic compounds by the solid-phase microextraction device. J Chromatogr A 892(1–2):455–467. doi:10.1016/s0021-9673(00)00295-8
30. Marin P, Periago JF, Prado C (2013) A new approach for diffusive sampling based on SPME for occupational exposure assessment. J Occup Environ Hyg 10(3):132–142. doi:10.1080/15459624.2012.755091
31. Lin L-T, Tsai AL, Tseng YM et al (2015) Passive sampling of airborne furan indoors by solid-phase microextraction. Int J Environ Anal Chem 95(1):45–56. doi:10.1080/03067319.2014.983438
32. Huang Y, Ho SSH, Ho KF et al (2012) Optimization of solid-phase microextraction (SPME) to determine airborne biogenic volatile organic compounds (BVOCs): an application for measurement of household cleaning products. Anal Methods 4(1):277–283. doi:10.1039/c1ay05490b
33. Gibson LT, Ewlad-Ahmed A, Knight B et al (2012) Measurement of volatile organic compounds emitted in libraries and archives: an inferential indicator of paper decay? Chem Cent J 6:42. doi:10.1186/1752-153x-6-42
34. Hyspler R, Crhova S, Gasparic J et al (2000) Determination of isoprene in human expired breath using solid-phase microextraction and gas chromatography-mass spectrometry. J Chromatogr B 739(1):183–190. doi:10.1016/S0378-4347(99)00423-5
35. Phillips M, Herrera J, Krishnan S et al (1999) Variation in volatile organic compounds in the breath of normal humans. J Chromatogr B 729(1–2):75–88. doi:10.1016/s0378-4347(99)00127-9
36. Cao WQ, Duan YX (2006) Breath analysis: potential for clinical diagnosis and exposure assessment. Clin Chem 52(5):800–811. doi:10.1373/clinchem
37. Yu H, Xu L, Wang P (2005) Solid phase microextraction for analysis of alkanes and aromatic hydrocarbons in human breath. J Chromatogr B 826(1–2):69–74. doi:10.1016/j.jchromb
38. Amorim LCA, Cardeal ZDL (2007) Breath air analysis and its use as a biomarker in biological monitoring of occupational and environmental exposure to chemical agents. J Chromatogr B 853(1–2):1–9. doi:10.1016/j.jchromb
39. Buszewski B, Ulanowska A, Ligor T et al (2009) Analysis of exhaled breath from smokers, passive smokers and non-smokers by solid-phase microextraction gas chromatography/mass spectrometry. Biomed Chromatogr 23(5):551–556. doi:10.1002/bmc.1141

40. Filipiak W, Ruzsanyi V, Mochalski P et al (2012) Dependence of exhaled breath composition on exogenous factors, smoking habits and exposure to air pollutants. J Breath Res 6(3):21. doi:10.1088/1752-7155/6/3/036008
41. Ligor M, Ligor T, Bajtarevic A et al (2009) Determination of volatile organic compounds in exhaled breath of patients with lung cancer using solid phase microextraction and gas chromatography mass spectrometry. Clin Chem Lab Med 47(5):550–560. doi:10.1515/cclm
42. Martin AN, Farquar GR, Jones AD et al (2010) Human breath analysis: methods for sample collection and reduction of localized background effects. Anal Bioanal Chem 396(2):739–750. doi:10.1007/s00216-009-3217-7
43. Ma HY, Li X, Chen JM et al (2014) Analysis of human breath samples of lung cancer patients and healthy controls with solid-phase microextraction (SPME) and flow-modulated comprehensive two-dimensional gas chromatography (GC × GC). Anal Methods 6(17):6841–6849. doi:10.1039/c4ay01220h
44. Grote C, Pawliszyn J (1997) Solid-phase microextraction for the analysis of human breath. Anal Chem 69(4):587–596. doi:10.1021/ac960749l
45. Li Y, Xu H (2015) Development of a novel graphene/polyaniline electrodeposited coating for on-line in-tube solid phase microextraction of aldehydes in human exhaled breath condensate. J Chromatogr A 1395:23–31. doi:10.1016/j.chroma
46. Deng CH, Li N, Zhang XM (2004) Development of headspace solid-phase microextraction with on-fiber derivatization for determination of hexanal and heptanal in human blood. J Chromatogr B 813(1–2):47–52. doi:10.1016/j.jchromb
47. Mochalski P, King J, Klieber M et al (2013) Blood and breath levels of selected volatile organic compounds in healthy volunteers. Analyst 138(7):2134–2145. doi:10.1039/c3an36756h

Chapter 4
Solid Phase Microextraction for Sensing Freely Dissolved Analytes in Complex Water Sample

Ruifen Jiang, Jianqiao Xu and Gangfeng Ouyang

Abstract Ever since its invention, solid phase microextraction (SPME) has been widely used for aqueous sampling. In recent years, this technique has received increasing attention for the analysis of complex water sample because of its capability of sensing freely dissolved analytes in the complex matrix. In this chapter, the development of negligible depletion SPME (nd-SPME) for the determination of freely dissolved analytes in complex water samples is reviewed. The fundamentals and sampling conditions of nd-SPME are first discussed. Then, the application of nd-SPME for measurement of sorption coefficients in complex matrixes is summarized. In addition, the calibration methods of nd-SPME are discussed in detail, especially the effect of complex matrix on the sampling kinetic. For the first time, the advances of complex matrix effects including retarded and enhanced effects on the SPME sampling kinetic are fully explored.

Keywords SPME · Freely dissolved concentration · Complex matrix · Sorption coefficient · Fouling · Enhanced SPME kinetics

4.1 Abbreviations

K_F	Freundlich constant
BFRs	Brominated flame retardants
BSA	Bovine serum albumin
CPZ	Chlorpromazine

R. Jiang
School of Environment, Guangzhou Key Laboratory of Environmental Exposure and Health, and Guangdong Key Laboratory of Environmental Pollution and Health, Jinan University, Guangzhou 510632, China

J. Xu · G. Ouyang (✉)
MOE Key Laboratory of Aquatic Product Safety/KLGHEI of Environment and Energy Chemistry, School of Chemistry, Sun Yat-sen University, Guangzhou 510275
People's Republic of China
e-mail: cesoygf@mail.sysu.edu.cn

G. Ouyang and R. Jiang (eds.), *Solid Phase Microextraction*,
DOI 10.1007/978-3-662-53598-1_4

DI	Direct immersion
DMPC	Dimyristoylphosphatidylcholine
DOC	Dissolved organic carbon
DOM	Dissolved organic matter
ELISA	Enzyme-linked immunosorbent assay
EqC	Equilibrium calibration
ExC	External calibration
FA	Fulvic acid
FQT	Fluorescence quenching technique
HA	Humic acid
HOCs	Humic organic carbons
HS	Headspace
HSA	Human serum albumin
IS	Internal standard
LAS	Linear alkylbenzenesulfonate
LLE	Liquid–liquid extraction
LLP	Liquid–liquid partition
OC	Organic carbon
OCCs	Organochlorine compounds
OCPs	Organochlorine pesticides
OCs	Organic compounds
OPEs	Organophosphate esters
PA	Polyacrylate
PAHs	Polycyclic aromatic hydrocarbons
PBDEs	Polybrominated diphenyl ethers
PBS	Phosphate-buffered saline
PDMS	Polydimethylsiloxane
SOM	Sediment organic matter
SPs	Suspended particulates
SUVA	Specific UV absorbance
TECAM	Triolein-embedded cellulose acetate membrane
UWL	Unstirred water layer

4.2 Introduction

The presence of complex matrixes such as dissolved organic matter (DOM) and suspended particulates (SPs) in environmental water can significantly affect the partitioning behavior of organic compounds (OCs), thus determining the fate of many pollutants present in the aquatic matrix [1–3]. Several studies have showed that the DOM increases the solubility [4], decreases the volatilization [5], and also affects the bioavailability and toxicity of the OCs [6, 7], as well as controls the leaching of pesticides in soil [8–11]. In addition, the presence of DOM enhances the photodegradation of pollutants through the production of reactive species such as hydroxyl radical or singlet oxygen [12], and inhibits the degradation of pesticides [13, 14]. The mechanisms lie behind the above phenomenon depends on the interaction between DOM and OCs and their physicochemical properties such as sorption coefficient.

Sorption coefficient describes the sorption capacity of the complex matrixes and equals the ratio of freely dissolved concentration to matrix-bound concentration of the analytes in the sample matrix. The freely dissolved compound is believed to be the driving force for the transportation, distribution, and bioaccumulation of a contaminant, and thus, its concentration is a key parameter controlling the bioavailability and toxic effect of the OCs. Thus, it is necessary to measure the freely dissolved concentrations of OCs when study the sorption capacity of the complex matrix presents in the water sample. Several methods including equilibrium dialysis, ultrafiltration, protein precipitation, chromatographic method, and micro-extraction technique have been developed to measure freely dissolved concentrations of OCs in water sample. Some of these techniques require the physical separation of the DOM-bound chemicals from the freely dissolved chemicals, while others do not need the phase separation. All of the methods have both demerits and merits. The limitation of a method can lead to large uncertainties in the quantification of the freely dissolved concentrations of the analytes. An overview of these methods is presented in the literature [15]. Among the techniques used, SPME seems to be the most promising technique since it is simple, fast, and sensitive compared to other methods [16–19]. In addition, SPME appears to have smaller biases since it causes the least disruption to the sample, and is capable of determining freely dissolved compounds in complex matrixes without additional phase separation steps in the workflow [20, 21].

4.3 Negligible Depletion SPME (nd-SPME)

In the 1990s, SPME had been established as a powerful tool in solventless trace analysis [22]. The fundamental of this analytical technique is that only the freely dissolved analyte fraction is sampled by the SPME fiber, whereas the analyte fraction bound to both solid sorbents and dissolved/suspended matter is not. In addition, only a small fraction of the freely dissolved compounds is extracted due to the small volume of the fiber coating, leading to a negligible depletion of the freely dissolved fraction and thus a negligible shift in the equilibrium between the bound and the free fraction. The equilibrium between the freely dissolved compound and the fraction bound to the matrix will not be perturbed significantly. Vaes et al. first introduced this method to measure the freely dissolved concentrations of four polar compounds in biological samples [23] and called it as nd-SPME.

4.3.1 The Mechanism and Operation of nd-SPME

Because fundamentally SPME is a partition or equilibrium extraction process, the amount of analytes accumulates on the fiber can be described by a one-phase association system and the first-order kinetics [24] using Eq. 4.1.

$$n_t = n_e[1 - \exp(-at)] \tag{4.1}$$

where n_t is the amount extracted onto the fiber when the sampling time is t, n_e is the amount extracted when sampling reaches equilibrium and can be described by Eq. 4.2. a is the sampling time constant indicating how fast a sampling process reaches equilibrium.

$$n_e = \frac{K_{ff} V_f V_s}{K_{ff} V_f + V_s} C_0 \tag{4.2}$$

where K_{ff} is the sorption coefficient of the analyte between fiber and water sample. V_f and V_s are the volume of the fiber coating and the water sample, respectively. C_0 is the original concentration of the analyte in the water sample. To achieve negligible extraction, the extracted amount at equilibrium should be less than 10% of the original amount of the target compound in the sample matrix, which can be described by Formula 4.3.

$$\frac{K_{ff} V_f V_s}{K_{ff} V_f + V_s} C_0 < 10\% C_0 V_s \tag{4.3}$$

The following formula can be obtained by rearranging the Formula 4.3.

$$\frac{K_{ff} V_f}{V_s} < \frac{1}{9} \approx 0.11 \tag{4.4}$$

Regarding the kinetic parameter, a, it may be different from the one in the matrix-free sample. The presence of the complex matrix in the water sample could also influence the kinetic sampling process. More details will be discussed in Sect. 4.4.

nd-SPME can be applied by exposing the fiber in the headspace (HS) above the sample or directly in the sample (direct immersion, DI). The advantage of headspace nd-SPME is that the matrix in the sample cannot interfere with the fiber, but it is only applicable for volatile compounds. For both HS and DI sampling modes, the procedure can be subdivided into four steps [23]: (i) preparation of the sampling matrix by spiking the standards into the sample with and without complex matrix, and incubation of the spiked sample for a sufficient long period of time till the compounds reach distribution equilibrium; (ii) exposure of the SPME fiber in the headspace or directly in the standard spiked sample for sampling; (iii) quantification of the fiber-extracted amount with analytical instrument; (iv) selection of a proper calibration method to determine the freely dissolved concentration in the sample matrix.

4.3.2 The Conditions of nd-SPME

Three conditions should be fulfilled to achieve the nd-SPME sampling for determination of the freely dissolved analytes in complex water samples.

- Equilibrium between the bound and free fraction of analyte should be achieved before sampling;

This condition should be followed mainly for the spiked sample because analytes distributed between different phases in real sample commonly reach equilibrium before sampling, and it is not necessary to incubate them before analysis. On the other hand, the laboratory-spiked sample usually requires incubation for a sufficient period of time to ensure the equilibrium between different phases in water sample. The equilibration time between the complex matrixes and the water correlates with the diffusion coefficient of analytes and the structure of complex matrixes as well as the environmental temperature. Generally, small volatile compounds diffuse faster than the large molecule and less volatile compounds. Hu et al. [25] found that the equilibration time for the low-molecular weight compounds (phenanthrene and Fluoranthene) was 72 h, whereas it increased to approximately 120 h for the high-molecular weight compound (chrysene). For highly volatile compounds such as odor-active carbonyls [26], the incubation time can be shorten to 12 h in wine. In addition, the incubation time for DOM is usually shorter than the SPs or other solid matrixes. Poerschmann et al. [27] found that the sorption of organotin to DOM took a few minutes while the sorption onto particulate organic matter took longer time due to the different diffusion pathways. Liu et al. also found that the sorption of polycyclic aromatic hydrocarbons (PAHs) [25, 28] and organochlorine compounds (OCCs) [29] to the C60 took much longer time than to the humic acid (HA), and a 7-day incubation was applied during the study. Other reported works applied an incubation period of 24 h for organophosphate esters (OPEs) sorbed to 4 sources of HA, 16 h for pyrethroid in wastewater [30], 24 h for anion-surfactant LAS in sediment pore water [31], and 10 min for estradiol with human serum albumin (HSA) [32]. Furthermore, it is worth noting that pretreatment for the spiked sample to inhibit the growth of bacteria or microorganism should be done to prevent the biodegradation of the spiked compounds during the incubation period. NaN_3 solution is commonly used in this application [4, 30, 33, 34].

- Depletion of the free fraction caused by the SPME extraction should be negligible.

An essential prerequisite to apply nd-SPME for describing the sorption phenomena is that the amount of analyte extracted by the fiber is negligible compared to the freely dissolved amount in the sample. As the depletion amount will never be 0%, a limit amount should be set to consider the depletion as negligible. Researchers have set the different limits at 5% [23], 1% [35], and 10% [27, 36]. According to Eq. 4.4, when the limit was set at 1%, the depletion percentage can be determined by Eq. 4.5

$$\text{depletion}\,\% = 100K_{\text{ff}}V_{\text{f}}/V_{\text{s}} \tag{4.5}$$

which indicates that the depletion ratio relates to the K_{ff} value and the volume ratio of the fiber coating to the volume of the water sample, $V_{\text{f}}/V_{\text{s}}$. When the K_{ff} values of the target compounds are big, the $V_{\text{f}}/V_{\text{s}}$ ratio should be small. Therefore, small coating volume to large sample volume is always preferred when sensitivity is allowed. Different thicknesses of the coatings have been reported in the literatures (Table 4.1). Among the selected fibers, 7 μm polydimethysiloxane (PDMS) coating is the most commonly used one due to the small volume of coating. For a PDMS coating with a thickness of 7 μm, the coating volume is approximately 0.026 μL. When the sample volume is 1L, the K_{ff} value of the compound can be up to 10^6. However, when the fiber volume is 100 μm ($V_{\text{f}} = 0.65\,\mu\text{L}$), the K_{ff} value of the compound should be lower than 10^5 to ensure the nondepletion condition.

Selection of a fiber coating depends on the properties of the compounds and the coatings. PDMS coating is preferred by nonpolar compounds such as PAHs [37], polybrominateddiphenyl esters (PBDEs) [34], OCCs [29], and organochlorine pesticides (OCPs) [38, 39] while polyacrylate (PA) coating is usually used for relatively polar compounds such as phenols [37], alcohols [32, 40, 41], organophosphate ester [30], surfactants [25, 42–44], and anilines [45]. On the other hand, when the volume ratio of the fiber coating to the sample is constant, fibers with low K_{ff} value can be chosen as an alternative option when the sensitivity is allowed. For example, PDMS can be used in the nd-SPME to determine the freely dissolved concentration of phenols due to the low K_{ff} value [46].

Equation 4.5 is only applicable to the equilibrium sampling. However, pre-equilibrium can be also used for nd-SPME. In pre-SPME, the sampling amount is lower than the extracted amount at equilibrium, which can in turn minimize the depletion amount and achieve the negligible condition. However, when using the pre-equilibrium SPME, one should pay much attention to the calibration process because the presence of the complex matrixes may influence the sampling kinetics. More details regarding to the calibration process can be referred to Sect. 4.4.

In literature, another application of SPME where negligible depletion condition is not required, but freely dissolved concentrations can nevertheless be measured, is matrix-SPME. The basic idea of matrix-SPME is to expose the fiber coating in the sample matrix for a sufficient long period of time till the analytes distributed

Table 4.1 Application of nd-SPME for determination of sorption coefficient

Compounds	Fibers	Matrix	Calibration method	Measurement endpoint and their relationship	References
40 priority compounds	PDMS, PDMS/DVB	Nordic reference fulvic acid	ExC	K_{DOM}, k log K_{DOM} versus log K_{ow}	[53]
Pesticide and trifluralin	PA	DOM	ExC	log K_{DOM} versus DOM properties	[49]
PAHs, phenols	PA, PDMS	Plant root, DOM, SOM	ExC	K_{DOM}, K_{OC}, K_{root} log K_{oc} versus log K_{ow}	[37]
PAHs	30 μm PDMS	Fullerene, C60 Aldrich HA		k_2, K_{HA}, K_{C60}, K_{HA+C60} log k_2 versus log K_{ow}	[28]
Odor-active carbonyls	65 μm PDMS/DVB, 50 μm DVB/CAR/PDMS, 85 μm PA HS-SPME	Wine	IS for total conc. ExC for free conc.	C_{free} C_{bound} K_a: apparent equilibrium constants	[26]
9 OPEs	85 μm PA, 30 μm PDMS	4 sources of HA, biota	EqC	K_{DOM} LC50 C_{DOM} versus death rate	[30]
PBDEs	PDMS disk	Liposome protein	EqC	K_{lipw}, K_{BSAw} log K_{lipw} versus log K_{ow} log K_{BSAw} versus log K_{ow}	[54]
Polar and ionic compounds	35 μm PA	DOC	EqC	K_{DOC} log K_{DOC} versus log K_{ow} Effect of pH and ion	[43]

(continued)

Table 4.1 (continued)

Compounds	Fibers	Matrix	Calibration method	Measurement endpoint and their relationship	References
Pyrethroid insecticides	7 μm PDMS	Secondary wastewater effluent	Pre-SPME was used ExC for free con. IS for total conc.	K_{id}, K_{DOM}	[55]
Cationic surfactant	7 μm PA	Homologues, micelles HA	ExC	Effect of pH, salinity Sorption isotherm Ion-exchange	[42]
PBDEs BFRs	Disposable 35 μm PDMS	SOM in pore water DOM in pore water	EqC	K_{fiber} K_{OC} K_{DOM} obtained from LLE and SPME	[34]
Chlorpromazine	34.5 μm and 7.5 μm PA	BSA	EqC	K_{BSA}	[17]
OCCs	30 μm PDMS	Fullerene, biota	EqC	K_{fiber}, $K_{lipid-PDMS}$ k_1, k_2, C_{60} versus C_{free} C_{fiber} versus C_{fish}	[29]
Anion-surfactant LAS	75 μm PA	Sediment pore water	IS	Sorption isotherms for sediment-water K_F, K_d, ion effect $\log K_f$ versus $\log K_d$	[31]
PAHs	Disposable 28.5 μm PDMS	Presence of DOM in pore water from different origins	EqC, LLE for total	K_{DOC}, temperature effect, ΔH_{sorp}, ΔS_{fus} $\log K_{DOC}$ versus $\log K_{ow}$ $\log K_{DOC}$ versus SUVA	[56]

(continued)

Table 4.1 (continued)

Compounds	Fibers	Matrix	Calibration method	Measurement endpoint and their relationship	References
8 pyrethroids	7 μm PDMS	4 sediment pore water, runoff water	IS for total conc. ExC LLE for total conc.	K_{DOC}	[18]
OCPs, BkF	TECAM	HA, biota	Pre-equilibrium ExC	K_{DOC}, K_{TECAM}, BAF, log K_{DOC} versus log K_{ow} EROD activity versus log C_{free}	[38, 39]
PAHs	30 μm PDMS	Aldrich HA, Fullerene, HA + C60	EqC, ExC	$K_{C_{60}}$, K_{DOM} Effect of pH and NaCl	[25]
Alcohol ethoxylates	35 μm PA	Marine sediment pore water	EqC	Sorption isotherm K sediment	[40, 41]
Pyrene	28.5 μm PDMS	BSA	EqC	K_{BSA}	[57]
Octylphenol	7 μm PDMS	Biological sample, Protein, BSA	Equilibrium at different C_{BSA}	K_{BSA} ELISA	[47]
PAHs (log K_{ow} > 6)	PDMS	Aldrich HA	Equilibrium at different DF	K_{DOM}	[58]
PAHs Alkyl PAHs	7 μm PDMS	Pore water with DOM (4–27 mg/L)	Internal calibration External calibration curve	K_{DOM}	[59]
Bifenthrin, permethrin isomers	30 μm PDMS	DOM and OM from sediment	Pre-SPME, ExC	K_{DOM}, K_{OM}	[60]

(continued)

Table 4.1 (continued)

Compounds	Fibers	Matrix	Calibration method	Measurement endpoint and their relationship	References
Halogenated compounds, Narcotic compounds	85 μm PA	Biota Midge Larve	N/A	LC50 C_{biota} versus C_{fiber} C_{biota} versus K_{fiber}	[61]
Radiochemicals, 3[H] Estradiol	Disposable 7 μm PA	HSA	ExC	K_{HSA}	[32]
PCB, PAHs, alkanes	PDMS, PAA ester membrane	DOM	ExC	K_{DOM} Analytes Solubility BCF	[5]
PAHs	Poly(acrylic acid) ester (PAAE)	DOM, OM	ExC	K_{DOM}, S_{DOM} ln K_{DOM} versus ln K_{ow}, S_{DOM}	[62]
HOC, LogK_{ow} 5.2–7.5	PDMS-coated glass fiber	Sediment, soil	EqC	Matrix-SPME C_{free}	[63]
OCPs, PCBs	Disposable 7 μm PDMS Automated SPME	HA from Aldrich	ExC	K_{DOM} BCF with and without HA LLE measure the total Conc.	[20]
Phenols PAHs	7 μm and 100 μm PDMS	DOM from contaminated water and artificial water	ExC for free conc. IS for total conc.	K_{DOM} log K_{DOC} versus log K_{ow}	[46]
Alcohols, Benzenes, Phenols, Anilines	PA	Phospholipid DMPC; Trout liver, Rat Hepatocytes	ExC	K_{DMPC}	[45]
Organotin compounds	7, 100 μm PDMS 85 μm PA	FA was isolated from a water plant	ExC	K_{OC} and K_d K_{OC} and K_{DOM}	[27]

All abbreviations can be found in the abbreviation form at the end of the chapter between the fiber, water, and complex matrix reach equilibrium. The fiber-extracted amount may deplete the original freely dissolved concentration. However, if at a steady state, net desorption from the matrix can re-establish and reach equilibrium, and the concentration on the fiber is still proportional to the initial freely dissolved concentration. Matrix-SPME has been applied by Mayer et al. to measure freely dissolved concentrations in sediment pore water [37]. It is preferable to apply matrix-SPME at equilibrium, but kinetic measurements can be performed when the equilibrium between matrix-bound and freely dissolved analytes is established much more quickly than the extraction equilibrium, so the equilibrium condition is met.

- Binding matrix should not affect the extraction process.

Ideally, nd-SPME only senses the freely dissolved compounds when used for complex sample analysis [38, 39]. However, a binding matrix can theoretically adsorb to the fiber surface (fouling), which possibly influences the extracted amount. Fouling effect, on one hand, could block the uptake of analyte onto the fiber coating and underestimate the fiber-extracted amount. On the other hand, fouling could also lead to an overestimation of the concentration in the fiber coating as the matrix-bound analyte is measured along with the fiber-extracted amount. Heringa et al. [47] tested the potential fouling effect by measuring the amount of HSA on the exposed fibers with a competitive enzyme-linked immunosorbent assay (ELISA). Result showed that about 68 ng of the HSA was present on each of the exposed fibers and the fouling by protein adsorption did happen. Oomen et al. [48] also tested the fouling of chyme on the fiber surface by a Bradford assay; however, result showed there was no matrix adsorbed onto the fiber coating. Similar result was obtained by Broeders et al. who tested the possible effect of albumin on the total extraction of chlorpromazine (CPZ), and found that the addition of BSA did not cause fouling or influence the uptake kinetics of CPZ onto the fiber [17]. Bondarenko et al. [18] and Steven et al. [41] also found no fouling of matrix on the fiber coating in their research work.

Some researchers also claimed that short extraction time can be used to avoid the effect of complex matrix. Caupos et al. [49] and Ohlenbusch et al. [50] selected short extraction times for the SPME sampling and claimed that the fouling effect and the enhancement effect on the uptake kinetic resulted from the complex matrixes were maximally minimized.

Another effect of the complex matrix is on the sampling kinetics of SPME, and it is caused by the net desorption of analytes from the binding matrix in the aqueous diffusion layer around the fiber coating. The desorbed analytes speeds up the uptake of analytes onto the fiber in comparison with the samples without binding matrix. This phenomenon has been observed and reported to influence the quantification results when using pre-equilibrium SPME [51, 52]. Detailed information regarding the enhancement mechanism and calibrationapproach will be discussed in Sect. 4.4.

4.3.3 The Calibration Methods for nd-SPME

As nd-SPME is a nonexhaustive sample preparation technique that only extracts a small amount of analytes present in the sample, proper calibration method is required to quantify the concentration in the sample based on the fiber-extracted amount. Traditional SPME calibration methods including external calibration, standard addition, internal calibration, and equilibrium calibration, as well as on fiber-standard calibration can be utilized for nd-SPME. However, because the presence of complex matrix in the sample could affect the sampling kinetics, a proper calibration method should be selected.

External calibration (ExC) method is the simplest and the most common one applied in nd-SPME (Table 4.1), which involves the preparation of several standard samples using analyte-free samples. The analyte-free sample can be matrix-matched blank sample or the matrix-free blank sample like pure water. If a matrix-matched blank sample is used to prepare the calibration curve, the calibrated concentration is referred to the total concentration of the analytes. On the other hand, if a matrix-free blank sample is selected, the free concentration of the analytes is quantified. In this case, one assume no matrix effect is occurred during the sampling process. If fouling or kinetic enhancement effect of the complex matrix is occurred, external calibration method could not be accurate, and more discussion can refer to Sect. 4.4.

Internal standard method is also used in nd-SPME calibration by adding a known concentration of internal standard that is not present in the sample matrix, and can mimic the behavior of the analytes bound to the complex matrix during the extraction and analysis process. Calibration is done by comparing the ratio of the analytes signals to the internal standard with a known concentration [18, 26, 31, 55, 59, 64]. However, it is worth noting that the calibration concentration obtained by internal standard method is the total concentration in the sample not the freely dissolved concentration.

Equilibrium calibration (EqC) is based on the relationship amount and the described by Eq. 4.2, which indicates that the analyte concentration in the sample (C_0) is correlated with thc amount of analyte extracted onto the coating (n_e) when other parameters are known. Distribution coefficient of the analytes between the fiber coating and the water can be obtained by sampling from the spiked pure water. EpC can minimize the matrix effect on the enhancement of sampling kinetic.

4.3.4 Determination of Sorption Coefficient Using nd-SPME

One of the most important applications of nd-SPME is for determination of the sorption coefficient of organic compounds to the complex matrixes, which is defined as the ratio of freely dissolved concentration to matrix-bound concentration of analytes in the sample as shown in Eq. 4.6.

$$C_{\text{free}} = \frac{C_{\text{DOM}}}{K_{\text{DOM}}} = \frac{C_{\text{SPs}}}{K_{\text{SPs}}} = \frac{C_{\text{biota}}}{BCF} \tag{4.6}$$

Rearranging Eqs. 4.6 and 4.7 can be obtained for determination of the sorption coefficient, K_{matrix}

$$\text{free fraction} = \frac{C_{\text{free}}}{C_{\text{total}}} = \frac{1}{1 + K_{\text{matrix}} C_{\text{matrix}}} \tag{4.7}$$

K_{matrix} can be obtained by knowing the freely dissolved concentration (C_{free}) in samples containing different concentrations of complex matrix. The freely dissolved concentration in different samples can be determined using nd-SPME, while the total concentration (C_{total}) equals to the spiked concentration if the original sample does not contain target compounds. The total concentration can also be determined by liquid liquid extraction (LLE) method. Then, the K_{matrix} value can be obtained by plotting the free fraction of analytes in the sample with the concentration of the complex matrix based on Eq. 4.7 (as shown in Fig. 4.1).

The complex matrixes present in the aqueous sample can be DOM or SPs. In literatures, the sorption coefficients of a variety of organic contaminants to binding matrixes including commercial humic acid [25, 58, 65], surfactant [4], protein [17, 57], liposome [54], and DOM from pore water/wastewater [18, 30, 31, 34, 55, 56, 64] have been investigated and determined. Sorption coefficients of suspended particles including the C60 [29], latex [66], and particles with different sizes separated from sediment [55] were also measured with similar procedure described above. Moreover, Liu et al. measured the sorption coefficients of PAHs [25, 28] onto the mixture of C60 and humic acid, and investigated the combined effect of both complex matrixes. The outstanding advantages of the nd-SPME technique for studying sorption coefficient are as follows: (i) The sorption behavior of both

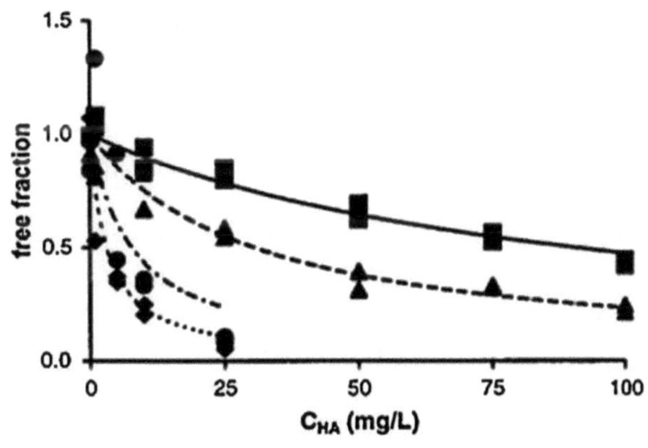

Fig. 4.1 Free fraction determined using SPME vs the humic acid concentration in water. Reprinted with the permission from Ref. [20]. Copyright 1998 American Chemical Society

dissolved and suspended particulate matters can be investigated with the same experimental method, and more consistent results can be obtained and compared; (ii) in contrast to traditional methods such as fluorescence quenching technique (FQT), the SPME allows the investigation of multicomponent sorbate mixtures when coupled with high-resolution analytical instrument such as GC or HPLC; and (iii) the SPME method can easily achieve the measurement of sorption coefficients at different temperatures, which is considered as the most significant and relevant parameter for nonionic humic organic carbons (HOCs) in environmental aquatic system [56, 67]. Haftka et al. applied the nd-SPME method to [56] study the structural and thermodynamic aspects of PAH sorption to natural DOM as well as other environmental processes such as enhanced diffusion phenomena in pore water, and then provided further insight into the site-specific exposure to freely dissolved PAH concentrations in soil and sedimentary pore water.

The accuracy of the obtained sorption coefficients measured by nd-SPME has been evaluated by comparing them with the ones obtained by other methods. Lee et al. compared the sorption coefficients of bifenthrin and permethrin isomers in creek and filed sediment measured by liquid–liquid partition (LLP) and nd-SPME. Results showed that the K_{DOM} obtained by nd-SPME gave 0.6- to 4.4-fold greater than those obtained by LLP as shown in Table 4.2. The authors stated that the underestimation by LLP method was attributed to the DOM-bound compounds which were not excluded from the aqueous phase by centrifugation. In addition, the degree of underestimation depended on the source and the amount of DOM, and may be generally significant for compounds that have DOM adsorption coefficient larger than 10^4 [60].

The nd-SPME method for sorption coefficient measurement was also compared with FQT [62]. As shown in Fig. 4.2, the FQT values for PAHs are consistently higher by a factor of 3–5 than the SPME ones. The authors claimed that two methods detected different types of interactions, and obtained concentration-base and activity-base coefficients for FQT and SPME method, respectively.

Table 4.2 Sorption coefficients of bifenthrin and permethrin isomers in creek and field sediments measured by LLP or SPME

Sediment	Bonita creek		San Diego creek		Field sediment	
	K_d	r^2	K_d	r^2	K_d	r^2
Bifenthrin						
LLP	3.4×10^4	0.88	1.2×10^4	0.87	3.6×10^3	0.77
SPME	6.5×10^4	0.78	4.3×10^4	0.9	5.8×10^3	0.89
cis-Permethrin						
LLP	1.4×10^4	0.78	1.3×10^4	0.85	2.6×10^4	0.9
SPME	7.5×10^4	0.83	5.1×10^4	0.86	7.1×10^4	0.88
trans-Permethrin						
LLP	1.5×10^4	0.89	1.4×10^4	0.85	1.7×10^4	0.81
SPME	3.7×10^4	0.78	4.4×10^4	0.88	3.2×10^4	0.88

Reprinted with the permission from Ref. [60]. Copyright 2003 American Chemical Society

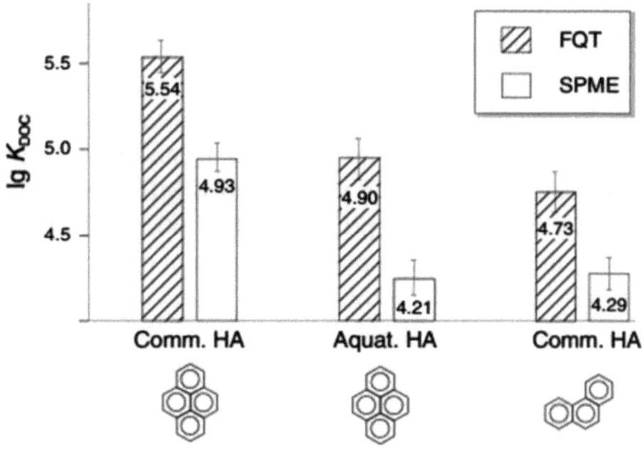

Fig. 4.2 Comparison of sorption coefficients determined by FQT and SPME for commercially available Roth HA and aquatic HA (HO10 HA) for pyrene and phenanthrene. Reproduced from Ref. [62] with the permission of John Wiley & Sons Ltd

Besides DOM and SPs, environmental aquatic system also contains different types of biota. OCs can partition into the biological lipid membrane and accumulate in biota [30]. Since the uptake of OCs from the aquatic phase to the biota is a passive and diffusion-driven process, a chemical partitioning technique such as nd-SPME can be used to simulate this process, offer a very useful estimation of the potential accumulation by biota from the water phase, and predict the baseline toxic effect. This approach was initially developed as an extraction method with solid-phase extraction disks (C18 Empore disks) [68]. The application of SPME for this purpose was introduced by Parkerton and Stone [69], and it was further developed by Verbruggen et al. [36] as a tool to mimic the accumulation in lipid membranes. This approach assumes that the amount of hydrophobic organics retained by the SPME fiber is a good mimic for the amount that can accumulate in lipid membranes. The advantage of biomimetic SPME is that only the bioavailable fraction of the toxicants partitions the fiber and is measured. In complex water sample, only a small amount of the freely dissolved fraction of OCs is extracted from water samples by SPME fiber. In contrast to exhaustive extraction, biomimetic SPME is similar to the field situation where uptake by biota does not affect environmental concentration.

Biomimetic SPME has been widely applied to study the bioconcentration toxicity effect and uptake kinetics of OCs in aquatic biota [5, 20, 29, 30, 38, 39, 61]. Leslie et al. [61] used an 85 μm polyacrylate (PA) coating to mimic the uptake of 2,4,5-trichloroaniline and 1,2,3,4-tetrachlorobenzene by midge larvae. Results found that the SPME fiber–water partition coefficient (K_{ff}) was related to the toxicant concentration found in midge larvae. In the complex sample containing different sources of HA, the bioavailability of OCs can be affected by the concentration of HA. Chen et al. studied the bioaccumulation of benzo[k]fluoranthene on Japanese medaka in the water sample containing different concentrations of HA and found that the bioaccumulation of BkF was reduced with an increase in HA concentration in exposure medium. Similar results also obtained by Pang et al. [30] who utilized nd-SPME to evaluate the impact of 4 representative HAs on the aquatic invertebrate *Daphnia magna*, and concluded that the interaction of HA with 9 organophosphate esters (OPEs) was corroborated by the effects of HA on the acute toxicity of OPEs to *D. magna*. The effect of SPs (nC60) in water sample on the bioaccumulation of 8 OCCs to biota was also investigated by Hu et al. [29] Results indicated that the OCCs concentrations in medaka at kinetic uptake regime linearly correlated with that in nd-SPME fiber without nC60, but this correlation deteriorated with the presence of nC60. These results suggested that in solution containing nC60, the uptake mechanism of OCC to medaka is different from that to nd-SPME fiber. In addition, the presence of nC60 slightly enhanced the bioaccumulation of the less hydrophobic OCCs. More discussions regarding the enhancement of uptake kinetics can be referred to Sect. 4.4.

4.4 Passive Dosing for Determination of Sorption Coefficient

4.4.1 SPME Fiber Dosing Method

As discussed in Sect. 4.1, by applying nd-SPME, a sample is first spiked with selected target compounds, and then a blank SPME fiber is exposed to the spiked sample for a predetermined period of time. Last, the extracted fiber is quantified, and the concentration of the freely dissolved organic compounds is calibrated. Nd-SPME method has been reported for the determination of the freely dissolved concentration of a variety of organic pollutants in the complex sample [20, 23, 27, 32, 33, 45, 46, 48, 60, 70, 71]. However, several issues are raised when applying this extraction strategy for more hydrophobic compounds. The first issue pertains to the complicated spiking procedure due to the low aqueous solubility of the hydrophobic compounds. In addition, spiking process normally introduces organic solvent into the samples and may influence the partition of OCs in a given aqueous system. Furthermore, with the SPME extraction method, either negligible depletion sampling [23, 45] or equilibrium sampling [70] process should be fulfilled to minimize the disruption of partition equilibrium in the sample.

On the contrary, fiber depletion method (also called solid-phase dosing method) can be employed to overcome the limitations of extraction strategy [64, 72–74]. In SPME depletion procedure (as shown in Fig. 4.3), the SPME fiber is first loaded with targeted compounds. Then, the analyte-loaded fiber is exposed to the solutions that contained different amounts of complex matrix. Analytes present in the sample is distributed between fiber coating, water, and complex matrix till equilibrium, and the following relationship can be applied to describe the amount of analytes between different phases.

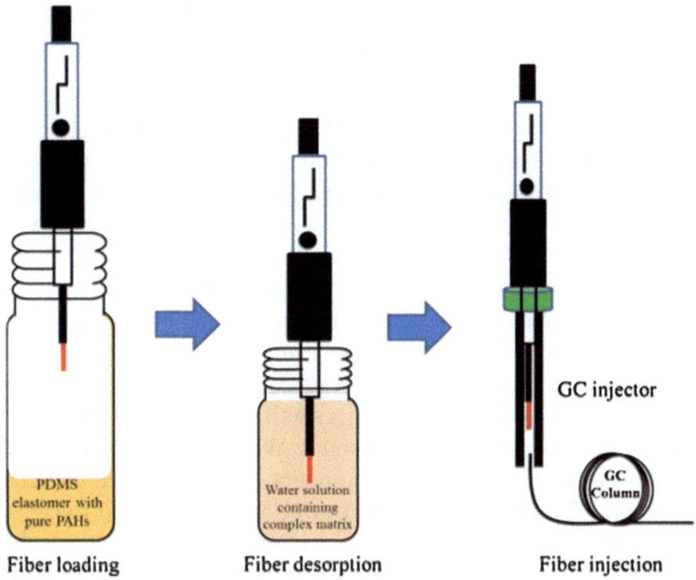

Fig. 4.3 Schematic of the experimental procedure on the autosampler. Reprinted from Ref. [75], Copyright 2015, with the permission from Elsevier

$$q_{\mathrm{f,e}} + q_{\mathrm{e}}^{\mathrm{u}} + q_{\mathrm{e}}^{\mathrm{b}} = q_0 \tag{4.8}$$

$$K_{\mathrm{matrix}} = \frac{C_{\mathrm{m}}^{\mathrm{b}}}{C_{\mathrm{m}}^{\mathrm{u}}} = \frac{\frac{q_{\mathrm{e}}^{\mathrm{b}}}{n_{\mathrm{matrix}}}}{\frac{q_{\mathrm{e}}^{\mathrm{u}}}{V_{\mathrm{s}}}} = \frac{q_{\mathrm{e}}^{\mathrm{b}}}{q_{\mathrm{e}}^{\mathrm{u}} \times C_{\mathrm{matrix}}} \tag{4.9}$$

$$K_{\mathrm{ff}} = \frac{q_{\mathrm{f,e}} \times V_{\mathrm{s}}}{V_{\mathrm{f}} \times q_{\mathrm{e}}^{\mathrm{u}}} \tag{4.10}$$

where $q_{\mathrm{f,e}}$, $q_{\mathrm{e}}^{\mathrm{u}}$, and $q_{\mathrm{e}}^{\mathrm{b}}$ are the amounts of analytes left on the fiber coating, freely dissolved in solution, and bound to the matrix when the desorption reaches equilibrium, respectively; q_0 is the initial loaded amount; K_{matrix} is the partition coefficient of analyte between the matrix and the water; K_{ff} is the partition coefficient of analyte between the fiber coating and the water; n_{matrix} is the amount of additional matrix, while C_{matrix} is the concentration; and V_{f} and V_{s} are the volumes of fiber coating and sample, respectively. By integrating the above equations, one can obtain the following equation:

$$\frac{q_0}{q_{\mathrm{f,e}}} = 1 + \frac{V_{\mathrm{s}}}{K_{\mathrm{ff}} V_{\mathrm{f}}} + \frac{V_{\mathrm{s}} K_{\mathrm{matrix}} C_{\mathrm{matrix}}}{K_{\mathrm{ff}} V_{\mathrm{f}}} \tag{4.11}$$

where $\frac{q_0}{q_{f,e}}$ is linearly proportional to the C_{matrix} when the other parameters in the equation are constant. From the intercept and slope of the curve, one can determine the K_{ff} and K_{matrix}, respectively, by knowing the value of $\frac{V_s}{V_f}$. Equation 4.11 indicates that the K_{matrix} can be obtained by plotting the $q_0/q_{f,e}$ versus C_{matrix}.

The main challenge associated with depletion method pertains to the fiber dosing process, which should be reproducible and the dosing amount should be easily quantified. In the reported research [58], the blank fibers are directly exposed to the solution spiked with high concentration of targeted compounds. This method usually requires a long exposure time, and a large number of fibers are needed to be prepared at the same time. Additionally, the experimental procedure for the reported depletion method is labor intensive, and full automation is difficult to achieve with the currently available autosampler. In our group [51, 52, 75], we have proposed a fully automated loading method by using a standard gas generation vial, which can provide continuously constant loaded amount. When coupled with an autosampler, the fiber-loading process can achieve automation, and the SPME depletion method for determination of sorption coefficient can be fully automated too.

With fully automated SPME dosing technique, we measured the sorption coefficients (K_{matrix}) of 4 PAHs between HA/hydroxypropyl-β-cyclodextrin (β-HPCD) and aqueous sample [75]. In addition, the temperature effect on the sorption coefficient was also investigated. Finally, the proposed method was applied for the real biological fluid analysis. The sorption coefficients of 6 PAHs between the complex matrixes in the fetal bovine serum and water were determined and compared to those obtained by SPME extraction method. The result demonstrated that the proposed method can be applied to determine the sorption coefficients of hydrophobic compounds between complex matrix and water in a variety of samples.

For the SPME fiber dosing technique, ter Laak et al. [72] proposed another strategy also based on Eq. 4.11. Instead of measuring the $q_0/q_{f,e}$ values in complex samples with different concentrations of complex matrix, the researchers plotted the $q_0/q_{f,e}$ against the $\frac{V_s}{V_f}$ while kept the concentration of the complex matrix constant. The partition coefficients of six PAHs with Aldrich humic acid in aqueous sample were determined, and the result was in great agreement with literature data.

4.4.2 Coated Vial Passive Dosing Method

The passive dosing vial method for determination of sorption coefficients of OCs with complex matrixes was introduced by Mayer et al. [65]. It is based on the principles of passive dosing developed for the control of freely dissolved concentrations in samples. Different from fiber dosing method, vial dosing preloads the analytes onto a polymer-coated vial (Fig. 4.4). Then the analyte dosing vial is loaded with samples with and without complex matrix. The vial is incubated till the

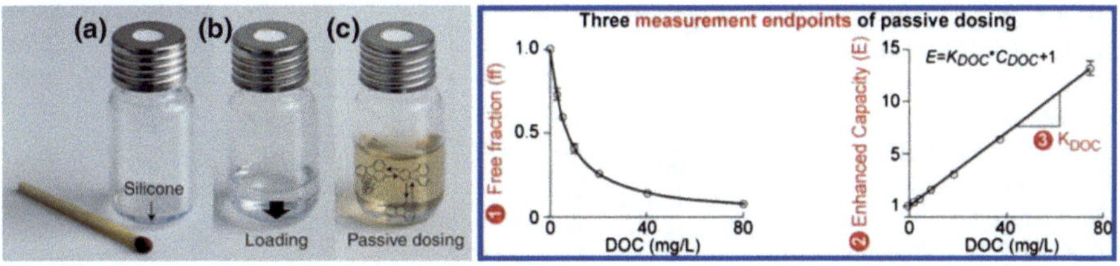

Fig. 4.4 Coated vial passive dosing procedure. Reprinted with the permission from Ref. [64]. Copyright 2010 American Chemical Society

analyte distribution equilibrium is reached between different phases. Last step is to measure the total concentration of the analytes in the solution with (C_{total}) and without complex matrix (C_{free}) using traditional liquid–liquid extraction method. According to Eq. 4.7, the K_{matrix} can be obtained from the plot of the $C_{\text{total}}/C_{\text{free}}$ value versus C_{DOM}.

This technique has been applied to determine the sorption coefficients of OCs in a variety of samples [3, 54, 64, 65]. In comparison with fiber dosing method, which achieves the speciation by the depletion of analytes concentration in the fiber coating under a complete mass balance assumption, the coated vial passive dosing method applies a much larger and thicker polymer phase to dominate the partitioning between polymer and sample. The analyte concentration in the polymer is not depleted during the speciation process. The freely dissolved concentration in samples is constant and independent of the complex matrix. Maenpaa et al. [76] compared the application of HS-SPME with coated glass jar for measurement of freely dissolved concentrations of PCBs in contaminated soil and sediment. Results obtained from the two methods were in good agreement with each other. Both methods combined high analytical performance, reduced equilibrium time, and new calibration possibilities, which makes them suited for bioaccumulation research and environmental monitoring.

4.5 SPME Kinetics in Aqueous Samples Containing Mobile-Binding Matrixes

It has been nearly two decades since the initiation of the studies on the influences of mobile-binding matrixes on SPME kinetics in aqueous samples, when nd-SPME in nonequilibrium mode was explored as a tool for the determination of freely dissolved concentrations in various aqueous samples which contained mobile-binding matrixes such as dissolved humic substances and proteins [23, 32, 77]. At the very beginning, the influence of mobile-binding matrixes on SPME kinetics was not well recognized by the scientific community. Scientists thought the sorption equilibrium of the target analyte bound to the binding matrix would not be disturbed under the investigation of nd-SPME, and only the free analytes would partition to the fiber

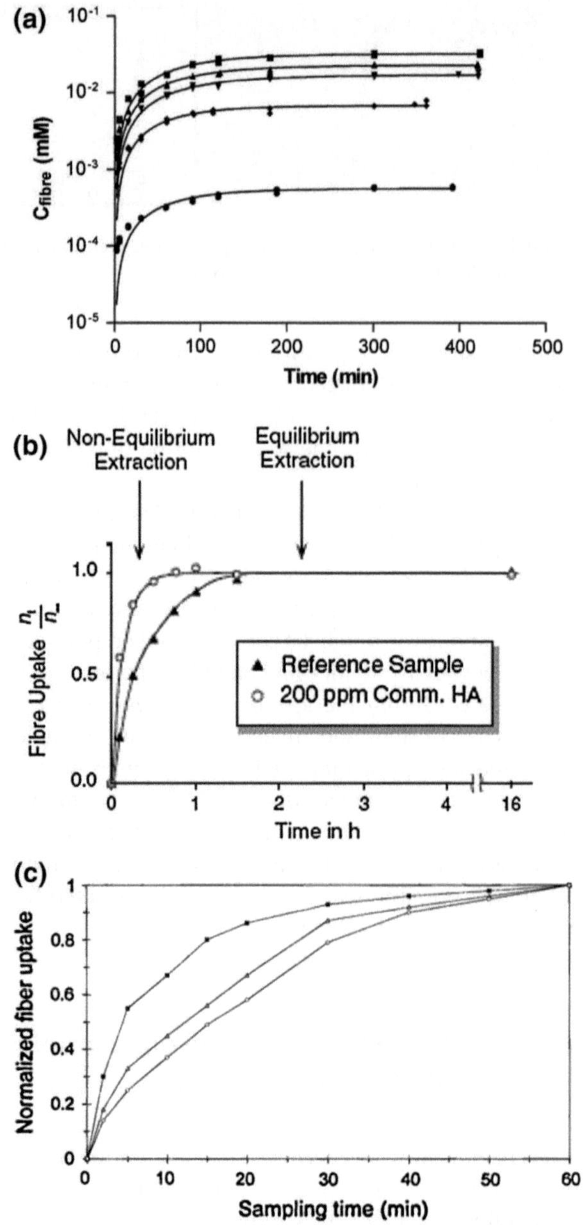

Fig. 4.5 **a** Adsorption profiles of [3H]estradiol to the 7 μm polyacrylate fiber in the presence of various concentrations of BSA: none (*filled square*), 6.5 × 10^{-6} M (*filled triangle*), 1.6 × 10^{-5} M (*filled inverted triangle*), 6.4 × 10^{-5} M (*filled diamond*), 1.0 × 10^{-3} M (*filled circle*). Reprinted with the permission from Ref. [32]. Copyright 2002 American Chemical Society. **b** Kinetics of direct immersion SPME of fluoranthene from pure water (reference sample) and from a humic acid (HA) solution (7 μm polydimethylsiloxane fiber). Reproduced from Ref. [62] with permission of John Wiley & Sons Ltd. **c** Kinetics of direct immersion SPME of tetrabutyltin from pure water (*filled square*), fulvic acid solution (*open triangle*) and humic acid solution (*open circle*) (100 μm polydimethylsiloxane fiber). Reprinted with the permission from Ref. [27]. Copyright 1997 American Chemical Society

coatings. Therefore, calibration curves established in pure water at a fixed time (not reach the extraction equilibrium) were convinced to be capable of determining freely dissolved concentrations in aqueous samples containing mobile-binding matrixes [23]. Actually, in the early study of Vaes et al. [23], the extracted amounts of 4-chloro-3-methylphenol by nonequilibrium SPME in phosphate buffer saline

(PBS) solutions outside dialysis membrane (containing bovine serum albumin, BSA) and PBS solutions inside dialysis membrane (no BSA) were nearly the same, which demonstrated that the SPME kinetics was not influenced by BSA. The same phenomenon was also observed by Heringa et al. in their study of SPME of [^{3}H] estradiol from BSA solutions (Fig. 4.5a) [32].

However, later studies revealed that the influences of the mobile-binding matrixes on SPME kinetics could be much complicated and ambiguous, as the presence of the binding matrixes such as humic organic matters could possibly enhance SPME kinetics (Fig. 4.5b) [77], as well as retard SPME kinetics (Fig. 4.5c) [46] in aqueous samples. Kopinke, Pörschmann and Georgi explained these phenomena in their early book chapter by hypothesizing two opposite effects, i.e., the barrier effect and the shuttle effect [78], which indicated that the mass transfer toward the fiber might be hampered by forming a thin film of the binding matrix at the surface of the fiber, or the depletion of the free analyte adjacent to the fiber because extraction could be compensated by the release of the bound analyte from the binding matrix. Since then, these two effects have been widely adopted to explain the distorted SPME kinetics in complex aqueous samples containing mobile-binding matrixes, and they are also termed as "fouling effect" and "diffusion layer effect," respectively, in the literature [15].

4.5.1 Retarded SPME Kinetics

The fouling effect or the barrier effect is unwanted in SPME practices, not only because the retarded SPME kinetics can sacrifice analytical sensitivity, but also because the sorption of binding matrixes on SPME fibers can shorten the lifetimes of the fibers and lead to erroneous results [15]. In previous studies, the occurrence of the fouling effect seemed quite common in the presence of mobile-binding matrixes [46, 79, 80]. Even though, in some studies, no obvious retardation effects on SPME kinetics were observed, the possible attachment of binding matrixes on fibers should not be fully excluded. For example, the presence of silica nanoparticles with a mean radius of 7.5 nm was proven to be capable of enhancing SPME kinetics of diclofenac [81]. However, silica nanoparticles with a mean radius of 10 nm were observed to aggregate on the surface of PDMS coating and even entering the coating [82].

The sorption of binding matrixes on fiber coatings is probably dependent on the nature of the coatings and the nature of the mobile-binding matrixes, as well as the exposure time and agitation conditions. Mobile-binding matrixes in aqueous samples attach to SPME fibers mostly due to the hydrophobic interactions between binding matrixes and fiber coatings, as binding matrixes such as dissolved humic substances possess surfactant-like nature, while fiber coatings generally act as reverse-phase sorbents in aqueous samples. From this perspective, modifying the surfaces of fiber coatings to be hydrophilic can possibly make fibers fouling-resistant [83]. In addition, it might be expected that short-time exposure under agitated conditions could probably avoid serious attachment of binding matrixes on fiber coatings.

Except for the fouling effect, Alam et al. proposed another two possible reasons for the retarded SPME kinetics [84], i.e., (i) the local depletion due to the relatively large analyte diffusivity in fiber coating while the ever-increasing diffusion distance in sample medium, and (ii) SPME kinetics controlled by dissociation kinetics of the analyte from the binding matrix. However, the authors hypothesized totally static conditions in their modeling. Actually, aqueous samples are difficult to be kept totally static due to mechanical agitation or uneven heat distribution. In addition, when the aqueous sample was hypothesized totally static, SPME in it was no longer a first-order kinetic process.

4.5.2 Enhanced SPME Kinetics

As for the enhanced SPME kinetics in the presence of certain binding matrixes, even though the shuttle effects of the binding matrixes have long been proposed, studies were focused on the following issues: (i) under what circumstances the shuttle effect would happen, and (ii) how significantly the shuttle effect would enhance the SPME kinetics.

Kopinke, Pörschmann, and Georgi [78] proposed the concept of "shuttle effect" and put forward two preconditions for the occurrence of shuttle effect; i.e., (i) diffusion through the stagnant water layer surrounding the fiber is the rate-limiting step, and (ii) the association kinetics of the analyte to the binding matrix is in the timescale the binding matrix diffusing through the stagnant water layer. In addition, Kopinke, Pörschmann, and Georgi claimed that the more analytes the binding matrixes were loaded with, the more significant the shuttle effect would be. However, the second precondition was in contrast to the considerations of Jeannot and Cantwell [85], who claimed that the association–dissociation kinetics should be fast compared with the mass transfer rate through the stagnant water layer.

The following studies inherited the thoughts of Jeannot and Cantwell, probably owing to the progresses on the understanding of the analogous processes during electrochemical analysis, in which ligands also influence the kinetics of electrochemical reactions on the surfaces of the electrodes [86]. Oomen et al. for the first time conceptually visualized the shuttle effect in their study, as shown in Fig. 4.6 [48]. In their mass transfer model, it also proposed that diffusion through the unstirred water layer (UWL) adjacent to the fiber was the rate-limiting step, the release of analyte from the mobile-binding matrix was fast, and the bound concentration was large. However, Oomen et al. just discussed the total flux when the association–dissociation equilibrium between the free analyte and bound analyte was kept everywhere in the UWL, as shown in Eq. 4.12:

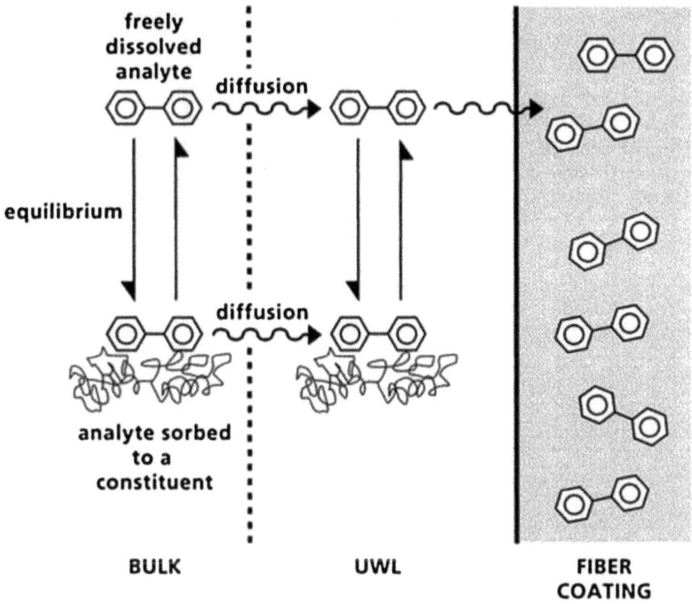

Fig. 4.6 Conceptual representation of the uptake model for analyte fluxes toward the SPME fiber coating. Both the freely dissolved analytes and the sorbed analytes diffuse into the unstirred water layer (UWL). Only the freely dissolved analytes in the UWL directly partition into the fiber coating. If diffusion through the UWL is the rate-limiting step for the entire uptake process, a concentration gradient in the UWL is formed. Depending on the concentration of sorbed analytes in the UWL and on their desorption kinetics, desorbed analytes contribute to the flux toward the fiber coating. Reprinted with the permission from Ref. [48]. Copyright 2000 American Chemical Society

$$F_{tot} = \frac{A}{l}\left(D_{bound}Z_{bound}\upsilon_b + D_{free}Z_{free}\upsilon_w\right) \tag{4.12}$$

where F_{tot} is the total conductivity through the UWL; A and l are the surface area and the thickness of the UWL, respectively; D_{bound} and D_{free} are the diffusivities of the bound analyte and free analyte, respectively; Z_{bound} and Z_{free} are the fugacity capacities of the bound analyte and free analyte, respectively; and υ_b and υ_w are the relative volumes of the binding matrix and water, respectively.

The air-bridge system established by Oomen et al. is useful for demonstrating the shuttle effects of different binding matrixes (Fig. 4.7) [48]. Chyme solution, water, or artificial soil solution was added to the liquid vial, and the two vials were equilibrated via the glass tube and then the analyte concentrations in the headspace vial should be in certain proportion to the analyte concentrations in water, the free analyte concentrations in chyme solution, and the freely dissolved concentrations in artificial soil solution. If the ratio of the extracted amount from the aqueous phase to that from the headspace vial for the binding matrix-containing solution was larger than the ratio for the matrix-free water, then the shuttle effect occurred. In this

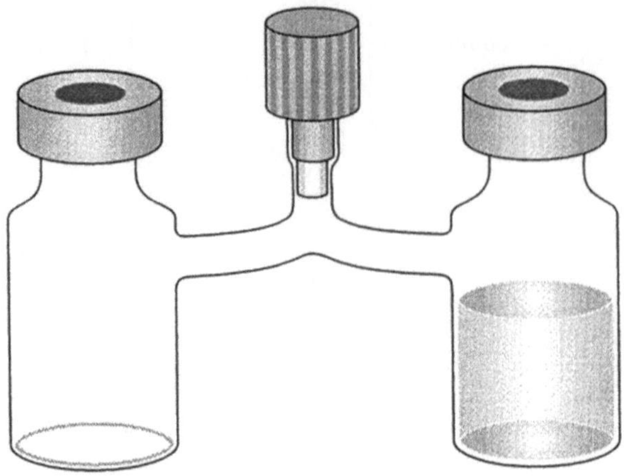

Fig. 4.7 Schematic representation of an air-bridge system consisting of two vials connected by a glass tube. Reprinted with the permission from Ref. [48]. Copyright 2000 by American Chemical Society

study, the shuttle effect was observed for chyme while no shuttle effect was observed for the artificial soil. The authors demonstrated that the dissociation of the analyte from the artificial soil might not be as fast as required.

Kramer et al. studied the desorption kinetics of pyrene preloaded in polydimethylsiloxane fiber in BSA solution [57]. The authors assumed that the time for achieving analyte partition equilibrium at the fiber–sample interface and the time for reaching association–dissociation equilibrium between the free and bound analytes were both significantly shorter than the characteristic time for the depletion of the analyte from the fiber. Furthermore, continuity of fluxes over both the fiber coating–boundary layer interface and boundary layer-bulk medium interface was assumed. In this study, the boundary layer was equal to the UWL in the model of Oomen et al. [48], both of them represented the bottlenecks of the mass transfer processes. Then, under the quasi-steady assumption, the total flux of the analyte $J(t)$ at time point t through the boundary layer was written as the combined mass transfer of the free analyte and the bound analyte:

$$J(t) = -\left[m_{\text{free}}\left(C_{\text{i,free}}(t) - C_{\text{b,free}}(t)\right) + m_{\text{bound}}\left(C_{\text{i,bound}}(t) - C_{\text{b,bound}}(t)\right)\right] \quad (4.13)$$

where m_{free} and m_{bound} are the mass transfer coefficients of free analyte and bound analyte, respectively; $m_{\text{free}} = \frac{D_{\text{free}}}{\delta}$, $m_{\text{bound}} = \frac{D_{\text{bound}}}{\delta}$, δ is the thickness of the boundary layer. The subscripts "i" and "b" refer to the fiber–sample interface and bulk solution, respectively. Based on the assumption that the time required to reach association–dissociation equilibrium between the free and bound analytes was significantly shorter than the characteristic time for the depletion of the analyte from the fiber, the authors got

$$\frac{C_{\text{i,bound}}(t)}{C_{\text{i,free}}(t)} = \frac{C_{\text{m,bound}}(t)}{C_{\text{m,free}}(t)} = K_{\text{b}}C_{\text{b}} \tag{4.14}$$

where K_{b} is the association constant of the analyte to the binding matrix, and C_{b} is the concentration of the binding matrix, respectively. Then, the Prandtl boundary layer was adopted in the model, in other words, δ only depended on the fiber radius, fluid velocity, and fluid viscosity. By assuming that the thicknesses of the boundary layers were nearly unchanged after the incorporation of BSA in the aqueous media, the authors wrote the facilitated transport ratio (FTR) as follows:

$$\text{FTR} = \frac{m_{\text{free}} + m_{\text{bound}}K_{\text{b}}C_{\text{b}}}{m_{\text{free}}} = 1 + \frac{D_{\text{bound}}}{D_{\text{free}}}K_{\text{b}}C_{\text{b}} \tag{4.14}$$

It was the first time to evaluate the extent of shuttle effect; however, the assumption that the free analyte and bound analyte were in association–dissociation equilibrium everywhere in the boundary layer could certainly not be adopted in all binding matrix-analyte systems. For example, Oomen et al. demonstrated that the release of polychlorinated biphenyl from the artificial soil might not be as fast as required, as no shuttle effect was observed [48].

The subsequent studies were quite cautious about the dissociation kinetics of the analytes from the binding matrixes. ter Laak et al. [87] developed the model of Kramer et al. [57] by replacing the Prandtl boundary layer with the Nernst boundary layer, as well as introducing a so-called lability factor to correct for the incomplete association and dissociation of the analyte and binding matrix in the boundary layer. The thickness of the Nernst boundary layer could be calculated from the thickness of the Prandtl boundary layer:

$$\delta_{\text{N}} = \left(\frac{D}{\nu}\right)^{1/3} \delta_{\text{P}} \tag{4.15}$$

where δ_N and δ_P are the thicknesses of the Nernst boundary layer and the Prandtl boundary layer, respectively; D was the diffusivity of the analyte; and ν was the kinematic viscosity of the aqueous medium. Equation 4.15 demonstrates that the thickness of the Nernst boundary layer is dependent on the diffusivity of the analyte. The authors defined that the bound analyte will be completely labile in the boundary layer when the free analyte and bound analyte were in equilibrium everywhere in the boundary layer and then, the weighted average diffusivity $\overline{D}$ was expressed as:

$$\overline{D} = \frac{D_{\text{free}} + D_{\text{bound}} K_{\text{b}} C_{\text{b}}}{1 + K_{\text{b}} C_{\text{b}}} \tag{4.16}$$

By introducing the lability factor ξ, the authors rewrote Eq. 4.16 for the incompletely bound analyte in the boundary layer as:

$$\overline{D} = \frac{D_{\text{free}} + \xi D_{\text{bound}} K_{\text{b}} C_{\text{b}}}{1 + \xi K_{\text{b}} C_{\text{b}}} \tag{4.17}$$

Theoretically, the lability factor ranged from unity (completely labile) to zero (completely inert). Thereafter, the enhanced SPME kinetics could be evaluated by the average mass transfer coefficient $\overline{m}$:

$$\overline{m} = \frac{\overline{D}}{R_{\text{f}} \ln\left(1 + \overline{\delta}/R_{\text{f}}\right)} \tag{4.18}$$

where R_{f} is the radius of the fiber, $\overline{\delta}$ is the average thickness of the boundary layer, which could be calculated by replacing Eq. 4.17 in Eq. 4.15. This expression of the average mass transfer coefficient is corrected for the cylindrical geometry of the fiber and the adjacent boundary layer.

From the model of ter Laak et al., it is obvious that the shuttle effect occurs when $\xi > 0$; and the larger the ξ is, the more the bound analyte contributes to the overall mass transfer through the boundary layer.

Van Leeuwen, Town and colleagues [66, 81] provided a criterion to evaluate the lability of the bound analyte in the boundary layer by comparing the so-called kinetic flux J_{kin} and diffusive flux J_{dif}:

$$J_{\text{kin}} = k_{\text{d}} C_{\text{bound}} \mu \tag{4.19}$$

where k_{d} is the dissociation kinetic coefficient of the bound analyte from the binding matrix, and μ is the thickness of the reaction layer in which the equilibrium between free analyte and bound analyte was distorted, as shown in Fig. 4.8. $\mu = (D_{\text{free}}/k_{\text{a}} C_{\text{b}})^{1/2}$, k_{a} is the association kinetic coefficient of the free analyte bound to the binding matrix. This expression demonstrates that the thickness of the reaction layer is determined by the mobility of the free analyte (D_{free}) and its mean-free lifetime ($1/k_{\text{a}} C_{\text{b}}$).

$$J_{\text{dif}} = \frac{\overline{D} C_{\text{tot}}}{\overline{\delta}} \tag{4.20}$$

where C_{tot} was the total concentration in the bulk solution. Equations 4.19 and 4.20 demonstrate that the kinetic flux results from the desorption of analyte from the binding matrix in the reaction layer, while the diffusive flux results from the coupled diffusion of the free analyte and bound analyte through the boundary layer. van

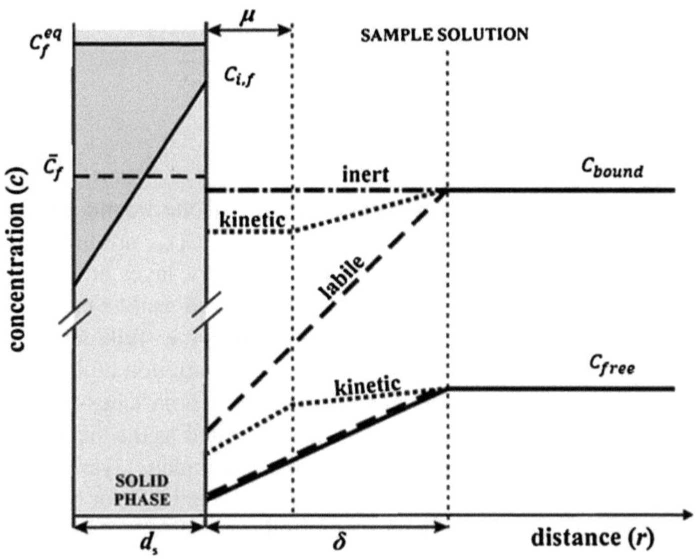

Fig. 4.8 Sketch of steady-state concentration profiles of the free analyte and bound analyte for the inert case (*dot with dashed lines*), the labile case (*dashed lines*), and the case controlled by desorption kinetics (*dashed lines*). $C_{i,f}$ is the concentration in the fiber coating at the water–solid interface, $\overline{C}_f$ is the mean concentration in fiber coating, C_f^{eq} denotes the final equilibrium concentration in the fiber coating, d_s is the thickness of the solid phase, δ is the diffusion layer thickness, and μ is the reaction layer thickness. Reprinted with a small modification with the permission from Ref. [81]. Copyright 2012 American Chemical Society

Leeuwen, Town, and colleagues demonstrated that the bound analyte in the boundary layer was labile when the ratio of the kinetic flux to the diffusive flux was significantly larger than unity [66, 81]. The authors provided the effective rate constant of SPME in aqueous samples containing binding matrixes as:

$$k_{\text{eff}} = \frac{A\overline{D}}{VK_{\text{fw}}\left[C_{\text{free}}/\left(C_{\text{free}} + C_{\text{lab, bound}}\right)\right]\overline{\delta}} \tag{4.21}$$

where k_{eff} is the effective rate constant of SPME, and $C_{\text{lab, bound}}$ is the labile fraction of the bound analyte.

Xu et al. [88] assumed a pseudo-steady state during SPME, as compared to the timescales for the free and bound analytes diffusing through the boundary layer (probably subsecond levels); the time to reach SPME equilibrium is quite long (often dozens of minutes to hours and even days). Based on the pseudo-steady-state assumption, the lability factor was demonstrated to be equal to the ratio of the normalized concentration gradients of the bound analyte to that of the free analyte across the boundary layer, as shown in Eq. 4.22:

$$\xi = \frac{1 - \frac{C_{\mathrm{i,bound}}}{C^*_{\mathrm{bound}}}}{1 - \frac{C_{\mathrm{i,free}}}{C^*_{\mathrm{free}}}} = \frac{\lambda - \tanh \lambda}{\lambda + \varepsilon K' \tanh \lambda} \tag{4.22}$$

where $\lambda = \sqrt{(k_\mathrm{d}\delta^2/D_{\mathrm{bound}})(\varepsilon K' + 1)}$, $K' = k_\mathrm{a}C_\mathrm{b}/k_\mathrm{d}$, $\varepsilon = D_{\mathrm{bound}}/D_{\mathrm{free}}$. As shown in Fig. 4.9, C^*_{bound} and C^*_{free} are the concentrations of the bound analyte and free analyte in the bulk solution, respectively; $C_{\mathrm{i,bound}}$ and $C_{\mathrm{i,free}}$ are the concentrations of the bound analyte and free analyte in the boundary layer at the fiber–sample interface, respectively. According to Eq. 4.22, the bound analyte can be subdivided into two fractions, of which $\xi K' C_{\mathrm{i,free}}$ can be treated as the "fully labile" fraction as it is always in proportion to the freely dissolved concentration at the same domain; on the contrary, $(1 - \xi)C^*_{\mathrm{bound}}$ remains unchanged at both sides of the boundary layer ($x = 0$ and $x = \delta$); thus, this fraction can be treated as the "inert" fraction that does not contribute to the mass transfer through the boundary layer. Both of the free fraction and the "fully labile" fraction can be treated available for SPME. Based on the interpretation of the lability factor, the authors defined a so-called effective average diffusivity as the weighted average diffusivities of the free analyte and the "labile" fraction of the bound analyte. The expression of the effective average diffusivity is the same as the average diffusivity provided in [85] (shown in Eq. 4.17). In the model of Xu et al., the effective average diffusivity was also used to calculate the Nernst boundary layer.

Then, comparing the rate constant of SPME when a portion of the bound analyte is labile in the boundary layer $\left(\left(\frac{\mathrm{d}n}{\mathrm{d}t}\right)_{\xi \neq 0}\right)$, with the rate constant when the bound

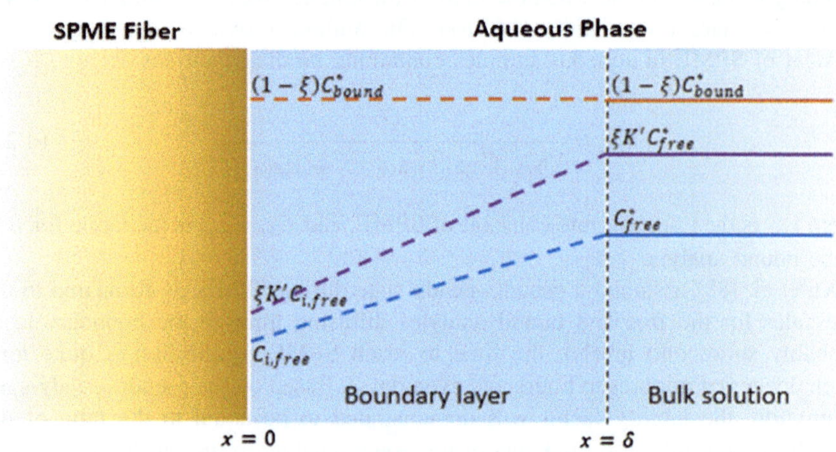

Fig. 4.9 Conceptual scheme of the mass transfer through the boundary layer from the aqueous phase to SPME fiber. Reprinted with a small modification from Ref. [88], Copyright 2016, with the permission from Elsevier

analyte is totally inert in the boundary layer $\left(\left(\frac{\mathrm{d}n}{\mathrm{d}t}\right)_{\xi=0}\right)$, the enhanced mass transfer ratio is expressed as:

$$R_{\mathrm{e}} = \frac{\left(\frac{\mathrm{d}n}{\mathrm{d}t}\right)_{\xi\neq 0}}{\left(\frac{\mathrm{d}n}{\mathrm{d}t}\right)_{\xi=0}} = (1+\varepsilon\xi K')^{2/3}(1+\xi K')^{1/3} \tag{4.23}$$

Based on Eq. 4.23, when R_{e}, ε, and K' are determined by experiment, the lability factor of the bound analyte in the boundary layer of a given system can be estimated.

The overall diffusivity $\left(\overline{D}_{\mathrm{overall}} = \frac{D_{\mathrm{free}} + \xi D_{\mathrm{bound}} K_{\mathrm{b}} C_{\mathrm{b}}}{1 + K_{\mathrm{b}} C_{\mathrm{b}}}\right)$ was adopted in the mathematical models to explain the enhanced mass transfer at nonaqueous liquid phase–aqueous phase interface [89] and water–air interface [90]. Xu et al. pointed out the model would be failed in the situations when $\xi = 0$[88]. The lability factors of bound PAHs in the boundary layers of sodium do decylsulphate solutions were

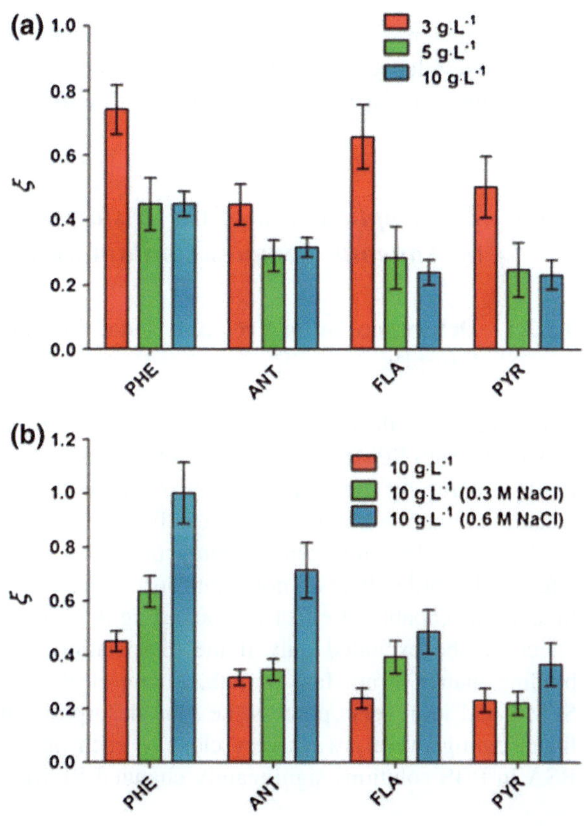

Fig. 4.10 Lability factors of the bound PAHs in DBLs under **a** different SDS concentrations and **b** different salinities. *PHE* phenanthrene, *ANT* anthracene, *FLA* fluoranthene, and PYR pyrene. *Error bars* represent the standard errors. Reprinted with a small modification from Ref. [88], Copyright 2016, with the permission from Elsevier

estimated according to Eq. 4.23. As shown in Fig. 4.10, the lability factors generally decreased with increasing SDS concentration and with decreasing salinity [88]. Based on the same model, the same group found that the lability factors increased with increasing the temperature of the sample solutions; higher dissociation kinetics under higher temperature should be the reason for the higher lability factors.

In the studies discussed above [48, 51, 57, 66, 81, 87, 88], SPME kinetics in aqueous samples containing binding matrixes was modeled by considering diffusion in boundary layers to be the rate-limiting step; therefore, the stagnant water layer (also termed as UWL [48], Prandtl boundary layer [57], and Nernst boundary layer [51, 66, 81, 87, 88] in the literatures) model was used in these studies. However, van Eijkeren et al. claimed that it might be difficult to precisely characterize the thicknesses of the stagnant layers; thus, they adopted a finite mass transfer coefficient instead of diffusion through a stagnant water layer surrounding the fiber in their model [91]. In addition, the diffusion of the analyte within the fiber coating was described as a nonsteady process, while in other studies [48, 51, 57, 66, 81, 87, 88], this process was thought much faster than the diffusion through the stagnant water layers and was not taken into consideration for the evaluation of the SPME kinetics in aqueous samples containing binding matrixes [48, 51, 57, 66, 81, 87, 88]. Alam et al. made an attempt to develop a general mathematical model to evaluate the effects of binding matrixes on SPME kinetics [84]. However, the authors excluded convection in the bulk solutions to explain the effects of the binding matrixes, which were actually not the real situations in aqueous samples.

4.5.3 The Implications of the Distorted SPME Kinetics in Aqueous Samples Containing Binding Matrixes

4.5.3.1 Determination of Free and Total Concentrations in Complex Samples

The revealing of the effects of the binding matrixes on SPME kinetics indicates that nonequilibrium SPME calibrated with the standard curves established in pure water is not suitable for the determination of freely dissolved concentrations in samples containing mobile-binding matrixes. It seems that equilibrium SPME should be used for the determination of freely dissolved concentrations unless the fouling effect and shuttle effect do not occur during the extraction process. Short extraction time seems capable of effectively avoiding the fouling effect [49], while the shuttle effect can be excluded only if the dissociation kinetics of the analyte from the binding matrix is not fast enough. Zhang et al. proved that the rate constants of SPME and its reverse process, i.e., the desorption of the analyte preloaded in the fiber coating, were always very close to each other, even though the presence of BSA in PBS solutions significantly changed SPME kinetics (Fig. 4.11) [92]. This

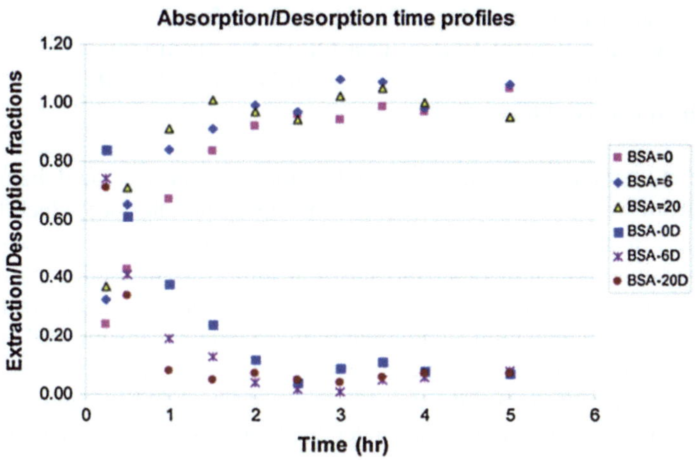

Fig. 4.11 Verification of the symmetry between extraction (absorption) of atrazine from the BSA solutions to C18-coated SPME fibers and desorption of preloaded standards from the fiber to the sample solutions (n = 6, RSDs 10–15%). Absorption and desorption (D) profiles were obtained in PBS solutions with different BSA concentrations (0, 6, and 20 μM). Reprinted with the permission from Ref. [92]. Copyright 2011 American Chemical Society

phenomenon is important as it demonstrates the extraction processes of the analytes can be calibrated by the desorption processes of their analogues. Then, the extracted amount at equilibrium n_0 can be determined according to:

$$\frac{n}{n_0} + \frac{q}{q_0} = 1 \tag{4.24}$$

where the extracted amount in the fiber coating n, the analogue of the analyte preloaded in the fiber q_0, and the residual analogue after extraction q all can be determined in the experiment. The freely dissolved concentration can be calculated by

$$C_{\text{free}} = \frac{n_0}{K_{\text{fw}} V_{\text{f}}} \tag{4.25}$$

where V_{f} is the volume of the fiber coating, and K_{fw} is the partition coefficient of the analyte between the fiber coating and pure water, which can be determined by equilibrium SPME of the analyte in pure water.

When nonequilibrium SPME is used to determine the total concentration of the analyte in a complex sample, the calibration should be conducted in a matrix, of which the composition is very close to the investigated sample. From this perspective, establishing the standard curve in blank samples, or utilizing standard addition or internal standard calibration methods can be ideal for obtaining accurate data.

4.5.3.2 Mass Transport of Organics at the Interfaces of Environmental Waters and Other Environmental Compartments, and the Bioavailability of Pollutants in the Environment

Ramus et al. demonstrated the dissolved humic substances could hamper mass transport from water to air by forming a thin film at the interface which hampered surface renewal of the water; on the other hand, dissolved humic substances could also act as shuttles which facilitate the mass transport [88]. In another study [93], ter Laak et al. also observed that the presence of dissolved organic matter could enhance the bioaccumulation of pollutants in aquatic organisms. All these phenomena were very similar to the processes of SPME in complex aqueous samples containing binding matrixes. The investigations into the SPME kinetics in complex aqueous samples can be helpful for understanding these environmentally relevant processes.

Acknowledgement We acknowledge financial support from the project of National Natural Science Foundation of China (Grant No. 21407184) and the NSF of Guangdong Province (Grant S2013030013474).

References

1. Leenheer JA, Croue JP (2003) Peer reviewed: characterizing aquatic dissolved organic matter. Environ Sci Technol 37:18A–26A
2. Delle Site A (2001) Factors affecting sorption of organic compounds in natural sorbent/water systems and sorption coefficients for selected pollutants. A review. J Phys Chem Ref Data 30:187–439. doi:10.1063/1.1347984
3. Escher BI, Cowan-ellsberry CE, Dyer S et al (2011) Protein and lipid binding parameters in rainbow trout (*Oncorhynchus mykiss*) blood and liver fractions to extrapolate from an in vitro metabolic degradation assay to in vivo bioaccumulation potential of hydrophobic organic chemicals. Chem Res Toxicol 24:1134–1143. doi:10.1021/tx200114y
4. Schacht VJ, Grant SC, Escher BI et al (2016) Solubility enhancement of dioxins and PCBs by surfactant monomers and micelles quantified with polymer depletion techniques. Chemosphere 152:99–106. doi:10.1016/j.chemosphere.2016.02.122
5. Poerschmann J, Kopinke FD (2001) Sorption of very hydrophobic organic compounds (VHOCs) on dissolved humic organic matter (DOM). 2. Measurement of sorption and application of a Flory-Huggins concept to interpret the data. Environ Sci Technol 35:1142–1148
6. Bejarano AC, Chandler GT, Decho AW (2005) Influence of natural dissolved organic matter (DOM) on acute and chronic toxicity of the pesticides chlorothalonil, chlorpyrifos and fipronil on the meiobenthic estuarine copepod *Amphiascus tenuiremis*. J Exp Mar Biol Ecol 321:43–57. doi:10.1016/j.jembe.2005.01.003
7. Lundqvist A, Bertilsson S, Goedkoop W (2012) Interactions with DOM and biofilms affect the fate and bioavailability of insecticides to invertebrate grazers. Ecotoxicology 21:2398–2408. doi:10.1007/s10646-012-0995-z

8. Worrall F, Fernandez-Perez M, Johnson AC et al (2001) Limitations on the role of incorporated organic matter in reducing pesticide leaching. J Contam Hydrol 49:241–262. doi:10.1016/S0169-7722(00)00197-2
9. Spark KM, Swift RS (2002) Effect of soil composition and dissolved organic matter on pesticide sorption. Sci Total Environ 298:147–161
10. Thevenot M, Dousset S, Hertkorn N et al (2009) Interactions of diuron with dissolved organic matter from organic amendments. Sci Total Environ 407:4297–4302. doi:10.1016/j.scitotenv.2009.04.021
11. Ding Q, Wu HL, Xu Y et al (2011) Impact of low molecular weight organic acids and dissolved organic matter on sorption and mobility of isoproturon in two soils. J Hazard Mater 190:823–832. doi:10.1016/j.jhazmat.2011.04.003
12. Auger JP, Richard C, Andreux F (1999) Effect of light on humic substances: production of reactive species: humic substances. Analusis 27:387–390
13. Dimou AD, Sakkas VA, Albanis TA (2004) Trifluralin photolysis in natural waters and under the presence of isolated organic matter and nitrate ions: Kinetics and photoproduct analysis. J Photochem Photobiol A Chem 163:473–480. doi:10.1016/j.jphotochem.2004.02.001
14. Bachman J, Patterson HH (1999) Photodecomposition of the carbamate pesticide carbofuran: kinetics and the influence of dissolved organic matter. Environ Sci Technol 33:874–881. doi:10.1021/es9802652
15. Heringa MBB, Hermens JLMLM (2003) Measurement of free concentrations using negligible depletion-solid phase microextraction (nd-SPME). TrAC Trends Anal Chem 22:575–587. doi:10.1016/S0165-9936(03)01006-9
16. Kramer NI, Krismartina M, Rico-Rico Á et al (2012) Quantifying processes determining the free concentration of phenanthrene in basal cytotoxicity assays. Chem Res Toxicol 25:436–445. doi:10.1021/tx200479k
17. Broeders JJW, Blaauboer BJ, Hermens JLM (2011) Development of a negligible depletion-solid phase microextraction method to determine the free concentration of chlorpromazine in aqueous samples containing albumin. J Chromatogr A 1218:8529–8535. doi:10.1016/j.chroma.2011.09.064
18. Bondarenko S, Gan J (2009) Simultaneous measurement of free and total concentrations of hydrophobic compounds. Environ Sci Technol 43:3772–3777
19. De Perre C, Le Ménach K, Ibalot F et al (2014) Development of solid-phase microextraction to study dissolved organic matter—polycyclic aromatic hydrocarbon interactions in aquatic environment. Anal Chim Acta 807:51–60. doi:10.1016/j.aca.2013.11.026
20. Urrestarazu Ramos E, Meijer SN, Vaes WHJ et al (1998) Using solid-phase microextraction to determine partition coefficients to humic acids and bioavailable concentrations of hydrophobic chemicals. Environ Sci Technol 32:3430–3435. doi:10.1021/es980274a
21. Burkhard LP (2000) Estimating dissolved organic carbon partition coefficients for nonionic organic chemicals. Environ Sci Technol 34:4663–4668
22. Pawliszyn J (2012) Handbook of solid phase microextraction. Elsevier, London
23. Vaes WHJ, Urrestarazu Ramos E, Verhaar HJM et al (1996) Measurement of the free concentration using solid-phase microextraction: binding to protein. Anal Chem 68:4463–4467. doi:10.1021/ac960337c
24. Ai J (1997) Solid phase microextraction for quantitative analysis in nonequilibrium situations. Anal Chem 69:1230–1236. doi:10.1021/ac9609541
25. Hu X, Liu J, Mayer P, Jiang G (2008) Impacts of some environmentally relevant parameters on the sorption of polycyclic aromatic hydrocarbons to aqueous suspensions of fullerene. Environ Toxicol Chem 27:1868–1874. doi:10.1897/08-009.1
26. Bueno M, Zapata J, Ferreira V (2014) Simultaneous determination of free and bonded forms of odor-active carbonyls in wine using a headspace solid phase microextraction strategy. J Chromatogr A 1369:33–42. doi:10.1016/j.chroma.2014.10.004
27. Poerschmann J, Kopinke FD, Pawliszyn J (1997) Solid phase microextraction to study the sorption of organotin compounds onto particulate and dissolved humic organic matter. Environ Sci Technol 31:3629–3636. doi:10.1021/es970377d

28. Hu X, Li J, Chen Q et al (2014) Combined effects of aqueous suspensions of fullerene and humic acid on the availability of polycyclic aromatic hydrocarbons: evaluated with negligible depletion solid-phase microextraction. Sci Total Environ 493:12–21. doi:10.1016/j.scitotenv.2014.05.107
29. Hu X, Liu J, Zhou Q et al (2010) Bioavailability of organochlorine compounds in aqueous suspensions of fullerene: evaluated with medaka (*Oryzias latipes*) and negligible depletion solid-phase microextraction. Chemosphere 80:693–700. doi:10.1016/j.chemosphere.2010.05.042
30. Pang L, Liu J, Yin Y, Shen M (2013) Evaluating the sorption of organophosphate esters to different sourced humic acids and its effects on the toxicity to *Daphnia magna*. Environ Toxicol Chem 32:2755–2761. doi:10.1002/etc.2360
31. Rico-Rico A, Droge STJ, Hermens JLM (2010) Predicting sediment sorption coefficients for linear alkylbenzenesulfonate congeners from polyacrylate-water partition coefficients at different salinities. Environ Sci Technol 44:941–947. doi:10.1021/es902453s
32. Heringa MB, Pastor D, Algra J et al (2002) Negligible depletion solid-phase microextraction with radiolabeled analytes to study free concentrations and protein binding: an example with [3H]estradiol. Anal Chem 74:5993–5997. doi:10.1021/ac0204552
33. Mackenzie K, Georgi A, Kumke M, Kopinke FD (2002) Sorption of pyrene to dissolved humic substances and related model polymers. 2. Solid-phase microextraction (SPME) and fluorescence quenching technique (FHT) as analytical methods. Environ Sci Technol 36:4403–4409
34. Jia F, Cui X, Wang W et al (2012) Using disposable solid-phase microextraction (SPME) to determine the freely dissolved concentration of polybrominated diphenyl ethers (PBDEs) in sediments. Environ Pollut 167:34–40. doi:10.1016/j.envpol.2012.03.025
35. Gorecki T, Pawliszyn J (1997) Effect of sample volume on quantitative analysis by solid-phase microextraction. Part 1. Theoretical considerations. Analyst (Cambridge, United Kingdom) 122:1079–1086. doi:10.1039/a701303e
36. Parkerton TF, Stone Ma, Letinski DJ (2000) Assessing the aquatic toxicity of complex hydrocarbon mixtures using solid phase microextraction. Toxicol Lett 112–113:273–282. doi:10.1016/S0378-4274(99)00237-4
37. Poerschmann J, Schultze-Nobre L (2014) Sorption determination of phenols and polycyclic aromatic hydrocarbons in a multiphase constructed wetland system by solid phase microextraction. Sci Total Environ 482–483:234–240. doi:10.1016/j.scitotenv.2014.03.004
38. Ke R, Wang Z, Huang S, Khan SU (2007) Accurate quantification of freely dissolved organochlorine pesticides in water in the presence of dissolved organic matter using triolein-embedded cellulose acetate membrane. Anal Bioanal Chem 387:2871–2879. doi:10.1007/s00216-007-1174-6
39. Chen S, Ke R, Zha J et al (2008) Influence of Humic Acid on Bioavailability and Toxicity of Benzo[k]fluoranthene to Japanese Medaka. Environ Sci Technol 42:9431–9436. doi:10.1021/es8014502
40. Droge STJ, Hermens JLM (2007) Nonlinear sorption of three alcohol ethoxylates to marine sediment: a combined Langmuir and linear sorption process. Environ Sci Technol 41:3192–3198. doi:10.1021/es062608z
41. Droge STJ, Sinnige TL, Hermens JLM (2007) Analysis of freely dissolved alcohol ethoxylate homologues in various seawater matrixes using solid-phase microextraction. Anal Chem 79:2885–2891. doi:10.1021/ac0620260
42. Chen Y, Droge STJ, Hermens JLM (2012) Analyzing freely dissolved concentrations of cationic surfactant utilizing ion-exchange capability of polyacrylate coated solid-phase microextraction fibers. J Chromatogr A 1252:15–22. doi:10.1016/j.chroma.2012.06.080
43. Haftka JJ, Scherpenisse P, Jonker MTO, Hermens JLM (2013) Using polyacrylate-coated SPME fibers to quantify sorption of polar and ionic organic contaminants to dissolved organic carbon. Environ Sci Technol 47:4455–4462. doi:10.1021/es400236a
44. Wang F, Chen Y, Hermens JLM, Droge STJ (2013) Evaluation of passive samplers with neutral or ion-exchange polymer coatings to determine freely dissolved concentrations of the

basic surfactant lauryl diethanolamine: measurements of acid dissociation constant and organic carbon-water sorption coefficie. J Chromatogr A 1315:8–14. doi:10.1016/j.chroma.2013.09.041
45. Vaes WH, Ramos EU, Hamwijk C et al (1997) Solid phase microextraction as a tool to determine membrane/water partition coefficients and bioavailable concentrations in in vitro systems. Chem Res Toxicol 10:1067–1072. doi:10.1021/tx970109t
46. Pörschmann J, Kopinke F-D, Pawliszyn J (1998) Solid-phase microextraction for determining the binding state of organic pollutants in contaminated water rich in humic organic matter. J Chromatogr A 816:159–167. doi:10.1016/S0021-9673(98)00525-1
47. Heringa MB, Hogevonder C, Busser F, Hermens JLM (2006) Measurement of the free concentration of octylphenol in biological samples with negligible depletion-solid phase microextraction (nd-SPME): analysis of matrix effects. J Chromatogr B Anal Technol Biomed Life Sci 834:35–41. doi:10.1016/j.jchromb.2006.02.009
48. Oomen AG, Mayer P, Tolls J (2000) Nonequilibrium solid-phase microextraction for determination of the freely dissolved concentration of hydrophobic organic compounds: matrix effects and limitations. Anal Chem 72:2802–2808
49. Caupos E, Touffet A, Mazellier P, Croue JP (2015) Partitioning of the pesticide trifluralin between dissolved organic matter and water using automated SPME-GC/MS. Environ Sci Pollut Res 22:4201–4212. doi:10.1007/s11356-014-3614-0
50. Ohlenbusch G, Kumke MU, Frimmel FH (2000) Sorption of phenols to dissolved organic matter investigated by solid phase microextraction. Sci Total Environ 253:63–74. doi:10.1016/S0048-9697(00)00376-4
51. Jiang R, Xu J, Lin W et al (2015) Investigation of the kinetic process of solid phase microextraction in complex matrix. Anal Chim Acta 900:111–116. doi:10.1016/j.aca.2015.09.010
52. Jiang R, Xu J, Zhu F et al (2015) Study of complex matrix effect on solid phase microextraction for biological sample analysis. J Chromatogr A 1411:34–40. doi:10.1016/j.chroma.2015.07.118
53. Ripszam M, Haglund P (2015) Automated method for determination of dissolved organic carbon-water distribution constants of structurally diverse pollutants using pre-equilibrium solid-phase microextraction. Environ Toxicol Chem 34:266–274. doi:10.1002/etc.2805
54. Endo S, Mewburn B, Escher BI (2013) Liposome and protein-water partitioning of polybrominated diphenyl ethers (PBDEs). Chemosphere 90:505–511. doi:10.1016/j.chemosphere.2012.07.069
55. Parry E, Young TM (2013) Distribution of pyrethroid insecticides in secondary wastewater effluent. Environ Toxicol Chem 32:2686–2694. doi:10.1002/etc.2347
56. Haftka JJH, Govers HAJ, Parsons JR (2010) Influence of temperature and origin of dissolved organic matter on the partitioning behavior of polycyclic aromatic hydrocarbons. Environ Sci Pollut Res Int 17:1070–1079. doi:10.1007/s11356-009-0263-9
57. Kramer NI, van Eijkeren JCH, Hermens JLM (2007) Influence of albumin on sorption kinetics in solid-phase microextraction: consequences for chemical analyses and uptake processes. Anal Chem 79:6941–6948. doi:10.1021/ac070574n
58. Ter Laak TL, Durjava M, Struijs J, Hermens JLM (2005) Solid phase dosing and sampling technique to determine partition coefficients of hydrophobic chemicals in complex matrixes. Environ Sci Technol 39:3736–3742. doi:10.1021/es048406p
59. Hawthorne SB, Grabanski CB, Miller DJ, Kreitinger JP (2005) Solid-phase microextraction measurement of parent and alkyl polycyclic aromatic hydrocarbons in milliliter sediment pore water samples and determination of K(DOC) values. Environ Sci Technol 39:2795–2803
60. Lee S, Gan J, Liu WP, Anderson MA (2003) Evaluation of K d Underestimation Using Solid Phase Microextraction. Environ Sci Technol 37:5597–5602
61. Leslie HAHA, Oosthoek AJPAJP, Busser FJM et al (2002) Biomimetic solid-phase microextraction to predict body residues and toxicity of chemicals that act by narcosis. Environ Toxicol Chem 21:229–234. doi:10.1897/1551-5028(2002)021<0229:BSPMTP>2.0.CO;2

62. Kopinke F-DD, Georgi A, Mackenzie K (2001) Sorption and chemical reactions of PAHs with dissolved humic substances and related model polymers. Acta Hydrochim Hydrobiol 28:385–399
63. Mayer P, Vaes WHJ, Wijnker F et al (2000) Sensing dissolved sediment porewater concentrations of persistent and bioaccumulative pollutants using disposable solid-phase microextraction fibers. Environ Sci Technol 34:5177–5183. doi:10.1021/es001179g
64. Birch H, Gouliarmou V, Lützhøft H-CH et al (2010) Passive dosing to determine the speciation of hydrophobic organic chemicals in aqueous samples. Anal Chem 82:1142–1146. doi:10.1021/ac902378w
65. Gouliarmou V, Smith KEC, de Jonge LW, Mayer P (2012) Measuring binding and speciation of hydrophobic organic chemicals at controlled freely dissolved concentrations and without phase separation. Anal Chem 84:1601–1608. doi:10.1021/ac2028497
66. Benhabib K, Town R, van Leeuwen H (2009) Dynamic speciation analysis of atrazine in aqueous latex nanoparticle dispersions using solid phase microextraction (SPME). Langmuir 25:3381–3386
67. Luers F, ter Hulscher TEM (1996) Temperature effect on the partitioning of polycyclic aromatic hydrocarbons between natural organic carbon and water. Chemsphere 33:643–657
68. Verhaar HJM, Busser FJM, Hermens JLM (1995) Surrogate parameter for the base-line toxicity content of contaminated water—simulating the bioconcentration of mixtures of pollutants and counting molecules. Environ Sci Technol 29:726–734. doi:10.1021/es00003a021
69. Parkerton TF, Stone MA (1995) Ecotoxicity on a stick: A novel analytical tool for predicting the ecotoxicity of petroleum contaminated samples. In: 17 annu. meet. soc. environ. toxicol. chem. Society of Environmental Toxicology and Chemistry, Pensacola, FL (United States), p 378
70. Mayer P, Vaes WHJ, Wijnker F et al (2000) Sensing dissolved sediment porewater concentrations of persistent and bioaccumulative pollutants using disposable solid-phase microextraction fibers. Environ Sci Technol 34:5177–5183. doi:10.1021/es001179g
71. Heringa MB, Hermens JLM (2003) Measurement of free concentrations using negligible depletion-solid phase microextraction (nd-SPME). TrAC Trends Anal Chem 22:575–587. doi:10.1016/S0165-9936(03)01006-9
72. Ter Laak TL, Durjava M, Struijs J, Hermens JLM (2005) Solid phase dosing and sampling technique to determine partition coefficients of hydrophobic chemicals in complex matrixes. Environ Sci Technol 39:3736–3742. doi:10.1021/es048406p
73. Mayer P, Karlson U, Christensen PS et al (2005) Quantifying the effect of medium composition on the diffusive mass transfer of hydrophobic organic chemicals through unstirred boundary layers. Environ Sci Technol 39:6123–6129
74. Ter Laak TL, Busser FJM, Hermens JLM (2008) Poly(dimethylsiloxane) as passive sampler material for hydrophobic chemicals: effect of chemical properties and sampler characteristics on partitioning and equilibration times. Anal Chem 80:3859–3866. doi:10.1021/ac800258j
75. Jiang R, Lin W, Wen S et al (2015) Development of a full automation solid phase microextraction method for investigating the partition coefficient of organic pollutant in complex sample. J Chromatogr A 1406:27–33. doi:10.1016/j.chroma.2015.06.018
76. Mäenpää K, Leppänen MT, Reichenberg F et al (2011) Equilibrium sampling of persistent and bioaccumulative compounds in soil and sediment: Comparison of two approaches to determine equilibrium partitioning concentrations in lipids. Environ Sci Technol 45:1041–1047. doi:10.1021/es1029969
77. Kopinke FD, Porschmann J, Remmler M (1995) Sorption behavior of anthropogenic humic matter. Naturwissenschaften 82:28–30
78. Kopinke F-D, Pörschmann J, Georgi A (1999) Application of SPME to study sorption phenomena on dissolved humic organic matter. In: Applications of solid phase microextraction. The royal society of chemistry

79. Poon KF, Lam PKS, Lam MHW (1999) Determination of polynuclear aromatic hydrocarbons in human blood serum by proteolytic digestion - Direct immersion SPME. Anal Chim Acta 396:303–308. doi:10.1016/S0003-2670(99)00447-X
80. Poon KF, Lam PKS, Lam MHW (1999) Determination of polychlorinated biphenyls in human blood serum by SPME. Chemosphere 39:905–912. doi:10.1016/S0045-6535(99)00033-8
81. Zielińska K, Van Leeuwen HP, Thibault S, Town RM (2012) Speciation analysis of aqueous nanoparticulate diclofenac complexes by solid-phase microextraction. Langmuir 28:14672–14680. doi:10.1021/la303143w
82. Zielińska K, van Leeuwen HP (2013) Role of nanoparticles in analytical solid phase microextraction (SPME). Environ Chem 10:120. doi:10.1071/EN13004
83. Xu J, Huang S, Wu R et al (2015) Bioinspired polydopamine sheathed nanofibers for high-efficient in vivo solid-phase microextraction of pharmaceuticals in fish muscle. Anal Chem 87:3453–3459. doi:10.1021/ac5048357
84. Alam MN, Ricardez-Sandoval L, Pawliszyn J (2015) Numerical modeling of solid-phase microextraction: binding matrix effect on equilibrium time. Anal Chem 87:9846–9854. doi:10.1021/acs.analchem.5b02239
85. Jeannot M, Cantwell F (1997) Solvent microextraction as a speciation tool: determination of free progesterone in a protein solution. Anal Chem 69:2935–2940
86. Leeuwenl HVAN, Cleven R, Buffle J (1989) Voltammetric techniques for complexation measurements in natural aquatic media. Role of the size of macromolecular liginds and dissociation kinetic of complexes. Int Union Pure Appl Chem 61:255–274
87. Ter Laak TL, Van Eijkeren JCH, Busser FJM et al (2009) Facilitated transport of polychlorinated biphenyls and polybrominated diphenyl ethers by dissolved organic matter. Environ Sci Technol 43:1379–1385. doi:10.1021/es802403v
88. Xu J, Huang S, Jiang R et al (2016) Evaluation of the availability of bound analyte for passive sampling in the presence of mobile binding matrix. Anal Chim Acta 917:19–26. doi:10.1016/j.aca.2016.02.039
89. Smith KEC, Thullner M, Wick LY, Harms H (2011) Dissolved organic carbon enhances the mass transfer of hydrophobic organic compounds from nonaqueous phase liquids (NAPLs) into the aqueous phase. Environ Sci Technol 45:8741–8747. doi:10.1021/es202983k
90. Ramus K, Kopinke FD, Georgi A (2012) Sorption-induced effects of humic substances on mass transfer of organic pollutants through aqueous diffusion boundary layers: the example of water/air exchange. Environ Sci Technol 46:2196–2203. doi:10.1021/es2038382
91. Van Eijkeren JCH, Heringa MB, Hermens JLM (2004) Modelling SPME data from kinetic measurements in complex samples. Analyst 129:1137–1142. doi:10.1039/b407926d
92. Zhang X, Oakes KD, Hoque E et al (2011) Pre-equilibrium solid-phase microextraction of free analyte in complex samples: correction for mass transfer variation from protein binding and matrix tortuosity. Anal Chem 83:3365–3370
93. Ter Laak TL, ter Bekke MA, Hermens JLM (2009) Dissolved organic matter enhances transport of PAHs to aquatic organisms. Environ Sci Technol 43:7212–7217. doi:10.1021/es903149f

Chapter 5
Application of Solid Phase Microextraction in Aqueous Sampling

Siming Huang, Juan Zheng and Gangfeng Ouyang

Abstract Various pollutants have been emitted due to the development of industry, which can cause serious problems for the production of drinking water and threaten the health of human. Aquatic monitoring has attracted more and more attention and needs to be developed urgently for the analysis of trace pollutants from complex matrix. Solid-phase microextraction (SPME), as a solventless extraction technique, is applicable for both off-site and on-site sampling. Here, two general strategies for on-site SPME sampling, including active and passive sampling, are introduced in detail. The main formats for active sampling include fiber, thin film, and stir bar, while the reported formats for passive sampling are fiber, polydimethylsiloxane (PDMS) rod, PDMS membrane, and PDMS tubing. To obtain better determination effects of emerging contaminants, new adsorptive materials have been employed gradually instead of commercial adsorbents to achieve the targets of high enrichment efficiencies and short analysis time. All in all, the SPME samplers have been diversified and developed over the past decade to obtain more information about the pollution status of water environment all over the world.

Keywords Solid-phase microextraction · Active sampling · Passive sampling · Fiber · Thin film · Stir bar · Polydimethylsiloxane

5.1 Introduction

After more than 25 years, solid-phase microextraction (SPME) has gained widespread acceptance as a flexible and well-automatable microextraction technology, for the analysis of pollutants or metabolites in water, air, sediment, plants, and animals [1–5]. With regard to water analysis, rapid sampling and sensitive detection

S. Huang · J. Zheng · G. Ouyang (✉)
MOE Key Laboratory of Aquatic Product Safety/KLGHEI of Environment and Energy Chemistry, School of Chemistry, Sun Yat-Sen University, Guangzhou 510275, PR, China
e-mail: cesoygf@mail.sysu.edu.cn

G. Ouyang and R. Jiang (eds.), *Solid Phase Microextraction*,
DOI 10.1007/978-3-662-53598-1_5

of contaminants—organic, inorganic as well as microorganisms—in drinking, recreational, ambient waters, wastewater, and seawater are critical for addressing pollution, providing adequate treatment solutions, and estimating potential public health risks upon exposure. Therefore, aquatic monitoring is still an ongoing challenge and a key issue in a plethora of researches.

Many analytical procedures can be used for water analysis, such as purge and trap, static headspace analysis, solid-phase extraction (SPE), liquid-phase extraction, dispersive liquid–liquid microextraction (DLLME), and SPME [6–8]. Purge and trap is an exhaustive extraction method and as such can extract all the purgeable compounds from the sample that may interfere with the quantification. SPE is the most popular means with the usage of organic solvents. DLLME is another new extraction technique that minimizes solvent usage and requires very small water samples. However, this method is not suitable for on-site monitoring. SPME, a solventless extraction technique, shows the advantages of automated, rapid, multimode, and applicable for off-site and in situ sampling.

SPME sampling in the aqueous samples can be operated in laboratory with direct immersion, headspace, membrane-protected or in-tube modes. In consideration that stirring and elevated temperature can improve the uptake kinetics and mass transfer, stir bar, agitator, and heat are usually used. SPME sampling also can be carried out in the field without the transportation and storage of a large number of water samples from the field to the laboratory, eliminating the possible losses of target analytes and providing more representative quantitative results of the original sample characteristics. And the sampling procedure requires very short extraction time. Thus, the on-site SPME sampling technique with designed samplers is of great interest.

There are two general strategies for the on-site SPME sampling, that is, active and passive sampling. Some new designed and developed on-site SPME samplers and calibration methods have been discussed in many reports. On-site SPME sampling devices which integrate sampling with sample isolation and sample enrichment are based on different calibration methods, including traditional, equilibrium extraction, and several diffusion-based approaches. And the diffusion-based approaches have been developed for passive sampling, because passive sampling is based on the completely free flow of analyte from the sample into the trap which follows Fick's first law of diffusion. And the sampling rate is controlled by the molecular diffusion coefficient of the analyte and the parameters of the sampler device. Passive sampling devices allow long-term monitoring of pollutant levels in aquatic systems. Compared with passive sampling, active sampling offers rapid sampling and sample preparation using electrically powered samplers. It becomes a preferable choice for on-site sampling because it allows for better control of the sampling rate.

5.2 SPME Active Sampling

Active sampling is based on the provision of a driving force such as pumping, forced stirring, or agitation. It performs more frequently in laboratory than on-site because the agitation or stirring in laboratory is easier to carry out. However, the on-site active sampling with active sampler increases the sampling rate and decreases the sampling time, which can control the agitation speed with a power supported device for complex field sampling. Therefore, active sampling has been a very much preferred approach in most sample extraction applications in recent time. For example, a portable electric drill and a battery-operated electric toothbrush as agitators in the field sampling are used for real sample analysis. The PDMS fiber or thin film is attached to the electric drill at a constant rotation speed for on-site extraction. The portable agitators provide a controllable rotation speed and accelerate the distributions of analytes in the coatings, leading to simple, rapid, on-site extraction.

5.2.1 *Fiber Format SPME Active Sampler*

For SPME sampling, the fiber configuration is the most commonly used mode for the analysis of environmental, biological, and food samples, and the most suitable one for automation and miniaturization. Among the various kinds of commercial SPME coatings, such as PA, CAR/PDMS, PDMS/DVB, and carbowax, PDMS is commonly used for the analytes with greater partition coefficients, such as all PAHs. A new active water sampling method using SPME fiber with PDMS coating for the analysis of environmental contaminants was proposed [9]. As shown in Fig. 5.1, two Teflon disks were attached to a metal shaft using small screws. The fiber could be screwed into the top disk. A small hole in the bottom disk was used to secure the position of the fiber. The top disk was movable and could be adjusted to expose or withdraw the fiber during or after sampling. When the drill was turned on, the fiber rotated and the water was agitated. The Teflon disks prevented the fiber from spinning outward during sampling. This method was compared with the thin-film microextraction. The results showed that using the SPME fiber required longer equilibrium time, and less extracted amount of analytes were detected due to the smaller volume of fiber. Therefore, this new designed active SPME sampler was not deployed for the on-site sampling in this work.

However, in another study, a similar but modified active sampler using a homemade SPME fiber tip with an electric drill was developed, for the sampling of 16 PAHs and 9 organochlorine pesticides (OCPs) in the seawater of Pearl River Estuary [10]. As shown in Fig. 5.2, a one-meter-long aluminum rod took place of the electric drill bit which was rotated with the portable electric drill, and then a waterproofing fabric stuck to the other side of the rod. The homemade PDMS fibers were uniformly inserted into the fabric. After sampling, the fibers could be removed

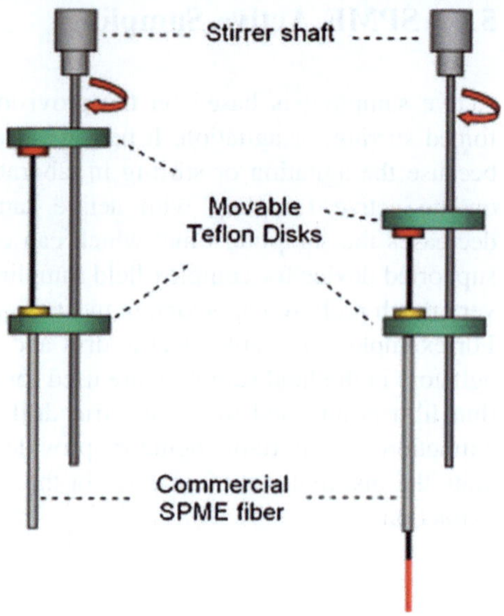

Fig. 5.1 New type of active SPME sampler showing the protected position of the fiber before and after sampling, and the exposed position of the fiber during sampling. Reprinted with the permission from Ref. [9]. Copyright 2009, with permission from Elsevier

and stored easily. In this study, the calibration method was called sampling-rate (SR) calibration, which was assumed that the rate of mass transfer or sampling rate remained constant in the sampling procedure. Sampling rates of PAHs and OCPs were determined in the laboratory, which were obtained at different levels of concentrations and extraction temperatures in 1-L spiked artificial seawater samples. The applicability and accuracy of the proposed technique were verified with traditional liquid–liquid extraction (LLE) method.

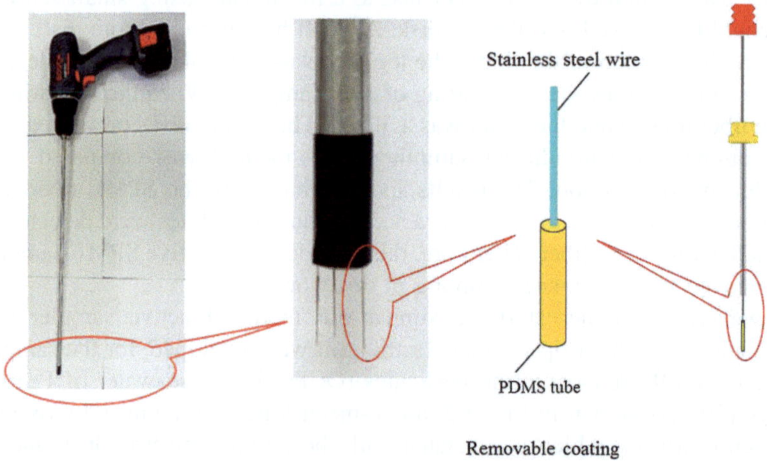

Fig. 5.2 Scheme of the developed on-site active sampler. Reprinted with the permission from Ref. [10]. Copyright 2015, with permission from Elsevier

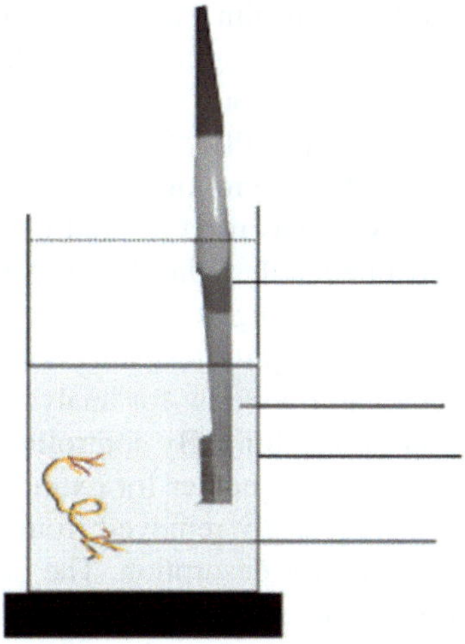

Fig. 5.3 Schematic of on-site extraction with a handheld battery-operated electric toothbrush as agitator. Reprinted with the permission from Ref. [11]. Copyright 2011, with permission from Elsevier

Apart from using an electric drill, a handheld battery-operated electric toothbrush as a novel on-site sampler with functional polymer-coated fibers was developed by Basheer et al. [11]. The toothbrush with a replaceable rotating power head was purchased from a local market without any modification (Fig. 5.3). A kind of functional conjugated polymer (2-(9,9-bis(6-bromo-ethylhexyl) 9-H-fluoren-2-yl)benzene-1,4-diamine) was coated on commercial Technora fibers (each strand consisted of 1000 filaments, each of diameter ca. 9.23 μm) which were then used for extraction. After extraction, the fibers were sealed in the vial and stored in an icebox until detection in the laboratory. We think it is not an absolute on-site sampling as the fibers and the toothbrush are separated, and the sampling procedures are operating off the shore, which the 10 mL seawater is collected in a beaker. Nevertheless, the method needs only a small amount of sample and solvent for extraction and desorption, respectively. Moreover, it avoids the transportation of water samples back to the laboratory.

5.2.2 Thin-Film Format SPME Active Sampler

Thin-film microextraction (TFME) with higher surface-to-volume ratio of extraction phase, which is typically 25–125 times larger than fiber geometry, has been developed in the last decades. It offers the analytical method with high extraction efficiency and sensitivity in a very short period of extraction time [12, 13]. The simple, durable TFME samplers have been successfully applied to the field water sampling, and it produces results similar to the traditional techniques which require extensive sample preparation [14, 15]. And it has been widely used for active and passive sampling of PAHs, pesticides, phenols, and nitrosamines in aqueous matrices [14–16].

Qin et al. utilized PDMS thin-film samplers coupled with an electric drill at a constant rotation speed for constant agitation in order to sample PAHs in aqueous samples rapidly [9, 17]. In a previous study, a prototype of active sampler with PDMS thin film was designed. The PDMS thin film was cut into a house-like shape as 2 cm × 2 cm square with a 1-cm-high triangle on top (5 cm^2 for one side). A piece of stainless steel wire was used to hold the thin film for easy movement. And in order to secure and flatten the thin film, a special framed holder as shown in Fig. 5.4 was built based on the design of the upper part of a snap pill jar. The thin film was fixed to the holder by tightly fastening the cap and jar part. A round hole was cut in the cap, providing a window for analyte to access the thin film. A rod was attached near the cap for handling. By controlling the movement of the rod, the thin film could be rotated with the holder for extraction with very flat shape. After sampling, the thin film can be easily removed from the holder by opening the cap and then pushed into the liner for desorption. The performance of TFME for PAH analysis including extraction rate, equilibration time, and sensitivity was investigated in 1L aqueous solution and a flow-through system. The results showed that the thin-film extraction reached equilibrium after 2 h which was dependent on the thickness of thin-film device (0.127 mm). Moreover, increasing agitation speed could improve the extraction efficiency. Although the thin film is flexible, the device with special designed holder could be operated under high rotation speed, such as 1400 rpm. Afterward, the device was deployed for the analysis of fluorene, anthracene, fluoranthene, and pyrene in Laurel Creek. The detected concentrations were similar to those determined by laboratory analyses using fiber SPME method, indicating a strong agreement between the two methods and demonstrating a feasibility of TFME method for field sampling.

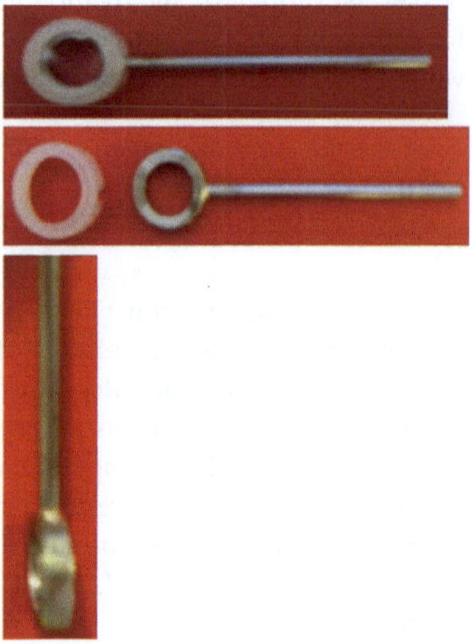

Fig. 5.4 Specially designed thin-film framed holder. Reprinted with the permission from Ref. [17]. Copyright 2008, with permission from Elsevier

In another study, a more convenient and effective TFME device as an active sampler for PAHs analysis in water sample was developed. The thin film was also cut into the shape of a house, as Ref. [17] mentioned, and positioned in the copper mesh pocket attached to the handing rod of the drill. The new designed thin film sampler and the application for on-site sampling were shown in Fig. 5.5. Step (a) depicted that a window was milled in the left part of a piece of copper mesh and another smaller window was milled in the right part. The left window was used to permit analytes' access to the thin film for extraction and to allow the thin film to go into the liner after extraction. The smaller window was just big enough to allow the passage of a small screw driver. It was smaller because it needed to support the thin film to prevent it from falling off during rotation and sampling. Step (b) involved folding the copper mesh along the dashed line and sealing the two edges by soldering. A pocket with one opening and two differently sized windows on two sides was thus formed. After placing the thin film in the pocket, the pocket was connected to a metal rod by two small screws. The rod could be extended or shortened depending on the specific requirements for on-site sampling. Step (c) represented the final thin-film field sampler, which resembled a small racket. When the extraction was completed, the thin film could be pushed from the small window through the big window into the liner by the small screw driver as in step (d). Finally, in step (e), capping two ends of the liner to seal the thin film provided an

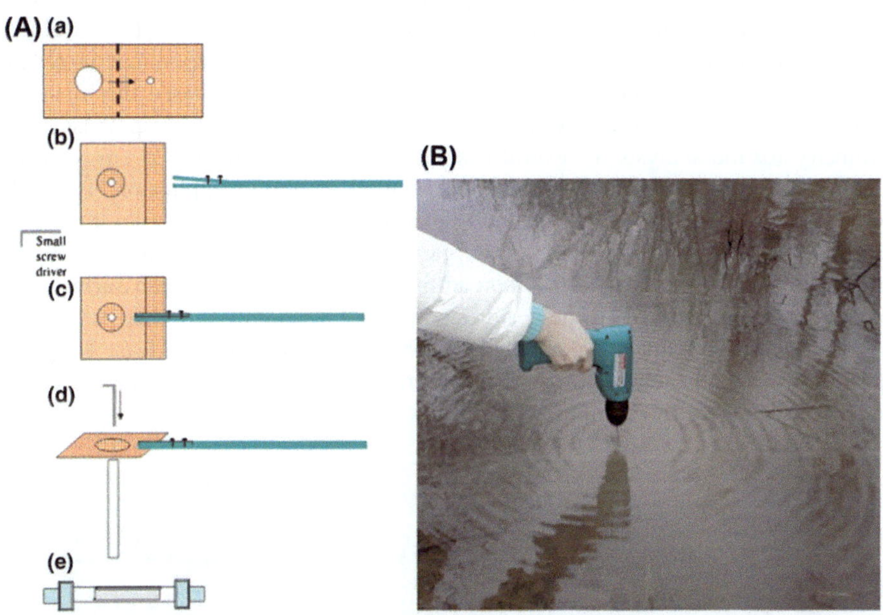

Fig. 5.5 **A** Schema of a thin-film copper mesh sampler: *a* Fold (*b*, *c*) fix on a rod (*d*) push into a liner (*e*) seal the liner for transportation; **B** on-site sampling of river water with the thin-film sampler. Reprinted with the permission from Ref. [9]. Copyright 2009, with permission from Elsevier

effective mechanism for preserving the integrity of the sample and for preventing contamination. The sampler was rotated vertically 0.2–0.5 m below the water surface with the portable drill at 600 rpm. And the sampling time was less than 6 min, which addressed the requirement of rapid sampling. The experimental values demonstrated that there was no significant difference in the concentration measured by the thin film and by the fiber geometry. In this study, sampling-rate calibration method was proposed to quantify the extraction of PAHs. The mass uptake predicted by the calibration method compares well with experimental mass uptake. This investigation confirmed that thin film could be a successful active sampler for determining the concentration of organic pollutants.

In summary, the advantages of the thin film are reflected in a shorter equilibration time and a higher extraction rate due to its thinner extraction phase and larger surface area. And it was successfully applied for the water sampling in the field. To reduce the extraction time, a handheld battery-operated drill was used simultaneously. Using this PDMS design, Qin et al. were able to demonstrate that a 10 cm^2 membrane provided approximately 10× the extraction efficiency for fluoranthene and pyrene compared to commercially available PDMS sorptive stir bar technology (area = 1 cm^2).

Despite the marked improvement, such preconstructed membranes are limited to PDMS in terms of available extraction solvent. Additionally, such large membranes require a firm supporting frame for direct sampling with agitation. Further works in the area have explored the in-house preparation of particle-loaded PDMS membranes and implementation of inert support materials. Grandy et al. [18] developed a thin-film membrane by the use of a high-density PDMS prepolymer in combination with DVB particles, which were spread onto a carbon-based mesh support coupled with hand-portable gas chromatography toroidial iron trap mass spectrometry for the analysis of several pesticides. An approximately 25 cm × 60 cm sheet of the carbon mesh was cut and secured to a film applicator. Then, the prepared coating mixture was manually placed in a thin strip along the top of the carbon mesh sheet. The coating bar gap was adjusted to the thinnest setting available and then used to slowly spread the mixture across the mesh surface. The coating was then cured inside a nitrogen-purged vacuum oven at a pressure of −15 mmHg (approximately), and at 190–200 °C for a period of at least 16 h. The procedure was repeated to spread the mixture onto the other side of the mesh. The device was shown in Fig. 5.6.

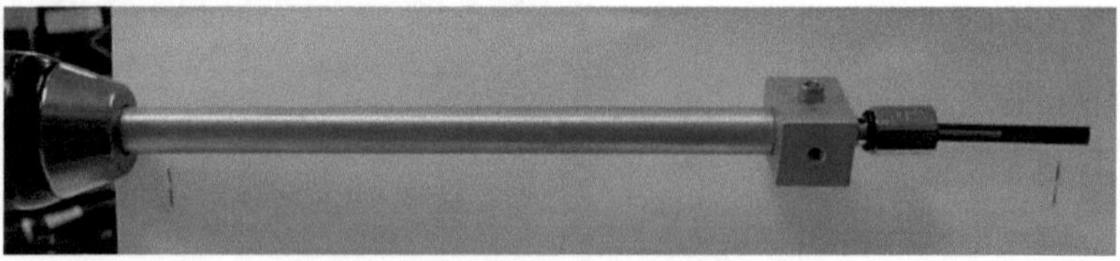

Fig. 5.6 4 cm × 4.85 cm DVB/PDMS/carbon mesh TME membrane couple with a modified power drill setup holding. Reprinted with the permission from Ref. [18]. Copyright 2016 American Chemical Society

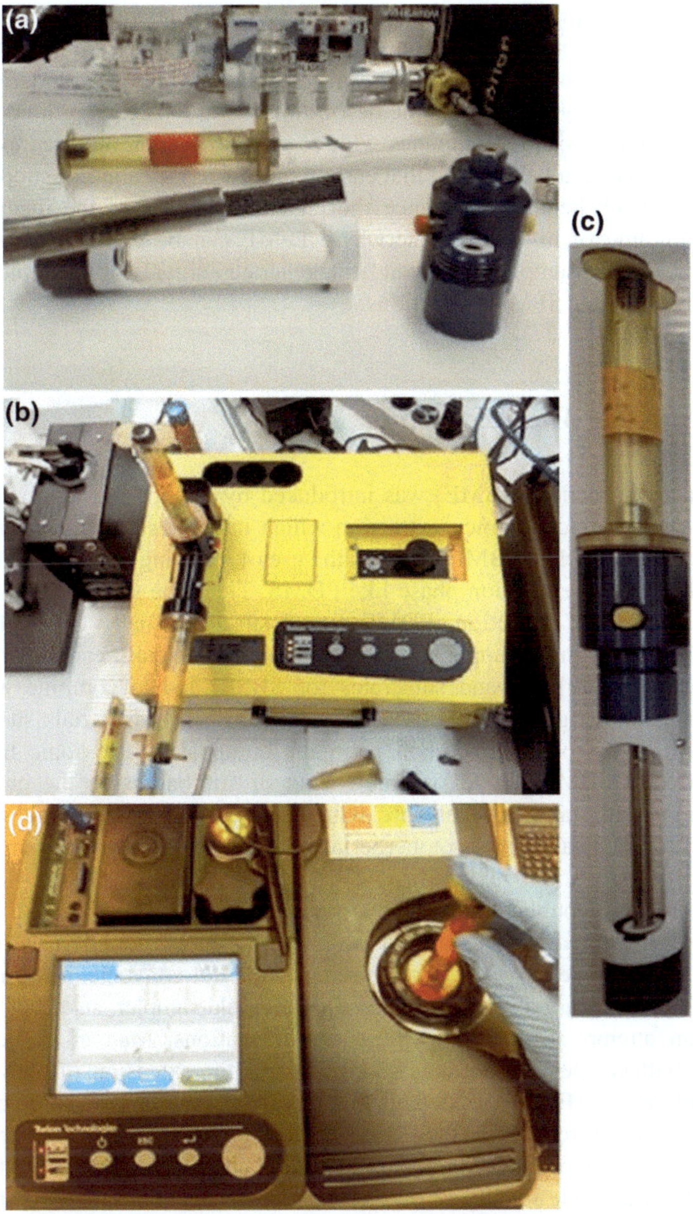

Fig. 5.7 Desorption of TFME membranes onto the portable desorption module: **a** Insertion of membrane into 3.5″ sorbent tube and into conventional trap holder; **b** Transfer of analytes from TFME membrane to the needle trap using SPS-3 desorption unit (in breakthrough test configuration); **c** Non-leaking linkage between the sorbent tube and 19-gauge needle trap device; **d** Desorption of needle trap onto portable GC-TMS for separation and analysis. Reprinted with the permission from Ref. [18]. Copyright 2016 American Chemical Society

The on-site sampling in lake was performed with a relatively short extraction time of 10 min under the constraint of battery power. After sampling, the membrane was inserted into a sorbent tube for desorption, which was undertaken with the use of the prototype high-volume desorption module (Fig. 5.7).In addition, the rigidity of different sized membranes with or without carbon mesh supported was investigated. The results indicated that carbon mesh was essential for supporting the membrane, which helped to enhance the physical stability of the membrane for reliable sampling of high-flow waterways. TFME can be used to perform much more rapid on-site sampling while it still generates signals comparable to those attained from longer SPME extractions.

5.2.3 Stir Bar Microextraction

Stir bar microextraction (SBME) was introduced by Sandra's group in 1999, presenting a special mode of microextraction which integrates the extraction element and the stirring unit [19]. SBME involves the use of a stirring bar that is coated with a relatively layer of extraction phase like PDMS. Then, the stir bar is added to the aqueous sample for extraction. PDMS-coated stir bar usually called Twister is widely employed to most applications due to its commercial availability [20–23]. To develop new sorbents and substrates for SBSE, and to combine with other extraction methods, great efforts have been made. Several materials such as graphene, hydrophilic polymer, carbon nanotube composites, and ionic liquids are introduced into SBME [24–27]. A new kind of stir bar using the cross-linked polymer-modified stainless steel wire with chemical resistance as substrate was developed for the analysis of protoberberine alkaloids in herbs and plasma samples [28].

There are fewer publications about on-site applications of SBME to real samples. The possible reasons might be as follows: (1) the commonly used large size magnetic stirrer powered with alternating current is not convenient to carry or use on-site; (2) stir bars are difficult to be fixed on the field sampler, such as the electric drill. In an attempt to overcome these two limitations, Mao et al. designed a miniature battery-operated portable electric stirrer for extension of SBME in the field sampling [29]. Figure 5.8 depicts that an electric stirrer was soldered together with a magnetic stir rod, where the PDMS-coated stir bar was fixed at the bottom by strong magnetic force. The stirring speed could reach 2000–6000 rpm with the use of an external portable intelligent digital tachometer. The device can be processed not only for off-site analysis in laboratory with different extraction modes including direct immersion, headspace, and continuous flow, but also for on-site field sampling. And the results obtained by on-site SBSE sampling were in good agreement with those obtained by off-site SBSE sampling. The LODs results of the target PAHs compared to other reported literatures using the methods of tube-SPME-HPLC-FLD, HS-SDME-HPLC-FLD, and PDMS-SBSE-TD-GC-MS showed that the proposed PDMS-SBSE-GC-FLD method was the most sensitive

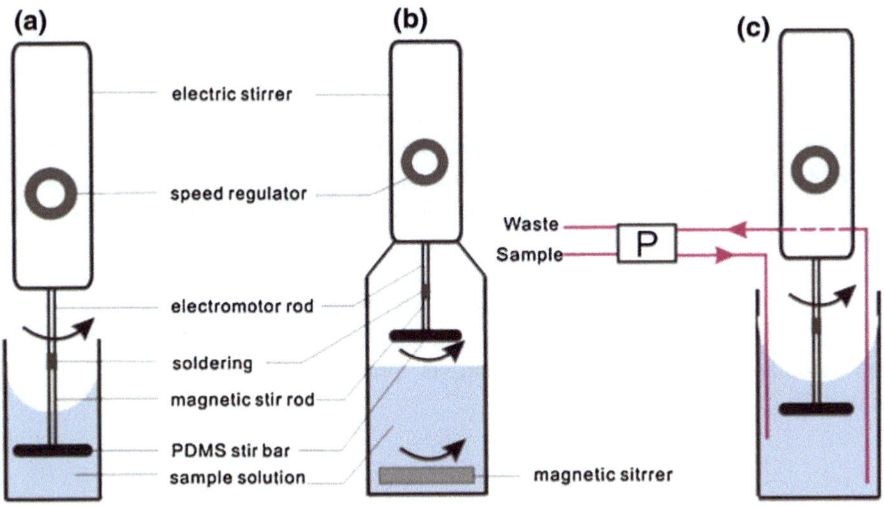

Fig. 5.8 Schematic diagram of the designed portable SBSE sampling device. Reprinted with the permission from Ref. [29]. Copyright 2012, with permission from Elsevier

method for PAHs determination. As for on-site sampling, the device at the speed of 4000 rpm was successfully employed for the sampling of PAHs in East Lake water in China. The homemade portable electric stirrer is simple, easy to operate, user friendly, and of low cost. And the stir bar is very convenient to fix and remove. Thus, it provides a great idea for the design device to be commercialized.

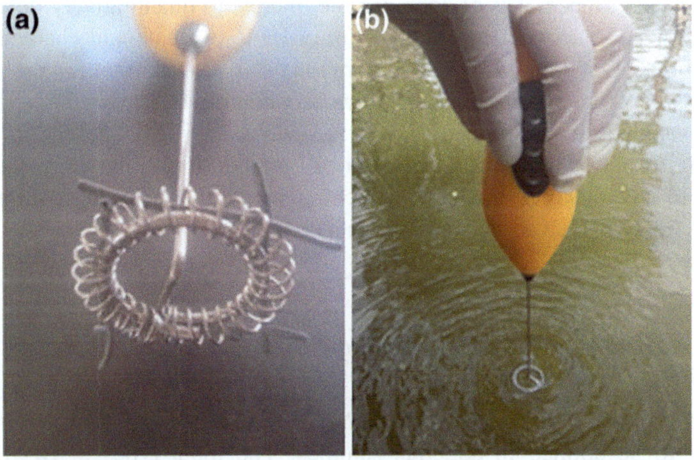

Fig. 5.9 Schematic of HF/GBs portable sampler (**a**) and on-site sampling and sample preparation with HF/GBs portable sampler (**b**). Reproduced from Ref. [30] by permission of John Wiley & Sons Ltd

A hollow fiber/graphene bar (HF/GB) was prepared for sampling four nitrobenzene compounds in lake water, with a handheld battery-operated electric egg beater as supporting frame by Xing et al. [30]. As shown in Fig. 5.9, four HF/GBs were fixed onto the electric egg beater. During on-site sampling, the sampler was placed in the lake water and the egg beater was started quickly, which lasted for 5 min. After sampling, the stir bars were placed in airtight vials, stored in an icebox, and brought back to laboratory for analysis. The approach is excellent for the following features: (1) the HF/GBs stir bar preparation is simple because only ultrasonic is required; (2) the handheld battery-operated egg beater is easy to obtain and cheap; (3) the device is small in size, easy to carry, and convenient to operate; (4) the sampling process has a short time, and the stir bars can be used repeatedly. In summary, this proposed method is practical for routine environmental monitoring.

5.3 SPME Passive Sampling

Persistent organic pollutants (POPs) have been increasingly emitted into the environment through wastewater, agricultural runoff, and industrial processes [31]. In consideration, POPs have caused serious problems for the production of drinking water, which could threaten the health of human [32]. More and more research works focusing on the detections and quantifications of POPs have been published in recent years. In the detection results, the freely dissolved concentrations and time-weighted average (TWA) concentrations have drawn great attention. The freely dissolved concentrations of POPs determine the main processes, such as transport, distribution, and bioaccumulation, and thus exhibit their potential toxic effects [33, 34]. Hence, it is of considerable interest to quantitatively determine the freely dissolved concentrations of POPs in order to evaluate and characterize their bioavailability. Obtaining TWA concentrations is also important for the ecological risk assessment process because chemical concentration is varying over time in aquatic systems. As mentioned above, passive sampling is based on free flow (according to Fick's first law of diffusion) of analyte molecules from the sampled medium to a collecting medium, which appears to meet the requirements of detecting freely dissolved and TWA concentrations.

5.3.1 *Formats of SPME Passive Samplers*

Among all kinds of passive samplers, SPME-based passive sampler is an ideal candidate to be applied in aqueous samples. The main types of SPME passive samplers are fiber, PDMS rod, PDMS membrane, and PDMS tubing (Table 5.1), which is given in detail as follows.

Table 5.1 Novel compounds sampled by passive samplers based on SPME

Analytes	Passive sampler	Detection	References
PAHs, PCBs	PDMS fiber/tubing	GC/MS	[35]
Lauryl diethanolamine	C18/SCX-SPME fiber	LC-MS/MS	[36]
PCBs, OCPs	PDMS	GC/MS	[37]
PBDEs, triclosan, methyltriclosan	PDMS	GC/MS	[38]
PAHs, PCBs	PDMS	GC/MS	[39]
PCBs	Silicone rubber	GC/MS	[40]
Pyrethroids	PDMS	GC/MS	[41]
PAHs	PDMS rod	GC/MS	[42]
PAHs	PDMS rod/membrane	GC/MS	[43]
Polar compounds, pesticides	PA	HPLC/UV, GC/MS	[44]
OCPs, PCBs, PAHs	membrane-enclosed sorptive coating sampler	GC/MS	[45]
Pharmaceuticals, industrial chemicals, hormones, pesticides	PA	HPLC/UV	[46]
PAHs	PDMS thin film	GC/MS	[9]
PAHs	PDMS thin film	GC/MS	[9]
Fragrances, phosphate flame retardants, pharmaceuticals, PAHs, PCBs, PBDEs, and pesticides	PDMS	GC/FID	[47]
PCBs, PAHs, OCPs	PDMS	GC/MS	[48]

5.3.1.1 Fiber Geometry

The passive sampler using traditional SPME fiber is exposed in aqueous samples directly for rapid sampling. The fiber coating is usually protected by a robust covering layer from physical collision with large objects, and the layer also slows bacterial growth (Fig. 5.10). Zeng et al. successfully developed the traditional SPME fiber-based sampling device employing PDMS-coated (100 μm) fiber as the sorbent phase, and field tests proved the feasibility of the passive device [49]. However, the sampling time required to reach equilibrium is of the order of several seconds and minutes for volatile organic compounds and of the order of several minutes up to half an hour for POPs. Thus, this type of sampling cannot be used for long-term monitoring, and the results obtained are only comparable with those obtained by grab sampling.

More and more works have appeared and emphasized the importance of SPME fibers as passive samplers in the detection of freely dissolved concentrations of POPs in the past decades due to remarkable advantages [50, 35, 36]. The SPME fiber-based passive sampler is easy to be deployed and operated by non-scientists.

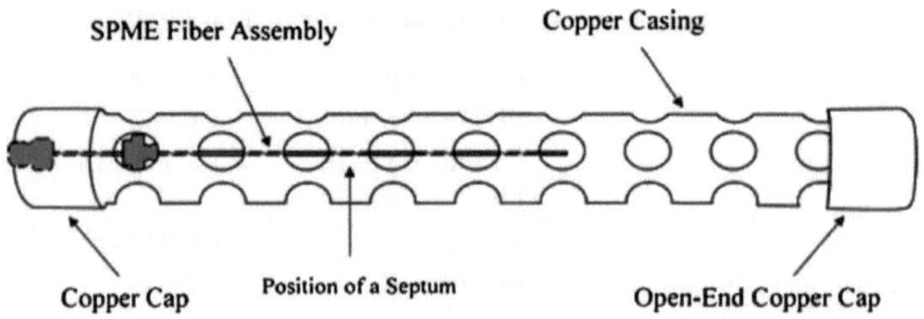

Fig. 5.10 Schematic of traditional SPME fiber-based sampling device. Reprinted with the permission from Ref. [49]. Copyright 2004 American Chemical Society

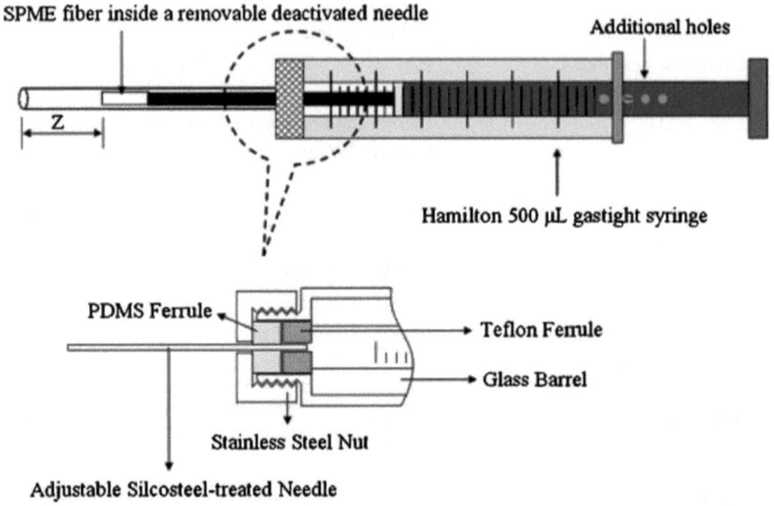

Fig. 5.11 Schematic diagram of the fiber-in-needle SPME device for TWA water sampling and the adjustable/removable needle. Reprinted with the permission from Ref. [51]. Copyright 2005 American Chemical Society

The use of SPME fiber-based passive sampler simplifies the analytical procedure at the most critical step of sampling, that is to say, sampling, extraction, and concentration are combined into one step. After concentration, the SPME fiber can be directly injected into a gas chromatograph (GC) or high-performance liquid chromatograph (HPLC) for analysis without further treatment. The cost of transport and storage is greatly reduced because of the tiny structure of SPME fiber.

The traditional SPME fiber-based passive device was modified by Ouyang et al. using a 500-μL gas-tight syringe and a removable needle (Fig. 5.11) for TWA water sampling, called fiber-in-needle SPME [51]. It is worth mentioning that the air in the needle can be removed easily by repeating exposure and retraction of the plunger for fiber-in-needle SPME to act as a diffusive sampler in water. Except for the same advantages of traditional SPME fiber-based passive device, fiber-in-needle

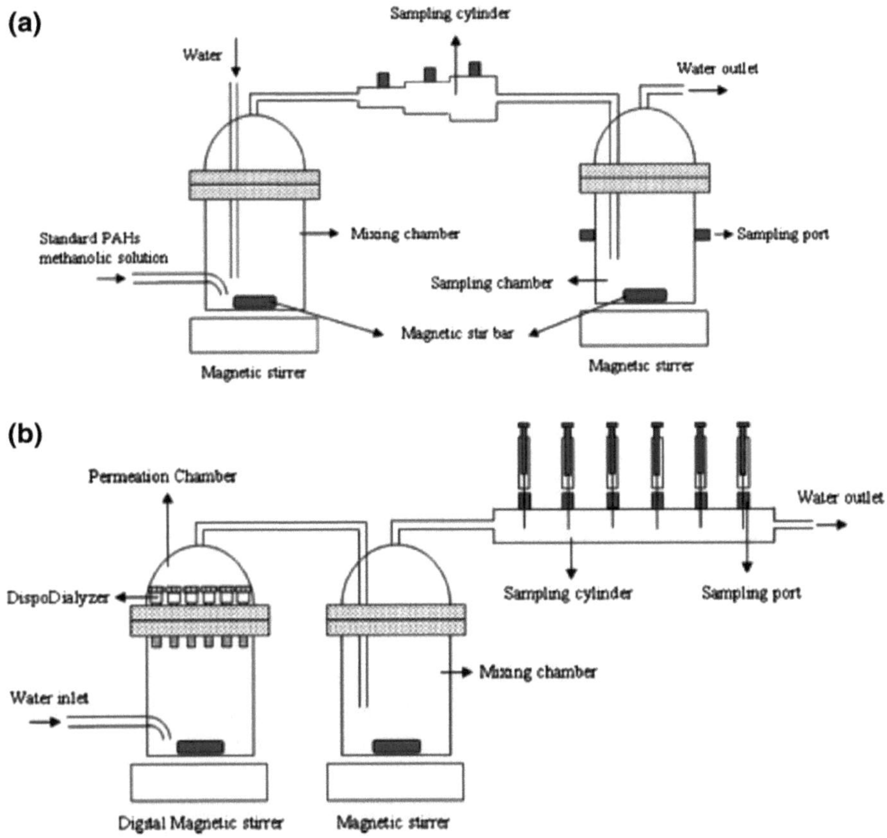

Fig. 5.12 **a** Schematic diagram of the flow-through system based on dilution. **b** Schematic diagram of the new flow-through system based on permeation. Reprinted with the permission from Ref. [51]. Copyright 2005 American Chemical Society

SPME device is independent of face velocity and affected only slightly by temperature, and the position of the SPME fiber. The obvious disadvantage of fiber-in-needle SPME device is the limited sampling rate, which leads to less extraction of analytes. However, rapid sampling is unessential to obtain TWA concentrations over weeks or months.

Subsequently, a flow system combined with SPME fiber for the TWA water sampling experiment was presented by Ouyang et al., as shown in Fig. 5.12, which consisted of a permeation chamber, a mixing chamber, and a long sampling cylinder with six sampling ports [9]. The key parts of the system were the DispoDialyzers (Spectrum Laboratories) in the permeation chamber, partially filled with pure standards (naphthalene, acenaphthene, fluorene, anthracene, fluoranthene, pyrene) and pure water. Then, the DispoDialyzers were deployed in the permeation

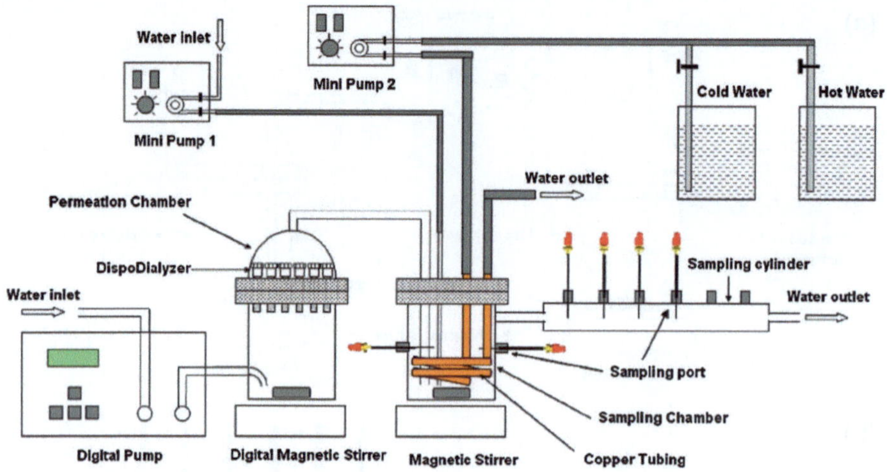

Fig. 5.13 Modified flow-through system for the generation of standard PAHs in aqueous solution. Reprinted with the permission from Ref. [52]. Copyright 2009 American Chemical Society

chamber, and the dissolved analytes were diffused from the membrane of the DispoDialyzer to the sampling cylinder.

The flow-through system is convenient, inexpensive, and solvent-free. In addition, the system possesses the advantages of long lifetime and production of high concentrations of target analytes, which are benefit to shorten the testing time of passive samplers.

To mimic the variety of the environment, such as temperature, turbulence, and the concentration of the analytes, the above mentioned flow-through system for the generation of standard aqueous polycyclic aromatic hydrocarbons (PAHs) solution has been modified by Ouyang et al. [52]. Figure 5.13 depicted that two minipumps were added to create concentration and temperature shifts, respectively. In detail, pure water was fed into the sampling chamber to create the shift of PAH concentration and flow velocity of water in the sampling chamber and cylinder, while hot or cold water was transferred by minipump 2 to create a temperature shift of the sampling system. After modification, the effect of environmental variables (temperature, turbulence, and concentrations of the analytes) to the sampling results was tested in this modified flow-through system, which was more similar to the real environment.

5.3.1.2 Thin-Film/Membrane and Rod Geometry

As for the fiber format SPME passive sampling, the sampling rate of analytes in water sample is low because of the small surface area/diffusion path ratio of the fiber in-needle, as well as low diffusivity and solubility of analytes in the diffusion medium (water) [37]. On the other hand, the utility of polymer-coated fiber with small capacity and inflexibility is limited in passive sampling, particularly in the measurements of hydrophobic organic compounds (HOCs) which are slightly dissolved in aquatic environments [38]. Thus, PDMS rod and membrane (Fig. 5.14) have been developed as passive samplers for field water analysis to overcome the shortcomings of polymer-coated fiber by Ouyang and Pawliszyn [53]. Compared with the volume of PDMS-coated fiber (100 μm, 0.61 μL), the volumes of PDMS rod (7.85 μL) and PDMS membrane (62.5 μL) are much larger, leading to higher sensitivity of the passive samplers. The flexibility of PDMS rod and PDMS membrane passive samplers have been demonstrated under laboratory as well as field conditions. The primary finding was that the PDMS membrane possessed highest sampling rates due to the highest surface-to-volume ratio and the largest volume at different sampling depths (Fig. 5.15). In detail, the sampling rate of membrane was about 10 times higher than that of PDMS rod and more than 1000 times than that of in-needle fiber. However, it was a little trouble for PDMS membrane analysis due to the requirement of large volume injector, while the PDMS-rod and the SPME fiber could be analyzed by a gas chromatograph/mass spectrometry (GC/MS) with a normal injector.

Another limitation for PDMS rod and PDMS membrane used in field sampling is that the deployed samplers must be retrieved before all of the preloaded standards are lost when in-fiber standardization is employed to calibrate the obtained data.

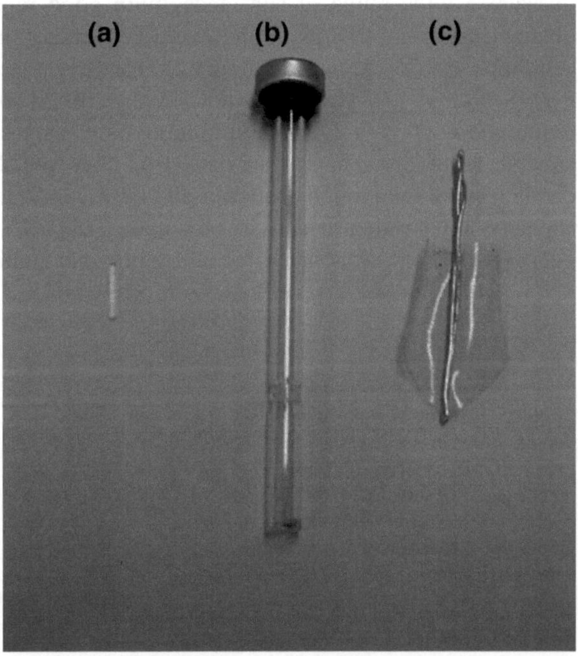

Fig. 5.14 SPME PDMS rod (**a**) PDMS membrane (**c**) passive samplers and inlet liner of GC (**b**) for automated injection with the ATAS system. Reprinted with the permission from Ref. [53]. Copyright 2007 American Chemical Society

5.3.1.3 PDMS Tubing

PDMS tubing has been developed as a good alternative for field monitoring because of easy operability and controllable thickness. Breedveld et al. selected PDMS tubing (core diameter, 750 μm; thickness, 200 μm), PDMS fiber (28 μm), polyoxymethylene (POM; thickness, 55 μm [POM-55] and 500 μm [POM-500]) as passive samplers to determine freely dissolved concentrations of HOCs in overlying surface water [4]. In consideration of fast equilibration and low detection limits, POM-55 was proved to be the optimal choice. If quick measurement was needed and low detection limits were not required, either SPME fibers or POM-55 could be employed, with PDMS tubing as a good alternative. In accordance with the comparison of results between different passive samplers, PDMS tubing was just suitable for field detection, without obvious advantages. Such a result is probably arisen from over-thick coating and small extraction volume. Generally, coating thickness of passive sampler is an important factor to influence the equilibration time, i.e., thicker coating makes equilibration time longer. While extraction volume determines the extraction capacity of sampler, the extraction capacity becomes stronger as the extraction volume increases.

Another form of PDMS tubing as passive sampler is shown in Fig. 5.16 [54]. The PDMS tubing with the length of 2 m was placed in the oil and analytes started to partition from the oil into the PDMS until equilibrium was established. Then, the PDMS tubing was filled with water (acceptor phase) and analytes diffused from PDMS to water. Finally, an equilibrium partitioning from the oil phase to the water phase via PDMS was established after a certain time. Such a passive sampler could measure a wide range of substances with good repeatability, and water samples obtained from the PDMS tubing could be directly measured by HPLC if the concentration of analyte was high enough. However, the passive sampler consisted of PDMS tubing, and water was not suitable for hydrophilic substances due to the requirement of very long equilibration time, as well as compounds with a low aqueous solubility. The polyacrylate (PA) fiber was also used to determine partition coefficients of compounds between oil and water since it exhibited a higher sorption capacity (thus higher sensitivity) for many substance classes, in particular, hydrogen bond donor compounds, including phenols, anilines, amides, and many drugs

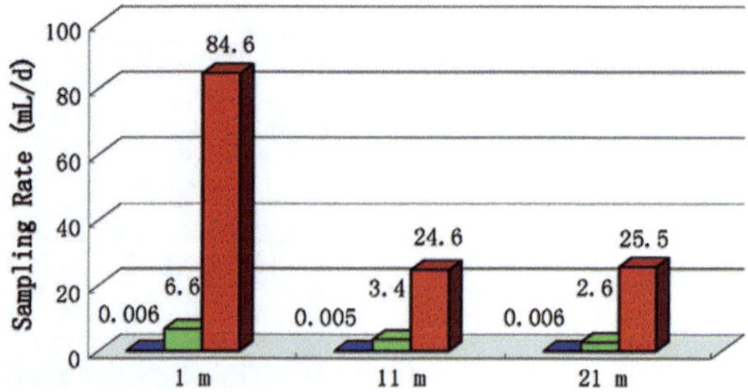

Fig. 5.15 Sampling rate of three types of SPME passive samplers at different sampling depths. *Blue*, in-needle fiber; *green*, PDMS rod; *red*, PDMS membrane. Reprinted with the permission from Ref. [53]. Copyright 2007 American Chemical Society

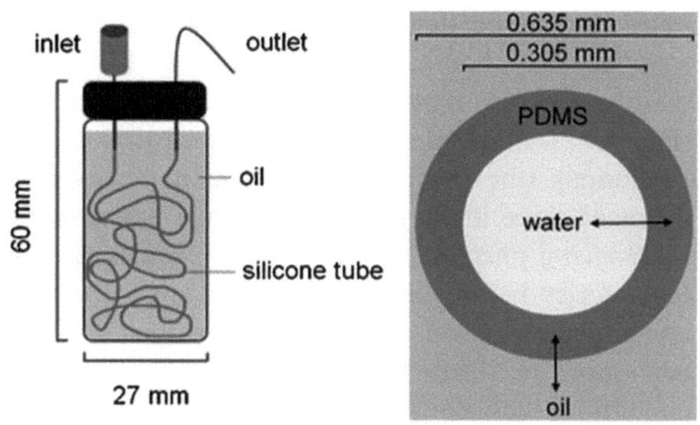

Fig. 5.16 Experimental setup with a PDMS tubing. Reprinted from Ref. [54], with kind permission from Springer Science + Business Media

and pesticides. Comparing the results obtained by PDMS tubing and PA fiber, the result from PDMS tubing was more reliable because partitioning between oil and water was surely in thermodynamic equilibrium and the influence of surrounding phase toward sorption properties for PDMS tubing was far less obvious as that for PA fiber. In addition, the sensitivity of PDMS tubing can be increased by lengthening or widening the PDMS tubing, with preconcentration of analyte before chromatographic analysis.

5.3.2 Chemical Information from Passive SPME Sampling

As more and more attention has been focused on the applications of SPME passive samplers to monitor pollutants in aqueous samples, the reliability of samplers in field deployment has been improved by calibration and comparison, as well as more information have been obtained to illustrate the distribution of different kinds of pollutants all over the world.

Determination of time-weighted average (TWA) concentration. In an on-site application, several passive samplers have been employed at the same location, and the agreement or disagreement between such independent measurements will give valuable novel information about the reliability of the samplers. Compared with other passive samplers, such as polyethylene (PE) strips [39] and polyoxymethylene (POM) sheets [40], the desorption process of SPME fiber is much easier, since there is no need for solvent desorption and the fiber can be coupled with GC injector to conduct thermal desorption. Except for simple operations, SPME fibers can facilitate fast mass transfer to shorten the equilibration time, as well as are small enough to reduce the cost of transport between laboratories and fields. Unfortunately, the sensitivity of SPME fibers deployed in fields to detect contaminants is incomparable with that of PE strip or POM sheet due to their small size; thus, large mass of adsorbent is needed to achieve analytical sensitivity. Besides, SPME fibers, without performance reference compound (PRC) corrections, produced values that are generally much lower (1–2 orders of magnitude) than those measured using PE and

POM, indicating that SPME may not been fully equilibrated with waters being sampled [55]. For instance, Burgess et al. investigated the differences in freely dissolved concentrations of *o,p'*- and *p,p'*-DDE deduced by PE and SPME fiber (Fig. 5.17), illustrating that PE-deduced dissolved *o,p'*- and *p,p'*-DDE water concentrations (the compounds for which the most comparable data are available) averaged three times greater than water concentrations deduced by SPME.

Deriving partition coefficients. It is necessary to derive partition coefficients in laboratories before field deployment, since the freely dissolved concentrations of contaminates in water samples are estimated from partition coefficients and concentrations in SPME fiber. If the partition coefficients are obtained under the same conditions, the values can be contrasted with those reported in other works to assess the performance of the used SPME fiber [54, 41, 42]. Table 5.2 summarizes the log SPME sampler water (log K_{SPME}) partition coefficient values. Generally, log K_{SPME} values are plotted against log K_{OW} values for further discussions. Pennell et al. found a good correlation between K_{SPME} and K_{OW} for most of the polybrominated diphenyl ethers (PBDE) congeners with the exception of 183 and 209, mainly because the Gibbs free energy required for the cavity formation in SPME fiber becomes much higher for larger chemicals, while the amount of energy needed for cavity formation in octanol changes little for the same chemicals [39].

Obtaining chemical activities and chemical activity ratios. Chemical activities and chemical activity ratios have been provided in the past few years, since they can be used to investigate discharge of pollutants downstream in the historically contaminated lake for assessing the distribution of the contamination. The chemical activity ratio also determines both the spatial and the temporal trends of target contaminants, which is the key factor for successfully remediating a contaminated site [35].

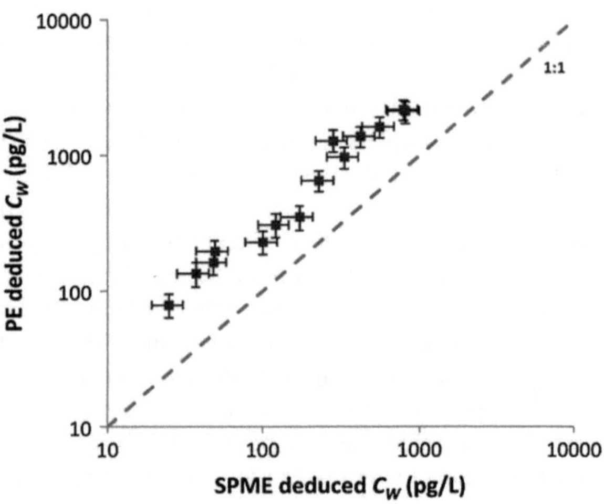

Fig. 5.17 Comparisons of *o,p'*-and *p,p'*-DDE PE deduced, C_W versus SPME (*squares*), in waters near the *bottom* at different stations. Reprinted with the permission from Ref. [55]. Copyright 2014 American Chemical Society

Table 5.2 Log SPME sampler water (log K_{SPME}) partition coefficient values derived in the laboratory

Compound	Log K_{SPME} (PDMS 100 μm, 22 ± 2 °C) [41]	Compound	Log K_{SPME} (PDMS 10 μm) [38]
Bifenthrin	5.87 ± 0.10	PBDE 17	5.92 ± 0.15
cis-Permethrin	6.43 ± 0.10	PBDE 47	6.61 ± 0.42
trans-Permethrin	6.33 ± 0.10	PBDE 71	6.52 ± 0.37
Cyfluthrin	5.93 ± 0.12	PBDE 99	7.16 ± 0.35
Cypermethrin	5.99 ± 0.12	PBDE 100	7.01 ± 0.35
Deltamethrin	5.89 ± 0.13	PBDE 183	7.63 ± 0.13
Esfenvalerate	6.33 ± 0.13	PBDE 209	6.79 ± 0.42
Fenpropathrin	5.73 ± 0.10	Triclosan	4.06 ± 0.13
λ-Cyhalothrin	6.12 ± 0.11	Methyltriclosan	4.90 ± 0.15
Compound	Log K_{SPME} (PDMS 28.5 μm, 20 °C) [35]	Compound	Log K_{SPME} (PDMS 10 μm) [39]
Naphthalene	3.20 ± 0.3	Fluoranthene	4.99 ± 0.04
Fluorene	3.76 ± 0.05	Pyrene	4.62 ± 0.24
Phenanthrene	3.95 ± 0.06	Benzo[ghi] perylene	5.60 ± 0.55
Anthracene	4.16 ± 0.14	PCB 77	5.85 ± 0.57
Fluoranthene	4.60 ± 0.2	PCB 118	6.13 ± 0.76
Benzo[a] anthracene	5.15 ± 0.14	PCB 153	6.37 ± 0.71
Chrysene	4.98 ± 0.13		
Benzo[b] fluoranthene	5.18 ± 0.19		
Benzo[k] fluoranthene	5.19 ± 0.18		
Benzo[e]pyrene	5.41 ± 0.15		
Indeno [1,2,3-cd] pyrene	5.33 ± 0.13		
Benzo[ghi] perylene	5.02 ± 0.07		
Pyrene	4.61 ± 0.16		
Benzo[a]pyrene	5.39 ± 0.16		

PCB—polychlorinated biphenyl, *PBDE*—polybrominated diphenyl ether

5.3.3 Recent Application of SPME Passive Sampling

Environmental problems have attracted worldwide attention, and all kinds of pollutants have appeared to threaten ecosystem balance and human health in different matrices, especially in aqueous matrices. In order to monitor the concentrations of pollutants, the deployment method coupled with SPME passive samplers has been successfully used for on-situ sampling.

Three main factors can affect the uptake of contaminates by SPME samplers, including the properties of the contaminant (e.g., *K*ow, molecular weight, and volatility), environmental parameters (e.g., temperature, flow rate, and the amount of biofouling present on the device surface), and SPME adsorbent types (e.g., polarity and thickness of adsorbents). According to contaminates need to be monitored, suitable SPME adsorbents can be chosen in laboratory for high sensitivity and accurate data before in situ sampling.

HOCs. It is a challenge to investigate the distributions of HOCs (such as PAHs and PCBs) in aqueous samples due to ultralow concentrations. SPME fibers, have been used to preconcentrate HOCs at ultralow concentrations for monitoring, assessment of bioavailability, fate and toxicity, and toxicity identification evaluation. For HOC extraction, the commercial PDMS fibers with different thicknesses are the most frequently used and the second is PDMS/DVB fiber. With the research development, the number of HOC congeners monitored by SPME fibers has increased, and calibration method has been perfected for wide field deployment [35, 39, 55, 48]. Investigations into freely dissolved concentrations of HOCs in sediment pore water is more important for predicting toxicity and uptake, since SPME fiber demonstrates obvious advantages compared with other passive samplers due to the difficulties for obtaining and preparing large volumes of pore water, as well as interferences from analytes associated with colloidal material [44].

Hydrophilic compounds. As more and more pharmaceuticals, industrial products, pesticides, and personal care products are used and emitted into different aquatic environments, and all kinds of hydrophilic compounds have appeared and needed to be quantified. PA is a moderate polar coating used as SPME fiber coating, which is believed to be a stronger hydrogen bond acceptor than PDMS or PE. Therefore, PA is a more ideal sampling phase for polar compounds with moderate hydrophobicity (log $K_{ow} > 2.5$) than other commercial SPME coatings. Currently, only a limited number of studies are available on the sorption of polar compounds in the aqueous samples with dissolved organic carbon (DOC) due to weak sorption and requirements of specific analytical method. Hermens et al. have used PA-coated fiber to correctly determine freely dissolved concentrations of hydrophilic compounds in the presence of DOC, considering the effect of concentration, pH, and ionic strength. However, it was difficult for PA-coated fiber to quantify the hydrophilic compounds (e.g., testosterone) with lower log K_{DOC} value [46].

Ionic pollutants. Apart from neutral compounds, there exist ionic pollutants in aquatic environment due to protonation or deprotonation of pollutants at different pH values. However, ionic pollutants are difficult to be extracted by common SPME coating because the affinity of the ionic species to neutral polymers is orders of magnitude lower than the neutral species. It is an obvious tendency for development of SPME fibers to meet the requirements of the diversification of pollutants. Hence, Droge et al. prepared new hydrophobic SPME coating with cation-exchange properties (C18/SCX), which extracted lauryl diethanolamine 100-fold stronger than PA-coated fiber [36].

5.4 Calibration

Unlike traditional extraction method, SPME is a non-exhaustive extraction technique in which only a small portion of analytes are extracted into the extraction phase. This unique feature indicates the requirement for careful calibration of SPME for quantitative analysis. Along with the development of SPME, a series of calibration methods were proposed which were based on the in-depth understanding of fundamental principles governing the mass transfer of analytes in multiple systems.

As for SPME aqueous sampling, in addition to external standard calibration method, equilibrium extraction, diffusion-based calibration, and kinetic calibration methods with several models were proposed.

In the rapid active sampling, the sampler worked in the linear uptake regime. The mass of analyte accumulated in the extraction phase after an exposure time (t) can be calculated as:

$$M_s(t) = C_w R_s t \tag{5.1}$$

where C_w is the concentration of analyte in the water phase, and R_s is the proportionality constant (sampling rate) and may be interpreted as the volume of water cleared of analyte per unit of exposure time by the sampler. When the sampler is used for on-site sampling, the sampling rate can be determined in the laboratory. If R_s is known, C_w can be calculated from the sampling rate (R_s), the exposure time (t), and the amount ($M_s(t)$) of the analyte trapped by the receiving phase. Also this calibration method is called sampling-rate (SR) calibration.

With regard to the passive sampling, the target analytes access the fiber coating by means of diffusion through the static water gap between the opening and fiber coating. Chen et al. [56] reported a physical model translated from heat transfer to a circular cylinder in cross flow to describe the rapid SPME extraction of benzene, toluene, ethylbenzene, and *o*-xylene using PDMS/DVB and CAR/PDMS fibers in aqueous samples. Parameters including extraction time, concentration, water velocity, and temperature were investigated. The model proposed for rapid water SPME sampling was used to predict the mass uptake for rapid sampling.

Unlike exhaustive sampling and equilibrium sampling methods, linear mass uptake to both sampling time and analyte concentration is crucial for the rapid SPME sampling method. To ensure this, feasible strategies include, first, the sample flow, and boundary layer mass-transfer condition must be maintained constant; second, the sorbent must be a zero sink for target analytes, which ensures $C_{sorbent}$ is zero; third, sampling time should be optimized according to analyte concentrations and sample velocities. Rapid SPME sampling requires controlled sample flow. Subsequently, the desorption kinetics of the analyte from a SPME fiber into an agitated sample matrix was investigated. This study facilitates the full integration of sampling, sample preparation, and sample introduction, especially for on-site or in vivo investigations, where the addition of standards to the sample matrix, or control of the velocity of the sample matrix, is very difficult [57]. The approach that allows the calibration of absorption using desorption is of great importance for the calibration of on-site, in situ, or in vivo analysis.

A standard-free kinetic calibration method was proposed for rapid on-site sampling by SPME, in which all extracted analytes could be calibrated with two samplings and did not need a standard to be preloaded to determine the desorption rate constant. However, the method requires that the conditions of the two samplings should be kept constant, and it is only suitable for rapid sampling [58].

Ouyang et al. proposed a new method, termed one-calibrate kinetic calibration technique, which used only one single standard to calibrate all extracted analytes. The theoretical considerations were validated by passive water sampling in laboratory and rapid water sampling in the field for polycyclic aromatic hydrocarbon (PAH) analysis [59].

5.5 Summary

This chapter focused on the development of SPME format for aqueous sampling. Various SPME configurations such as fiber, thin film, and stir bar with designed sampler have been widely applied to the analyses of environmental contaminations for on-site sampling. The on-site field sampling offered the advantages of requiring very short extraction time, eliminating the possible losses of target analytes during the transportation and storage procedures compared with traditional extraction technique, and providing more representative quantitative results of the original sample characteristics. The on-site active sampling could significantly increase the sampling rate and decrease the sampling time, in which the portable agitators provided a controllable rotation speed and accelerated the distributions of analytes in the coatings, leading to simple, rapid, on-site extraction.

SPME passive sampler have been successfully applied in monitoring all kinds of POPs from environmental aqueous samples, mainly focusing on freely dissolved and time-weighted average (TWA) concentrations. With research progress on passive sampler, adsorbents are so diversified for preconcentration so more and more pollutant congeners can be monitored, and the developed methods for quantification are more reliable with corrections and comparisons.

To develop better SPME passive sampler, the detection sensitivity of SPME fiber should be improved to meet the requirements of contaminates at ultralow concentrations. It is obvious that the extraction capacity of commercial adsorbents (PDMS, PA, PDMS/DVB et al.) with small volume is restricted. However, it is inconvenient to improve the sensitivity of SPME fiber by increasing the extraction volume since too large extraction volume may make it hard for direct thermal desorption. A good sign is that a large amount of adsorptive materials with much high surface area have been reported and they can take place of commercial adsorbents in SPME fibers for the target of higher sensitivity. In addition to sensitivity, the physical properties of SPME fibers are other points to be considered.

Suitable supporter and coating preparation method are necessary to make SPME fiber robust enough for complex matrix.

Acknowledgement We acknowledge financial support from the projects of NNSFC (Grants 21225731, 21377172, and 21477166) and the NSF of Guangdong Province (Grant S2013030013474).

References

1. George MJ, Marjanovic L, Williams DBG (2015) Solvent-assisted headspace sampling using solid phase microextraction for the analysis of phenols in water. Anal Chem 87:9559–9562. doi:10.1021/acs.analchem.5b02539
2. Stevens ME Jr, Tipple CA, Smith PA et al (2013) Application of a high surface area solid-phase microextraction air sampling device: collection and analysis of chemical warfare agent surrogate and degradation compounds. Anal Chem 85:8626–8633. doi:10.1021/ac401033a
3. Bianchi F, Bedini A, Riboni N et al (2014) Cavitand-based solid-phase microextraction coating for the selective detection of nitroaromatic explosives in air and soil. Anal Chem 86:10646–10652. doi:10.1021/ac5025045
4. Risticevic S, Souza-Silva EA, DeEll JR et al (2016) Capturing plant metabolome with direct-immersion in vivo solid phase microextraction of plant tissues. Anal Chem 88:1266–1274. doi:10.1021/acs.analchem.5b03684
5. Zhou SN, Oakes KD, Servos MR et al (2008) Application of solid-phase microextraction for in vivo laboratory and field sampling of pharmaceuticals in fish. Environ Sci Technol 42:6073–6079. doi:10.1021/es8001162
6. Voice TC, Kolb B (1994) Comparison of European and American techniques for the analysis of volatile organic compounds in environmental matrices. J Chromatogr Sci 32:306–311. doi:10.1093/chromsci/32.8.306
7. Voice TC, Kolb B (1993) Static and dynamic headspace analysis of volatile organic compounds in soils. Environ Sci Technol 27:709–713. doi:10.1021/es00041a014
8. Richardson SD, Kimura SY (2016) Water analysis: emerging contaminants and current issues. Anal Chem 88:546–582. doi:10.1021/acs.analchem.5b04493
9. Qin Z, Bragg L, Ouyang G et al (2009) Solid-Phase microextraction under controlled agitation conditions for rapid on-site sampling of organic pollutants in water. J Chromatogr A 1216:6979–6985. doi:10.1016/j.chroma.2009.08.052
10. Huang S, He S, Xu H et al (2015) Monitoring of persistent organic pollutants in seawater of the pearl river estuary with rapid on-site active SPME sampling technique. Environ Pollut 200:149–158. doi:10.1016/j.envpol.2015.02.016
11. Basheer C, Balaji G, Chua SH et al (2011) Novel on-site sample preparation approach with a portable agitator using functional polymer-coated multi-fibers for the microextraction of organophosphorus pesticides in seawater. J Chromatogr A 1218:654–661. doi:10.1016/j.chroma.2010.12.033
12. Kermani FR, Pawliszyn J (2012) Sorbent coated glass wool fabric as a thin film microextraction device. Anal Chem 84:8990–8995. doi:10.1021/ac301861z
13. Deng Z, Chen X, Wang Y et al (2015) Headspace thin-film microextraction coupled with surface-enhanced raman scattering as a facile method for reproducible and specific detection of sulfur dioxide in wine. Anal Chem 87:633–640. doi:10.1021/ac503341g
14. Bragg L, Qin Z, Pawliszyn J (2006) Field sampling with a polydimethylsiloxane thin-film. J Chromatogr Sci 44:317–323. doi:10.1093/chromsci/44.6.317

15. Ouyang G, Zhao W, Bragg L et al (2007) Time-weighted average water sampling in lake Ontario with solid-phase microextraction passive samplers. Environ Sci Technol 11:4026–4031. doi:10.1021/es062647a
16. Jiang R, Pawliszyn J (2014) Preparation of a particle-loaded membrane for trace gas sampling. Anal Chem 86:403–410. doi:10.1021/ac4035339
17. Qin Z, Bragg L, Ouyang G et al (2008) Comparison of thin-film microextraction and stir bar sorptive extraction for the analysis of polycyclic aromatic hydrocarbons in aqueous samples with controlled agitation conditions. J Chromatogr A 1196:89–95. doi:10.1016/j.chroma.2008.03.063
18. Grandy JJ, Boyaci E, Pawliszyn J (2016) Development of a carbon mesh supported thin film microextraction membrane as a means to lower the detection limits of benchtop and portable GC/MS instrumentation. Anal Chem 88:1760–1767. doi:10.1021/acs.analchem.5b04008
19. Baltussen E, Sandra P, David F et al (1999) Stir bar sorptive extraction (SBSE), a novel extraction technique for aqueous samples: theory and principles. J Microcolumn Sep 11:737–747. doi:10.1002/(SICI)1520-667X(1999)11:10<737::AID-MCS7>3.0.CO;2-4
20. Gomez MJ, Herrera S, Sole D et al (2011) Automatic searching and evaluation of priority and emerging contaminants in wastewater and river water by stir bar sorptive extraction followed by comprehensive two-dimensional gas chromatography-time-of-flight mass spectrometry. Anal Chem 83:2638–2647. doi:10.1021/ac102909g
21. Soini HA, Bruce KE, Klouckova I et al (2006) In situ surface sampling of biological objects and preconcentration of their volatiles for chromatographic analysis. Anal Chem 78:7161–7168. doi:10.1021/ac0606204
22. Benede JL, Chisvert A, Giokas DL et al (2016) Determination of ultraviolet filters in bathing waters by stir bar sorptive-dispersive microextraction coupled to thermal desorption-gas chromatography-mass spectrometry. Talanta 47:246–252. doi:10.1016/j.talanta.2015.09.054
23. Benet I, Ibanez C, Guardia MD et al (2015) Optimisation of stir-bar sorptive extraction (SBSE), targeting medium and long-chain free fatty acids in cooked ham exudates. Food Chem 185:75–83. doi:10.1016/j.foodchem.2015.03.102
24. Zhang W, Zhang Z, Zhang J et al (2014) Covalent immobilization of graphene onto stainless steel wire for jacket-free stir bar sorptive extraction. J Chromatogr A 1351:12–20. doi:10.1016/j.chroma.2014.05.038
25. Bratkowska D, Fontanals N, Cormack PA et al (2012) Preparation of a polar monolithic stir bar based on methacrylic acid and divinylbenzene for the sorptive extraction of polar pharmaceuticals from complex water samples. J Chromatogr A 1225:1–7. doi:10.1016/j.chroma.2011.12.064
26. Hu C, He M, Chen B et al (2015) Simultaneous determination of polar and apolar compounds in environmental samples by a polyaniline/hydroxyl multi-walled carbon nanotubes composite-coated stir bar sorptive extraction coupled with high performance liquid chromatography. J Chromatogr A 1394:36–45. doi:10.1016/j.chroma.2015.03.046
27. Zhou Q, Qian Y, Qian MC (2015) Analysis of volatile phenols in alcoholic beverage by ethylene glycol-polydimethylsiloxane based stir bar sorptive extraction and gas chromatography-mass spectrometry. J Chromatogr A 1390:22–27. doi:10.1016/j.chroma.2015.02.064
28. Zhang Z, Zhang W, Bao T et al (2015) Jacket-free stir bar sorptive extraction with bio-inspired polydopamine-functionalized immobilization of cross-linked polymer on stainless steel wire. J Chromatogr A 1407:1–10. doi:10.1016/j.chroma.2015.06.031
29. Mao X, Hu B, He M et al (2012) Stir bar sorptive extraction approaches with a home-made portable electric stirrer for the analysis of polycyclic aromatic hydrocarbon compounds in environmental water. J Chromatogr A 1260:16–24. doi:10.1016/j.chroma.2012.08.062
30. Xing R, Hu S, Chen X et al (2015) On-site sampling and sample-preparation approach with a portable sampler based on hollow-fiber/graphene bars for the microextraction of nitrobenzene compounds in lake water. J Sep Sci 38:368–373. doi:10.1002/jssc.201401129
31. Schwarzenbach RP, Egli T, Hofstetter TB et al (2010) Global water pollution and human health. Annu Rev Environ Resour 35:109–136. doi:10.1146/annurev-environ-100809-125342

32. Houtman CJ (2010) Emerging contaminants in surface waters and their relevance for the production of drinking water in Europe. J Integr Environ Sci 7:271–295. doi:10.1080/1943815X.2010.511648
33. Mackay D, Paterson S (1991) Evaluating the multimedia fate of organic chemicals: a level III fugacity model. Environ Sci Technol 25:427–436. doi:10.1021/es00015a008
34. Escher BI, Hermens JLM (2004) Peer reviewed: internal exposure: linking bioavailability to effects. Environ Sci Technol 38:455–462. doi:10.1021/es0406740
35. Cornelissen G, Pettersen A, Broman D et al (2008) Field testing of equilibrium passive samplers to determine freely dissolved native polycyclic aromatic hydrocarbon concentrations. Environ Toxicol Chem 27:499–508. doi:10.1897/07-253.1
36. Wang F, Chen Y, Hermens JLM et al (2013) Evaluation of passive samplers with neutral or ion-exchange polymer coatings to determine freely dissolved concentrations of the basic surfactant lauryl diethanolamine: measurements of acid dissociation constant and organic carbon-water sorption coefficient. J Chromatogr A 1315:8–14. doi:10.1016/j.chroma.2013.09.041
37. Zhao WN, Ouyang GF, Alaee M et al (2006) On-rod standardization technique for time-weighted average water sampling with a polydimethylsiloxane rod. J Chromatogr A 1124:112–120. doi:10.1016/j.chroma.2006.05.062
38. Duan CF, Shen Z, Wu DP et al (2011) Recent developments in solid-phase microextraction for on-site sampling and sample preparation. Trend Anal Chem 30:1568–1574. doi:10.1016/j.trac.2011.08.005
39. Fernandez LA, Lao WJ, Maruya KA et al (2012) Passive sampling to measure baseline dissolved persistent organic pollutant concentrations in the water column of the Palos Verdes shelf superfund site. Environ Sci Technol 46:11937–11947. doi:10.1021/es302139y
40. Perron MM, Burgess RM, Suuberg EM et al (2013) Performance of passive samplers for monitoring estuarine water column concentrations: 2. Emerging contaminants. Environ Toxicol Chem 32:2190–2196. doi:10.1002/etc.2248
41. Perron MM, Burgess RM, Suuberg EM et al (2013) Performance of passive samplers for monitoring estuarine water column concentrations: 1. Contaminants of concern. Environ Toxicol Chem 32:2182–2189. doi:10.1002/etc.2321
42. Mäenpää K, Leppänen MT, Figueiredo K et al (2015) Fate of polychlorinated biphenyls in a contaminated lake ecosystem: combining equilibrium passive sampling of sediment and water with total concentration measurements of biota. Environ Toxicol Chem 34:2463–2474. doi:10.1002/etc.3099
43. Lao WJ, Maruya KA, Tsukada D (2012) A two-component mass balance model for calibration of solid-phase microextraction fibers for pyrethroids in seawater. Anal Chem 84:9362–9369. doi:10.1021/ac302120m
44. Hawthorne SB, Grabanski CB, Miller DJ et al (2009) Solid-phase-microextraction measurement of 62 polychlorinated biphenyl congeners in milliliter sediment pore water samples and determination of K_{DOC} values. Anal Chem 81:6936–6943. doi:10.1021/ac901001j
45. Vrana B, Popp P, Paschke A et al (2001) Membrane-enclosed sorptive coating. An integrative passive sampler for monitoring organic contaminants in water. Anal Chem 73:5191–5200. doi:10.1021/ac010630z
46. Haftka JJH, Scherpenisse P, Jonker MTO et al (2013) Using polyacrylate-coated SPME fibers to quantify sorption of polar and ionic organic contaminants to dissolved organic carbon. Environ Sci Technol 47:4455–4462. doi:10.1021/es400236a
47. Alvarez DA, Maruya KA, Dodder NG et al (2014) Occurrence of contaminants of emerging concern along the california coast (2009–2010) using passive sampling devices. Mar Pollut Bull 81:347–354. doi:10.1016/j.marpolbul.2013.04.022
48. Burgess RM, Lohmann R, Schubauer-Berigan JP et al (2015) Application of passive sampling for measuring dissolved concentrations of organic contaminants in the water column at three marine superfund sites. Environ Toxicol Chem 34:1720–1733. doi:10.1002/etc.2995

49. Zeng E, Tsukada D, Diehl AW (2004) Development of a solid-phase microextraction-based method for sampling of persistent chlorinated hydrocarbons in an urbanized coastal environment. Environ Sci Technol 38:5737–5743. doi:10.1021/es049680m
50. Busser FJM, Hermens JLM (2008) Poly(dimethylsiloxane) as passive sampler material for hydrophobic chemicals: effect of chemical properties and sampler characteristics on partitioning and equilibration times. Anal Chem 80:3859–3866. doi:10.1021/ac800258j
51. Ouyang GF, Chen Y, Pawliszyn J (2005) Time-weighted average water sampling with a solid-phase microextraction device. Anal Chem 77:7319–7325. doi:10.1021/ac051035q
52. Ouyang GF, Cui SF, Qin ZP et al (2009) One-calibrant kinetic calibration for on-site water sampling with solid-phase microextraction. Anal Chem 81:5629–5636. doi:10.1021/ac900315w
53. Ouyang GF, Zhao WN, Bragg L et al (2007) Time-weighted average water sampling in lake Ontario with solid-phase microextraction passive samplers. Environ Sci Technol 41:4026–4031. doi:10.1021/es062647a
54. Oemisch L, Goss KU, Endo S (2013) Determination of oil–water partition coefficients of polar compounds: silicone membrane equilibrator vs. SPME Passive Sampler. Anal Bioanal Chem 405:2567–2574. doi:10.1007/s00216-012-6689-9
55. Fernandez LA, Lao WJ, Maruya KA et al (2014) Calculating the diffusive flux of persistent organic pollutants between sediments and the water column on the Palos Verdes shelf superfund site using polymeric passive sampler. Environ Sci Technol 48:3925–3934. doi:10.1021/es404475c
56. Chen Y, Koziel JA, Pawliszyn J (2003) Calibration for On-site analysis of hydrocarbons in aqueous and gaseous samples using solid-phase microextraction. Anal Chem 75:6485–6493. doi:10.1021/ac0349328
57. Chen Y, Pawliszyn J (2004) Kinetics and the on-site application of standards in a solid-phase microextration fiber. Anal Chem 76:5807–5815. doi:10.1021/ac0495081
58. Ouyang G, Cai J, Zhang X et al (2008) Standard-free kinetic calibration for rapid on-site analysis by solid-phase microextraction. J Sep Sci 31:1167–1172. doi:10.1002/jssc.200700495
59. Ouyang G, Cui S, Qin Z et al (2009) One-Calibrant kinetic calibration for on-site water sampling with solid-phase microextraction. Anal Chem 81:5629–5636. doi:10.1021/ac900315w

Chapter 6
Application of Solid-Phase Microextraction in Soil and Sediment Sampling

Ruifen Jiang and Gangfeng Ouyang

Abstract Solid-phase microextraction (SPME) has been widely applied in environmental analysis and received increasing attention in the sampling of soil/sediment due to its recognized advantages over the traditional methods. In this chapter, recent development of SPME for soil/sediment sampling is summarized. The method development including the selection of sampling formats, fiber coatings, extraction-phase geometries, and calibration methods is discussed in detail. Finally, the application of SPME for sensing of the freely dissolved organic compounds in pore water; the determination of the partition coefficients of organic pollutants between the sediment and pore water; the characterization of the bioavailability of organic pollutants in the soil/sediment; and the use as a biomimetic tool are reviewed.

Keywords Solid-phase microextraction (SPME) · Soil · Sediment · Sampling formats · Geometry · Calibration · Freely dissolved concentration · Biomimetic tool

Abbreviation

BCF	Bioconcentration factor
BFRs	Brominated flame retardants
BTEX	Benzene, toluene, ethylbenzene, xylene
CAR/PDMS	Carboxen/polydimethylsiloxane
CE	Capillary electrophoresis
CW/DVB	Carbowax/divinylbenzene

R. Jiang
School of Environment, Guangzhou Key Laboratory of Environmental Exposure and Health, and Guangdong Key Laboratory of Environmental Pollution and Health, Jinan University, Guangzhou 510632, People's Republic of China

G. Ouyang (✉)
MOE Key Laboratory of Aquatic Product Safety/KLGHEI of Environment and Energy Chemistry, School of Chemistry, Sun Yat-Sen University, Guangzhou 510275, People's Republic of China
e-mail: cesoygf@mail.sysu.edu.cn

G. Ouyang and R. Jiang (eds.), *Solid Phase Microextraction*,
DOI 10.1007/978-3-662-53598-1_6

CW	Carbon wax
DI	Direct immersion
DVB	Divinylbenzene
EC	Equilibrium calibration
ECC	External calibration curve
EVA	Ethylene-vinyl acetate
GC	Gas chromatography
HOCs	Hydrophobic organic compounds
HS	Headspace
IL	Ionic liquid
IMS	Ion mass spectrometry
IS	Internal standard
LC	Liquid chromatography
LOD	Limit of detection
LOQ	Limit of quantification
MDL	Method detection limit
MS	Mass spectrometry
MWA	Microwave-assisted
OCPs	Organochlorine pesticides
OS	Ultrasound
PA	Polyacrylate
PAHs	Polyaromatic hydrocarbon
PBDEs	Polybrominated diphenyl ethers
PCBs	Polychlorinated biphenyls
PCRs	Performance reference compounds
PDMS	Polydimethylsiloxane
PHW	Pressured hot water
PPy	Dodecylsulfate-doped polypyrrole
SBSE	Stir bar sorptive extraction
SPME	Solid-phase microextraction
TNT	Trinitrotoluene
USA	Ultrasonic-assisted

6.1 Comparison of SPME with EPA Method for Soil/Sediment Sampling

Soil/sediment is a kind of complex sample that contains several phases including solid particulate (SP), organic matter, dissolved organic matter (DOM), minerals, air, and water. After being discharged to the soil matrix, organic pollutants are distributed between these phases till equilibrium. Commonly, some organic compounds adsorb to the SP and DOM, while some of them dissolve freely in pore water. Based on different sampling methods, organic compounds from different phases can be extracted and quantified. Conventional sample preparation methods including soxhlet extraction (EPA 3540C), automated soxhlet extraction (EPA 3541), supercritical fluid extraction (EPA 3560, 3561, 3562), accelerated solvent extraction, ultrasonic extraction (EPA 3550C), and microwave-assisted extraction (MAE) (EPA 3556) use strong solvent or mild solvent to extract the organic pollutants from different phases of the soil sample. The sampling procedure is described in Fig. 6.1, which includes extraction, separation, cleanup, and preconcentration steps. Most of the steps are time-consuming and tedious and require large quantities of organic solvents and expensive devices such as microwave and ultrasonic device to facilitate the desorption of analytes from the matrix pores.

The development of SPME provides an alternative method for the soil/sediment sample preparation. During the sampling process, SPME can be either directly immersed in the solid sample (DI-SPME) or exposed to the headspace of the sample (HS-SPME). After sampling, the extracted analyst can be thermal-desorbed or solvent-desorbed from the fiber coating, and then quantified by the gas chromatography (GC) coupled with mass spectrometry (MS) or liquid chromatography (LC)/MS. The sampling procedure of SPME is shown in Fig. 6.2, which is much simpler compared with the traditional LLE methods shown in Fig. 6.1.

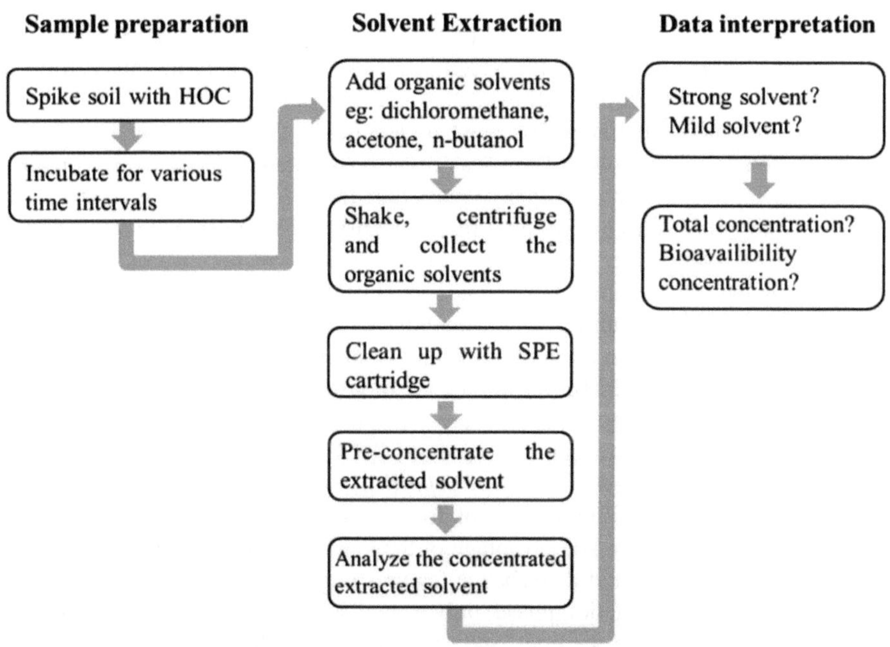

Fig. 6.1 The sampling procedure of tradition methods for soil sampling

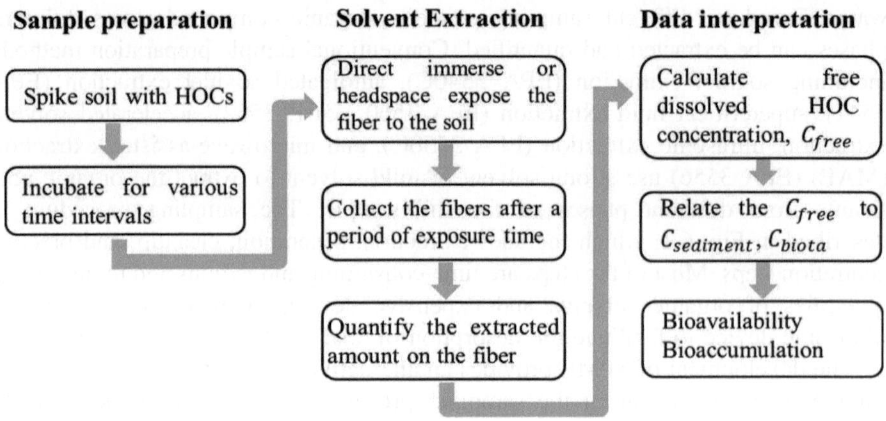

Fig. 6.2 The sampling procedure of SPME for soil/sediment sampling

6.2 Selection of Sampling Formats

Both DI and HS-SPME were reported for soil/sediment sampling (Table 6.1). Two protocols have been described for DI-SPME sampling. The most common one was by directly introducing the fiber to the solid matrices (Fig. 6.3a), which was also termed "matrix-SPME" [1] or "in situ SPME" [2]. During the exposure period, the analytes were distributed between the soil and the fiber coating till equilibrium. By determining the amount extracted by the fiber coating, the concentration in the soil matrix can be quantified. Since the SPME only sensed the freely dissolved form of the organic pollutants in the sediment, the quantified concentration of the organic pollutant presented the freely dissolved molecules in the pore water [1, 3]. The concentration of the dissolved form was often considered as the effective concentration that controlled several diffusive mass transfer processes such as evaporation, sorption, and uptake into macro- and microorganisms. Therefore, the matrix-SPME often used as biomimetic tool for investigating the bioavailability of the organic pollutants in the solid sample [4–8]. Conder et al. employed a disposable SPME fiber to measure the bioavailability of the explosive compounds, 2,4,6-trinitrotoluene (TNT), and its degradation products in sediment amended with activated carbon [4]. Fang et al. used matrix-SPME to estimate the bioavailability of DDT in soil to both earthworm and vegetables [7]. Maruya et al. compared the freely dissolved concentration of hydrophobic organic compounds (HOCs) measured by SPME with the bioaccumulation of *Macoma and Nereisspp* [8]. All results demonstrated that the SPME was a promise biomimetic tool for predicting bioavailable organic compounds in sediment. The main advantage of the matrix-SPME for soil sampling was the nondestructive, minimal-disturbance features during the sampling process. Due to the volume of the extraction phase, only a small amount of analytes in the sample matrix was extracted onto the fiber, and none or minimal disturbance was caused to the sample matrix. Stringer et al. [2] utilized the advantage of in situ SPME and

Table 6.1 Summary of SPME for complex soil matrix analysis

Samples	Analytes	Coatings	Separation methods	Calibration methods	Measurement endpoints and relationship	References
Sediment	PCBs and DDT	Disposable PDMS tube	Matrix-SPME	EC	$C_{fiber} C_{pw}$	[1]
Soil	BTEX and cumene	PDMS	USA-Water-HS-SPME	ECC (spiked in the extracted solution)	LOD: 1–2 ng/g Linear range: 2.5–288 ng/g	[35]
Sediment	2,4,6-trinitrotoluene and biodegradation products	Disposable PA	In situ SPME	EC	n_{fiber} versus $n_{sediment}$ $C_{sediment}$ versus log n_{biota} C_{fiber} versus C_{biota}, C_{fiber} versus log C_{biota}, $C_{sediment}$ versus C_{biota},	[4, 66]
Certified marine sediment	PAHs	PA	MWA-Micellar-DI-SPME	ECC (spiked in micellar solutions)	LODs: 0.28–7.66 ng/mL Recovery: 58.6–111.5% PA is better than PDMS for PAHs sampling	[19]
Soil	Chlorophenols	PA	MEA-Water-HS-SPME	ECC	LOD: 0.1–2.0 μg/kg Linear range: 0.5–300 μg/kg Recovery: 86.1–98.4%	[23]

(continued)

Table 6.1 (continued)

Samples	Analytes	Coatings	Separation methods	Calibration methods	Measurement endpoints and relationship	References
OECD soil	Chlorobenzenes	Disposable PDMS	Water-DI-SPME with shaking	/	C_{fiber} versus C_{biota}	[11]
Soil	PCBs and OCPs	PDMS fiber	Water-DI-SPME with shaking	EC	C_{fiber} versus C_{biota} k versus Log K_{ow} $C_{biota,cal}$ versus $C_{biota,exp}$ log BCF versus log K_{ow}	[12]
Soil	BTEX	CAR/PDMS	Multi-HS-SPME	ECC (spiked in water)	Linear range: 0.36–420 ng LODs: 0.2–1 ng Fiber type; temperature; addition of water; extraction time were evaluated	[27]
Marine sample	Energetic chemicals	CW/DVB	OS-water-DI-SPME	IS	MDL: 1–91 μg/L Effect of desorption temperature, NaCl concentration, adsorption time, desorption solution.	[56]
Soil	PAHs	PDMS	Water-DI-SPME with shaking	EC	C_{fiber}, C_{w}, $C_{sediment}$, K_{f}, k, log K_{fs} versus log k log K_{ow} versus log K_{oc}	[13]

(continued)

Table 6.1 (continued)

Samples	Analytes	Coatings	Separation methods	Calibration methods	Measurement endpoints and relationship	References
Estuarine soils and sediments	Nitrous oxides	CAR/PDMS	DI-SPME	ECC (standard gas)	Calibration curve: 4.1–2030 nM LOD: 4.1 nM (18 ppb) LOQ: 16 nM (72 ppb)	[81]
Soil	Degradation products of chemical warfare agent	PDMS	HS-SPME	ECC (spiked in soil)	Coupled with IMS; evaluation of fiber selection; extraction time; Linear range: 10–70 ppm LODs: 6.3–7.6 ppm	[26]
Soil	2-chloroethyl ethyl sulfide	Acrylate/silicone copolymer	Water-HS-SPME	ECC (spiked in soil)	Linear range: 0.1–10 ug/g LOD: 2.7 ng/g Recovery: 88–102%	[59]
Sediment	PAHs	Cold PDMS	HS-SPME	Exhaustive extraction	Linear curve: 0.0009–1000 ng/g LODs: 0.3–3 pg/g LOQs: 0.9–9 pg/g	[38]
Soil and sediment	Brominated flame retardants	PDMS	Water-HS-SPME	ECC (spiked in soil)	Reproducibility: 1.0–16% Recovery: 81.8–104.4% (sediment) 95.7–105.1% (soil) LODs: 5.2–214 pg/g (sediment) 14–625 pg/g (soil)	[28]

(continued)

Table 6.1 (continued)

Samples	Analytes	Coatings	Separation methods	Calibration methods	Measurement endpoints and relationship	References
Estuarine sediment	OCPs	PDMS	Water-HS-SPME	ECC (spiked in water)	Linear range: 0.2–4 ng/g Precision: 3.7–9.6% LODs: 0.034–0.301 ng/g Recovery: 71–121%	[30]
Landfill leachates and sediments	PAHs and PCB	PDMS	MEA-water-HS-SPME	ECC (spiked in leachates)	For PCB: LOD: 0.1–7 ng/L Recovery: 37–76% For PAHs: LOD: 5–926 ng/L for PAHs Recovery: 58–48%	[33]
Sediment	Organotin compounds	PDMS	OS-buffer-HS-SPME	Standard addition (IS for peak area calibration and standard addition for concentration quantification)	Recovery: >80% Repeatability: 2–30% LOD: 2.35–39.52 ng/Sng LOQ: 4.96–87.36 ng/Sng	[80]
Soil and clay	Chlorobenzenes	diglycidyloxycalix [4] arene yields, PDMS	OS-water-HS-SPME	Standard addition (spiked in soil)	Optimization of temperature; extraction time; solvent addition; Recovery: 64–109.3% Linear range: 53.3–533 ng/g LOD: 0.11–3.85	[58]

(continued)

Table 6.1 (continued)

Samples	Analytes	Coatings	Separation methods	Calibration methods	Measurement endpoints and relationship	References
Marine sediment	Alcohol ethoxylates	PA	Water-DI-SPME with shaking	ECC (spiked in water)	C_{fiber}, C_{w}, C_{sediment}, sorption parameters	[54]
Manufactured gas plant soils	PAHs	Disposable PDMS	Water-DI-SPME with shaking	EC	$C_{\text{fiber}} C_{\text{w}}$, $C_{\text{sediment}} C_{\text{biota}}$, K_{oc}, mortality, log $C_{\text{biota,cal}}$ versus log $C_{\text{biota, exp}}$	[9]
Soil	PAHs	PDMS	DI-SPME	EC	Compared different method for bioavailability assessment; BCF	[5]
Urban landfill soil	Global screening	PDMS/DVB	NaCl/Water-HS-SPME	/	Global screening	[29]
Soils	Tetracycline antibiotic residues	CW-DVB	Water-hollow membrane protect SPME	ECC	Coupled with CE Reproducibility: 5.3% LOD: 2.9–3.2 μg/kg Recovery: 54–98%	[57]
Estuarine sediments	OCPs	PDMS	MEA-OS-water-HS-SPME	Standard addition	Linear range: 0.1–1000 ng/g LOD: 0.0005–0.11 ng/g LOQ: 0.02–0.45 ng/g	[32]
Soil	Phenanthrene	PDMS fiber	HS-SPME	External calibration (spiked soil)	Identified degradation products	[82]

(continued)

Table 6.1 (continued)

Samples	Analytes	Coatings	Separation methods	Calibration methods	Measurement endpoints and relationship	References
Soil	PAHs	PDMS-coated vials	Water-DI-SPME with shaking	EC	Compared coated vial and SPME fiber; different A/V coating	[76]
Sediment	Permethrin	Disposable PDMS	DI-SPME	EC	f_{oc} versus C_{fiber} and C_{biota} log C_{fiber} versus log C_{biota} $C_{sediment}$ versus C_{biota} SPME showed better correlation with biota, PDMS as a surrogate	[6]
Soil	Pyrene	PDMS	DI-SPME	EC	Effect of organic matter and soil aging was investigated by EC50 for edaphic springtail	[83]
Certified reference sediment	PAHs	PDMS	MWA-ionic liquid-DI-SPME	ECC (spiked in ionic liquid extracted solution)	Three IL coatings were compared for sampling of PAHs Recovery: 84.6–101%	[21]
Soil	Atrazine and ametryn	Dodecylsulfate-doped polypyrrole fiber (PPy)	Water-cooled-HS-SPME	ECC (spiked in soil)	Coupled to IMS; cooled device Linear range: 200–4000 ng/g LOD: 200 ng/g	[60]

(continued)

Table 6.1 (continued)

Samples	Analytes	Coatings	Separation methods	Calibration methods	Measurement endpoints and relationship	References
Surface soil	PAHs	PDMS fiber	Hot water-HS-SPME USA-OS-HS-SPME	Standard addition	Compared to reference material	[84]
Sediment	Pyrethroids	Disposable PDMS	Water-DI-SPME with shaking	EC	Different soil types observe different uptake kinetic Non-agitated *versus* agitated Agitated log C_{fiber} *versus* non-agitated log C_{fiber}	[14]
Soil	16 PAHs	PDMS	USA-OS-water-DI-SPME	IS (spiked in the soil sample)	Five fibers were compared, and 100 μm PDMS was demonstrated to be the best Recovery: 79.3–88.6%	[22]
Marine sediment	Alkylbenzenesulfonate congeners	PA	DI-SPME	EC	Sediment sorption coefficient (K_d), K_F log K_f *versus* log K_d, log $C_{Na+ \text{ or } Ca2+}$ versus log K_d	[3]
Sediment	OCPs	PDMS-DVB	PHW-DI-SPME	ECC (spiked in water)	Extraction time, temperature, organic modifier	[41]

(continued)

Table 6.1 (continued)

Samples	Analytes	Coatings	Separation methods	Calibration methods	Measurement endpoints and relationship	References
Sediment	PBDEs	PA	Water-HS-SPME	Standard addition	Oxidative sample treatment to improve the sensitivity; effect of KMnO4, extraction time, temperature, reproducibility, recovery, and LOQs	[45]
Sewage sludge and sediment	Synthetic polycyclic musk	DVB/PDMS	MEA-Water-HS-SPME	Standard addition (spiked in soil)	Effect of MW power, extraction time, water addition, pH value, salt, and reproducibility LODs: 0.04–0.1 ng/g; RSD: 4–10%; Recovery: 85–96%	[34]
Soil	DDTs	PDMS	DI-SPME	/	C_{fiber}, $C_{sediment}$, C_{biota}, $C_{solvent}$, C_{water} C_{fiber} versus C_{biota}	[7]
Soil	Naphthalene	PDMS	DI-SPME	/	n_{fiber} versus $n_{degraded}$	[18]
Sediment DOM	PBDEs	PDMS fiber	Pore water separated from the sediment and DI-SPME	ECC (spiked pure water)	Compared the concentration of pore water obtained from LLE and SPME for pore water. C_{fiber}, $C_{sediment}$ $\log K_{ow}$ versus $\log K_{oc}$ and K_{DOM}	[46]

(continued)

Table 6.1 (continued)

Samples	Analytes	Coatings	Separation methods	Calibration methods	Measurement endpoints and relationship	References
Soil and sediment	PCBs	Disposable PDMS Coated PDMS	HS-SPME Coated vial	EC	Comparison of fiber and coated vial C_{fiber} versus C_{vial} $C_{olive\ oil}$ versus C_{biota} C_{fiber} versus C_{biota}	[42]
Mar analogue soil	PAHs	PDMS	HS-SPME	ECC (certified reference material)	Detected concentration: 1–60 ng/g;	[25]
Sediment	PAHs Petroleum hydrocarbon mixtures	Disposable PDMS	In situ SPME	EC	log K_{ow} versus BSAF $n_{carbon\ number}$ versus BSAF $C_{biota,cal}$ versus $C_{biota,\ exp}$, in situ SPME provide better prediction than OC prediction, log K_{ow} versus log BAF	[65]
Soil	PAHs	Large volume Silicone rod	DI-SPME	/	C_{rod} versus $C_{cyclodetrin}$ Large volume extraction to quantitatively extract the contaminants; demonstrate the carrier effect of cyclodextrin	[39]

(continued)

Table 6.1 (continued)

Samples	Analytes	Coatings	Separation methods	Calibration methods	Measurement endpoints and relationship	References
Sediment	PBC, PBDEs	Stir bar	OS-USA-centrifudge-Stir bar	Exhaustive extraction	Linear range: 0.5–50 ng/g Repeatability: 3.6–15.0% Recoveries: 62.8–91.5%	[85]
Soil	Thriazine and organophosphorus	PDMS	HS-SPME	ECC (spiked in soil)	The effect of organic matter and clay content was evaluated; compared the HS-SPME with DI-SPME and LSE	[24]
Sediment and soil	PCBs	PDMS	Water-DI-SPME with shaking	Stable isotope calibration	$C_{SI-SPME}$ versus $C_{Eq-SPME}$	[43]
Soil	Explosive compounds	Carbon ceramic copper nanoparticle	Water-DI-SPME	EC	Linear range: 2–160 ng/L LOD: 0.6 μg/g Reproducibility: 9.6% Effect of pH, time, stir speed, and NaCl content,	[61]
Sediment	PCBs		DI-SPME	EC	K_{AC}, k Used as a tool to study the effectiveness of sediment amendment	[44]
Sediment	PAHs	Disposable PDMS	In situ SPME	EC	Depth profiling; a special sampling device $\log K_{AC}$ versus $\log k$	[2]

(continued)

Table 6.1 (continued)

Samples	Analytes	Coatings	Separation methods	Calibration methods	Measurement endpoints and relationship	References
Soil	Aroclor 1260	CAR/PDMS	HS-SPME	IS addition	Recovery: 54.9–65.7% Repeatability <15% Linear range: 10–1000 ng/g	[86]
Marine sediment	Methanol, ethanol	CAR/PDMS	HS-SPME	/	Compared with DAJ and P&T, HS-SPME showed good agreement with P&T Effect of temperature, time, salt addition	[87]
Natural and artificial soil	DDT, PCB	Disposable PDMS	Water-DI-SPME with shaking	EC	$R = C_{fiber}/C_{sediment}$ $R_{natural}$ versus $R_{artificial}$ C_{biota} versus C_{soil}	[15]
OECD soil	C12-benzalkonium	PA	Water-DI-SPME	ECC	C_{fiber}, C_{pw}, C_{biota}, $C_{sediment}$ log C_{pw} versus survival of biota log C_{pw} versus log $C_{sediment}$	[16]
Soil	Aliphatic hydrocarbon	ZnO/polypyrrole	HS-SPME	/	Effect of extraction temperature, time, salt, stir rate LOD: 0.08–0.5 μg/L RSD: 5.4–7.6% and 8.6–10.4% Recovery: 85–108%	[62]

(continued)

Table 6.1 (continued)

Samples	Analytes	Coatings	Separation methods	Calibration methods	Measurement endpoints and relationship	References
Sludge and sediment	Chlorobenzenes	Nanostructure polysiloxane	HS-SPME	ECC	Effect of stir rate, salt, extract time, temperature Repeatability <6% Reproducibility: 3.3–9.7% LOD: 0.15–0.75 ng/L Recovery: 90–99%	[88]
Sand	PAHs	PDMS	Vacuum assisted HS-SPME	/	Extraction temperature, extraction time, modifiers Linear range: 1–400 ng/g LOD: 0.003–0.233 ng/g Repeatability: 4.3–10%	[89]
Sediment	HOCs	PDMS	DI-SPME	EC	C_{pw} versus C_{biota} serve as a proxy for bioaccumulation	[8]
Sediment	Permethrin	Disposable PDMS	DI-SPME	EC	C_{pw}	[64]
Soil	PAHs	Disposable PDMS	DI-SPME	EC	Chemical activity versus bioaccessibility, C_{pw}	[40]

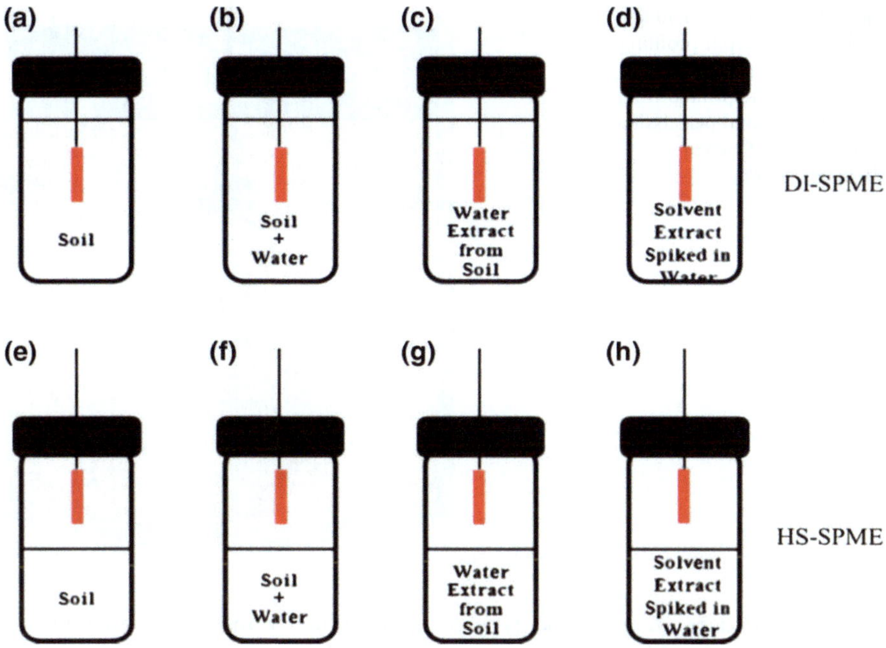

Fig. 6.3 Different sampling formats of SPME for soil and sediment sample. **a** DI-SPME in soil. **b** DI-SPME in soil water mixture.**c** DI-SPME in water that extractant from soil. **d** DI-SPME in solvent extractant spiked water. **e** HS-SPME in soil. **f** HS-SPME in soil water mixture. **g** HS-SPME in water that extractant from soil. **h** HS-SPME in solvent extractant spiked water

ied it for the depth profiling of the sediment treated with activated carbon. The SPME fiber was placed inside a perforated steel tube and directly inserted into the sediment with different depths as shown in Fig. 6.4. Since the volume of the SPME fiber was very small, only the PAHs around the coating were measured, and the concentration at different positions can be distinguished.

Another protocol of the DI-SPME method for soil/sediment sampling was done by adding water into the sample and then exposing the fiber into the suspension mixture (Fig. 6.3b) [9–16]. The added water did not only homogenize the sample, but also facilitated the release of analytes from the solid matrix. This procedure was quite effective if the analytes were not strongly bound to the sample. However, in the case that analytes were strongly adsorbed by the matrix, organic solvents [5, 17, 18] or other pre-extraction solvents such as surfactants [19] or ionic liquid [20] could be used to enhance the release of analytes from the solid sample. After pre-extraction, the organic solvent was spiked into the water prior to DI-SPME analysis (Fig. 6.3d). Besides the addition of water or organic solvent, other techniques such as microwave-assisted (MWA) [20, 21] or ultrasonic-assisted (USA) [22] were utilized to further improve the release of analytes from the solid sample prior to SPME and reduced the consumption of organic solvent. For example, microwave-assisted extraction may achieve 90% reduction in solvent

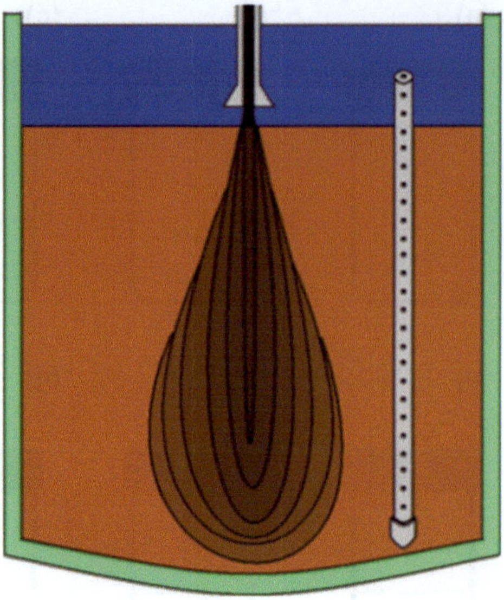

Fig. 6.4 Schematic of in situ SPME for the depth profiling of the organic pollutant in soil/sediment. Reprinted from Ref. [2], with kind permission from Springer Science +Business Media

consumption, depending on the type of application, or may be even higher if microwaves are combined with water extraction [23].

The major limitation of the DI-SPME for soil sampling was the complex matrix effect on the fiber coating. The solid particles may damage the coating during the agitation process, and the dissolved organic matter presented in the pore water may absorb onto the coating and modify the properties of the coating. Conversely, in HS mode, the fiber coating was exposed in the headspace of the solid sample (Fig. 6.3e) without directly contacting the sample matrix, and therefore, the matrix effect can be maximally reduced and can increase the lifetime of the fiber. In HS-SPME, the analytes should be released into the headspace of the sample before being trapped by the fiber [24–27]. As it is known, the analysts in the soil or sediment samples usually strongly bound to the matrix and strategies were required to release them into the HS. Similar to DI-SPME, common methods including adding modifiers such as water [28–31] (Fig. 6.3f, g) and organic solvent were reported in the literature. Salgado-Petinal et al. compared the extraction efficiency of brominated flame retardants (BFRs) from environmental solid sample with and without water addition [28], while Carvalho et al. investigated the effect of water and methanol addition on the sampling of organochlorine pesticides (OCPs) [32]. Results indicated that the addition of water or methanol significantly improved the release of organic pollutants from sediments, and the enhancement efficiency of methanol was better than water (Fig. 6.5). Besides the addition of modifier, microwave [23, 32–34] and ultrasonic wave [35] were often applied to facilitate the release of analytes into the HS too.

In fact, if the target compounds were thermostable, the simplest way to improve the desorption efficiency of analytes from the solid matrix was to use high sampling temperature, which can facilitate the release of analytes from the soil pore.

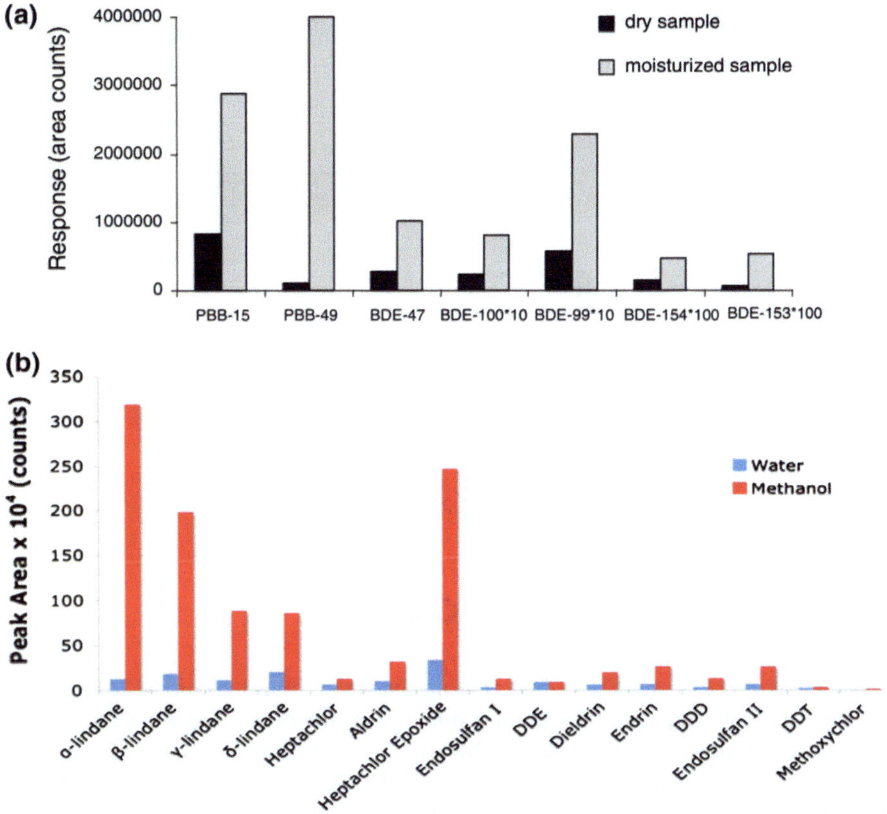

Fig. 6.5 **a** Comparison of the extraction efficiency of BFC with and without water addition [28] and **b** OCPs with water and organic solvent addition. Reprinted from Ref. [32], with kind permission from Springer Science+Business Media

However, according to the fundamentals of SPME, if the sampling temperature increases, the distribution coefficient of analytes between the fiber coating and the sample will decrease and so will the extraction efficiency. In order to simultaneously improve the HS concentration and the distribution coefficient of analytes between the fiber coating and the headspace, cold fiber SPME was introduced. In this approach, the fiber coating was cooled down while the sample matrix was heated up (Fig. 6.5) [36]. This strategy improved analytes' mass transfer rate due to the increase in sampling temperature, and concomitantly, the distribution coefficient was enhanced as the coating temperature was kept low. The cold fiber device has been applied to the headspace sampling of BTEX [36] and polycyclic aromatic hydrocarbons (PAHs) [37, 38] from soil/sediment. Rewarding results were obtained, showing that the extraction efficiency was significantly improved with the cooling effect, at the extent that exhaustive extraction was achieved in some cases (Fig. 6.6).

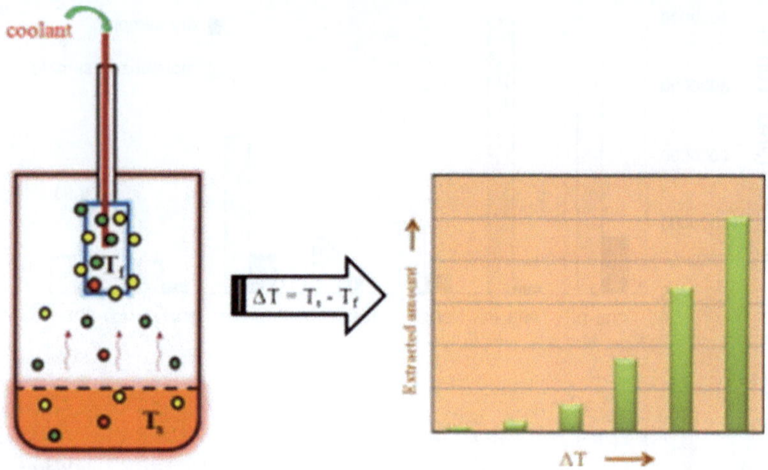

Fig. 6.6 Schematic diagram of cold fiber sampling. Reprinted from Ref. [37], Copyright 2012, with permission from Elsevier

6.3 Selection of SPME Fiber Coatings

The selection of a suitable fiber coating was the first step in SPME method development. Varied commercial and homemade fibers were available in the literature. Commercially available SPME coatings include polydimethylsiloxane (PDMS), polyacrylate (PA), carbon wax (CW)-PDMS, divinylbenzene (DVB)-PDMS, and Carboxen (CAR)-PDMS and can be classified as liquid sorbent and solid sorbent. Liquid coatings such as PDMS, PA, and CW compose primarily "liquid-like" polymers which retain analytes through absorption process. The analytes migrate in and out of the coating and absorb to the coating primarily by polarity. Solid coatings such as CAR-PDMS and DVB-PDMS are porous carbon or polymer which extracts analytes by surface interaction such as pi–pi binding, hydrogen binding, or van der Waals interaction. The retention of analytes depends upon the size of the analytes and the pore diameter.

The selection of a proper fiber coating for a particular application depends on the following three criteria: (i) molecular weight and size of an analyte, (ii) analyte polarity, and (iii) complexity of the matrix. The molecular weight of an analyte determines how rapidly it moves within the coating. Molecular with low weight moves faster and is difficult to be retained by liquid coating, especially the ones with low thickness. Generally, for small analytes (MW < 125), a solid coating is a better choice over a liquid coating and CAR-PDMS fiber is found to be the best choice for this type of compounds. Drescher et al. developed a SPME method using CAR-PDMS fiber to preconcentrate the nitrous oxide (N_2O) and achieved limit of detection of 4.1 nM. Zhang et al. also utilized the CAR-PDMS fiber for the enrichment of small alcohols in marine sediment. The result was comparable to the one obtained from direct aqueous injection, and lower LODs were obtained. On the other hand, for analytes with high molecular weight, liquid coating is a better option. PDMS and PA coatings have been widely used for sampling of PAHs [1, 5, 9, 13, 22, 25, 39, 40], OCPs [30, 41], PCB [12, 32, 33, 42–44], BFR [28], and PBDEs [45, 46]. In addition, DVB sorbent was found to be suitable for many of the large molecules due to the larger pores and has been widely used for sampling of OCPs in soil samples [47–49].

The second criterion for fiber selection is the analyte polarity. According to the "like dissolves like" principle, nonpolar compound tends to dissolve more in nonpolar coating, while polar compound prefers to retain in polar coating. PDMS is the most common nonpolar coating that is highly cross-linked and thermally stable. It has been widely reported for sampling of hydrophobic organic compounds in agriculture soils [47], estuarine sediment [30, 50–52], and landfill soil [28, 33, 53]. There are two polar fiber coatings among commercial fibers: CW-PDMS and PA coatings. PA coating, slightly less polar than CW-PDMS, is suitable for extraction of analyte that is poorly extracted by PDMS. It has been applied for sampling of polar compounds such as 2,4,6-trinitrotoluene [4], chlorophenols [23], alcohol ethoxylates [54], and alkylbenzenesulfonate congeners [3] from soil or sediment. However, when the sample matrix contained modifier such as surfactant, the compound affinity may be influenced by the modifier. Pino et al. [19] found that in the sample matrix containing micellar medium such as polyoxyethylene-10-laurylether, PA coating provided better extraction efficiency to the PAHs than the PDMS coating. In addition, PA has been reported for preconcentration of pesticides [55] and flame retardants [45] in river water and sediments, respectively. As another polar coating, CW-PDMS coating was reported for the sampling of explosives including nitroaromatics, 1,3,5-trinitro-1,3,5-triazine from marine sediment [56], and tetracycline antibiotic residues from soil [57].

The last criterion is the consideration of matrix complexity. This criterion is very important for the soil/sediment sampling whose matrix is usually very complex. For solid coating, competition for space in fiber coating can be challenging. Compounds with low affinity can be displaced by the one with high affinity and resulted in wrong quantification. In soil/sediment, lots of components presented in the sample which may displace the extracted analytes from the coating, and therefore, solid coatings are usually used in HS-SPME sampling.

In addition to commercial fibers, homemade fibers were widely investigated to overcome the limitations of commercial fibers regarding extraction efficiency and cost. Homemade coatings including diglycidyloxycalix[4]arene [58], acrylate/silicone copolymer [59], dodecylsulfate-doped polypyrrole fiber (PPy) [60], carbon ceramic copper nanoparticle [61], and ZnO/polypyrrole [62] were prepared to sample the chlorobenzenes, 2-chloroethyl ethyl sulfide, atrazine and ametryn, explosive compounds, and aliphatic hydrocarbon in soil sample, respectively. Results showed much higher extraction efficiency compared to the commercial PDMS and PA fibers.

6.4 Selection of Extraction-Phase Geometry

The originally developed geometry of SPME was the syringe-like fiber in which the extraction phase was coated on a fiber support such as fused silica fiber or metal wire which was attached to an assembly used to protect the fiber and penetrate the septum. The syringe-like SPME fiber can be mounted on a commercial holder for easy handling or coupled with an autosampler to achieve high throughput and precision. During sampling, the needle was pierced through the septum on the vial cap, and the whole device was placed on top of the vial containing the soil sample. Next, the fiber coating was exposed directly to the sample or the headspace of the sample for a period of time. After sampling, the fiber can be directly injected into an analytical instrument for analysis (Fig. 6.7a). The advantages of syringe-like SPME were related to the simple and fast operation process because it integrated the sampling, sample concentration, and sample introduction into one step. The syringe-like SPME can be easily integrated with analytical instrument such as GC-MS. It has also been reported to couple with IMS for the determination of degradation product of chemical warfare agent [26] and atrazine and ametryn [60]. Therefore, the syringe-like SPME fiber was often used for the in-lab sampling.

Disposable type SPME utilized SPME fibers but without the accessories, leading much increased flexibility and feasibility (Fig. 6.7b). The fiber was exposed directly to solid sample for a predetermined period of time, and then, the fiber was removed from the sample using tweezers, washed with Milli-Q water, and dried with a lint-free tissue. The extracted fibers can be thermally desorbed in the injector of conventional GCs [63], which however was not very time efficient. A more practical approach was to extract with a small amount of solvent, which then can be quantified not only by GC [15, 64] but also by HPLC [2, 40, 65]. A unique

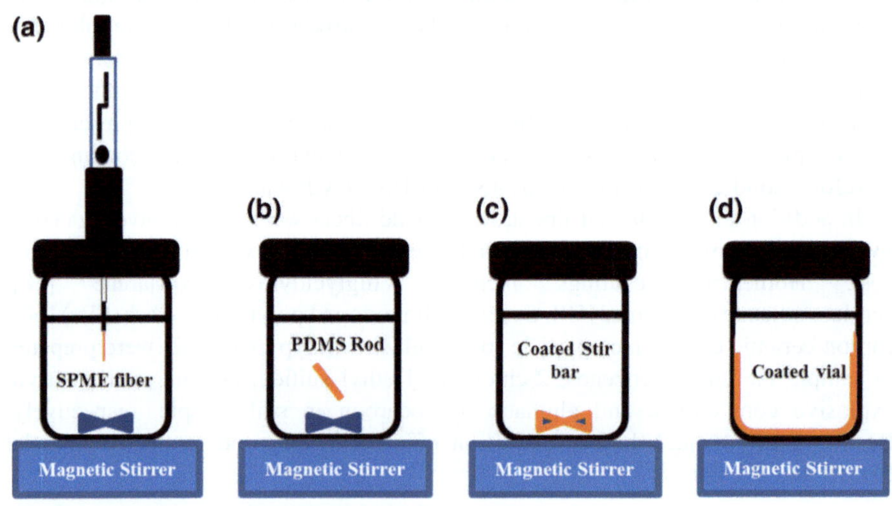

Fig. 6.7 Different sampling geometries of extraction phase

advantage of disposable SPME was its compatibility with bench-scale bioassays, largely because of the small size and flexibility of disposable fibers. The concurrent exposure of disposable fibers and organisms in the same bioassay chambers was considered to be more representative of organism exposure [4, 6, 11, 42, 65, 66]. Another advantage of the disposable type SPME was that different volume of the coatings can be chosen to improve the sampling sensitivity especially for the compounds that have trace amount in the environment [39].

Stir bar sorptive extraction (SBSE) was introduced by Baltussen et al. [67] based on the principle of SPME. A stir bar coated with 50–300 μL of PDMS was employed as extraction phase to preconcentrate analytes from aqueous sample. In general, the PDMS-coated stir bar was placed inside the water (Fig. 6.7c). After stirring for a period of time, the stir bar was removed from the solution and rinsed with deionized water, and dried and placed in an "injector tray." The compounds were thermo-desorbed in a programmable temperature vaporization injector [68–70] or solvent desorbed in a vial [71, 72] before introducing into GC instrument. SBSE presented advantages over traditional sample preparation techniques because no/less organic solvent was used and small volume of sample was required to achieve comparable sensitivity. With respect to SPME, lower detection limits and higher capacity can be obtained due to the larger volume of extraction phase. SBSE has been also reported for sampling the organic pesticides and phenolic organic pollutants in soil sample. Since the extraction phase itself was a magnetic stir bar, it was difficult to directly agitate in the soil sample. SBSE was usually coupled with other techniques such as solvent desorption [10], ultrasonic desorption [73], or pressurized subcritical water extraction [74] to extract the compounds from the soil matrixes.

Another geometry of the SPME technique was the coated vial. Wilcockson and Gobas first introduced the in-vial coating concept by depositing a thin layer of ethylene-vinyl acetate (EVA) on the internal wall of glass surface [75]. Reichenberg et al. extended this approach to coat the PDMS film on the wall of the inner vial [76]. The sampling procedure generally followed: (1) preparation of the coated vail; (2) equilibration of the targeted compounds between the coated vials and the investigated soil or sediment; (3) quantification of the targeted compounds extracted by the coated vial to determine the concentration on the polymer (C_{coating}); and (4) calculation of the free concentration in the soil or sediment (C_{free}) from C_{coating} and a predetermined partition coefficient. The general advantages of coated vials were as follows. First, the very thin coating thickness provided short equilibration time. A relatively large surface area-to-volume (A/V) ratio in the vial coating method accelerated extraction rate. For example, the A/V ratio of SPME fiber with 100 μm coating was 15.4 μL, while the A/V ratio for a coated vial with 2 μm coating was 500 μL [75]. Mäenpää et al. demonstrated that equilibrium could be reached within 2 weeks for PCB congeners, including PCB-209, when contaminated sediments were equilibrated with PDMS-coated jars under agitated conditions [42]. In addition, equilibrium may be further speed up by decreasing the thickness of polymer coating [75]. Reichenberg et al. developed such a method for contaminated soil, applying coated vials with different thicknesses ranging from 3 to 12 μm. The major advantage over other approaches was the built-in validation consisting of different polymer coating thicknesses that confirmed equilibrium partitioning between each sample and the polymer while at the same time showing that no sampling artifacts such as surface abrasion of the PDMS coatings were present [76]. Second, a sufficiently large polymer volume provided good sensitivity. Report found that by using a PDMS-coated jars, PCBs in Baltic Sea sediments were detected at levels as low as fg/L and pg/L in a recent study [77]. Last, agitation was allowed in this sampling format to ensure the fast sampling equilibrium.

6.5 Calibration Methods for SPME

SPME is a non-exhaustive sample preparation technique that requires proper calibration method to quantify the concentration in the sample matrix base on fiber extracted amount. Traditional analytical calibration methods including external calibration, standard addition, and internal calibration have been reported for the quantification of organic pollutants in soil/sediment sample when using SPME sampling (Table 6.1). External calibration involves the preparation of several standard samples using matrix-matched blank samples that can be uncontaminated soil or certified reference soil. External calibration is the simplest calibration method, but it may not be an appropriate method for the complex samples whose matrix-matched blank sample is difficult to obtain. Experimental result already demonstrated that the soil composition could significantly influence the slope of the calibration curves [24], which in turn indicated that the selection of matrix-matched sample was very important for external calibration method. In the case that the analytes were first extracted by water or the organic solvent, the matrix-matched blank sample for the preparation of the calibration curves was commonly used pure water or simulative solution.

Standard addition method was accomplished by adding known quantities of the targeted analytes to the sample matrix [32, 34, 58], which initially contained an unknown concentration of the analytes. A plot of the response versus the added concentration of the analytes is then developed, and the extrapolation of the plot to zero defined the original concentration of analytes in the sample. Compared to external calibration, this approach was able to compensate the matrix effects to some extent. However, it can be extremely tedious and time-consuming since it required to analyze large number of samples. In addition, a large amount of unknown samples was usually needed for preparation of the plot.

Internal standard approach involved adding a known concentration of internal standard that was not present in the sample matrix and can mimic the behavior of the analytes during extraction and analysis. Calibration then was done by comparing the ratio of the analytes signals to the internal standard with a known concentration [22]. This method showed significant advantages in terms of minimizing matrix effects and loss of target analytes during the sample preparation.

However, it was difficult to choose suitable compounds to use as internal standards. Isotopically labeled internal standards were preferable, but they were usually not easy to find and generally expensive.

Besides traditional calibration methods, equilibrium calibration (EC) and exhaustive calibration method were also reported for soil sample quantification [5, 9]. Equilibrium calibration was based on the relationship between the equilibrium extraction amount and the sample matrix concentration, which can be described using Eq. 6.1:

$$n_e = \frac{K_{fs} V_f V_s}{K_{fs} V_f + V_s} C_0 \tag{6.1}$$

where n_e is the extracted amount of the analyte, C_0 is the initial concentration of the target analyte in the sample, V_s is the sample volume, V_f is the fiber coating volume, and K_{fs} is the distribution coefficient of the analytes between the fiber coating and the sample. Equation 6.1 indicated that the analyte concentration in the sample (C_0) was related tithe amount of analyte extracted onto the coating (n_e), which was the analytical basis for quantification using SPME. When the sample volume was very large, such as in field sampling, then $V_s \gg K_{fs} V_f$ and Eq. 6.1 can be simplified to Eq. 6.2:

$$n_e = K_{fs} V_f C_0 \tag{6.2}$$

Equation 6.2 indicated that the equilibrium extracted amount was independent of the sample volume. In practice, there was no need to collect a defined sample prior to analysis, since the fiber can be exposed directly to the soil or sediment. The amount of extracted analyte will correspond directly to its concentration in the matrix, without depending on the sample volume. The SPME fiber was directly exposed in the soil or sediment for a period of time till it reached equilibrium. Then, amount extracted by the fiber was quantified and the concentration of the target analytes in the soil/sediment can be determined by knowing the distribution coefficients of the analytes between the fiber coating and the sample matrix. The limitation of this method was the long equilibrium time and the difficulty of obtaining an accurate distribution coefficient of analytes between the sample and the fiber. This method has been widely applied for the determination of the bioavailability of the analytes present in the soil sample. Since SPME only senses the freely dissolved analytes in the sample, if the K_{fs} value in Eq. 6.2 indicates the distribution coefficient of analytes between the pure water and the coating, the calibrated C_0 represents the freely dissolved concentration in the sample matrix, which usually correlates to the concentration of the analyte that is available for bioaccumulation.

In the case that $V_s \ll K_{fs} V_f$, Eq. 6.1 can be simplified to $n = V_s C_0$, indicating that all the analytes present in the sample matrix were extracted by the fiber coating. The extraction was an exhaustive extraction which rarely happened unless a high distribution coefficient was achieved. Cold fiber SPME have been reported for exhaustive sampling BTEX from the spiked sediment [38], and the concentration of the sample was determined by dividing the extracted amount to the volume of the sample.

In-fiber standardization method was one of the most popular calibration methods when SPME was used for in vivo or in situ sampling, in which other calibration methods were difficult to apply. The fundamental basis of in-fiber standardization method was the isotropy of absorption and desorption process of SPME. This method used the desorption of standards (also called performance reference compounds, PCRs) to calibrate the extraction of analytes (Fig. 6.8). The relationship between the extracted amount and the desorption amount can be described by Eq. 6.3:

$$\frac{n}{n_e} + \frac{Q}{q_0} = 1 \tag{6.3}$$

where n is the fiber extracted amount; Q is the amount of loaded standards left on the fiber; q_0 is the original loaded amount of standards; and n_e is the equilibrium extracted amount which for the field sampling which can be expressed by Eq. 6.2. Equation 6.3 can be rearranged to Eq. 6.4:

$$C_0 = \frac{q_0 n}{K_{fs} V_f (q_0 - Q)} \tag{6.4}$$

C_0 is the concentration of targeted compounds in the sample matrix. SPME in-fiber standard calibration method has been reported to measure the bioavailability of the HOCs in sediment. Compared to the equilibrium extraction method, the in-fiber

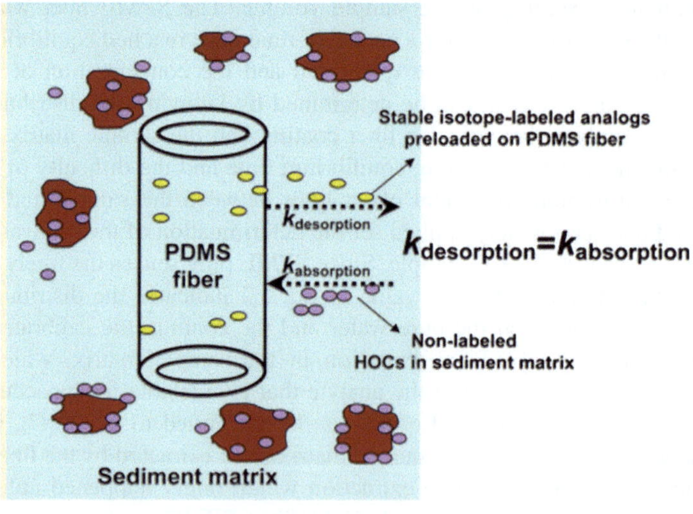

Fig. 6.8 Schematic diagram of the in-fiber standardization process. Reprinted with the permission from Ref. [43]. Copyright 2013 American Chemical Society

standard calibration method significantly shortens the sampling time and improves the accuracy of the results. However, the limitation of this method was the usage of deuterated compounds which were expensive and sometimes not available. In addition, reproducible loading procedure should be developed for the loading process.

6.6 Application

One of the most important applications of SPME for soil/sediment is the determination of bioavailability concentration of organic contaminants. The term "bioavailability" has been widely used in ecotoxicology and risk assessment. However, its definition seems to vary with discipline-specific designations. It has been indicated that the concept of bioavailability may be better inferred from the methods used for its measurement. Two endpoints have been commonly used by the researchers to measure the bioavailability of the organic pollutants, bioaccessibility, and chemical activity [76]. Bioaccessibility is defined as the measurement of the compound fraction that is weakly or reversibly absorbed and can undergo rapid desorption from the solid phase to the aqueous phase, while chemical activity refers to the potential of HOCs to partition into organism at equilibrium according to the equilibrium partitioning theory [78].

Chemical activity depicted the potential for a chemical to undergo spontaneous processes such as diffusion and partitioning. At equilibrium, chemical activity was the same in each matrix compartment (e.g., water, soil/sediment, SPME fiber, and biota). Consequently, at equilibrium, the HOC concentration in one compartment was proportional to that in another. When the relationship between two phases was known, the HOC concentration in one phase can be used along with a partition coefficient (e.g., K_{oc} or BCF) to predict the concentration in another phase. Therefore, by determining the chemical activity in pore water of the soil/sediment matrix, the chemical activity in other phases can be determined.

For HOCs present at environmentally relevant level, chemical activity was represented by the freely dissolved concentration. As a non-exhaustive extraction method, SPME has been increasingly adopted in analyzing the freely dissolved concentration of organic contaminants in soil/sediment. The relationship between the freely dissolved concentration of the contaminant and the extracted concentration on the fiber can be described using Eq. 6.5

$$K_{ff} = \frac{C_{fiber}}{C_{free}} \tag{6.5}$$

K_{ff} is the distribution coefficient between the fiber coating and the pure water. C_{fiber} is the concentration on the fiber coating, while C_{free} is the freely dissolved organic compounds in the sample matrix. By determining the C_{fiber} and knowing the K_{ff}, one can obtain the bioavailability concentration C_{free}. The matrix-SPME was based on the above-mentioned fundamental to determine the C_{free} in the sediment. Later, this method has been widely applied to monitor the free concentration of organic pollutants in soil/sediment after amended with activated carbon [2] or to investigate the effect of soil content on the organic pollutants [14]. More often, this technique was further applied to measure the distribution coefficient between the soil and pore water defined as the Eq. 6.6.

$$K_{\text{oc}} = \frac{C_{\text{soil}}}{C_{\text{free}} f_{\text{oc}}} \tag{6.6}$$

K_{oc} is the distribution coefficient between the soil sample and the pore water. C_{soil} is the concentration in the soil sample, while f_{oc} is the fraction of the organic carbon. If there is biota present in the soil/sediment, the organic pollutants also distribute into the biota, and the concentration in the biota and the pore water can be correlated using Eq. 6.7.

$$\text{BCF} = \frac{C_{\text{biota}}}{C_{\text{free}} f_{\text{lipid}}} \tag{6.7}$$

BCF is the bioconcentration factor. C_{biota} is the concentration in the biota, while f_{lipid} is the percentage of lipid in the biota. The values of K_{ff}, K_{oc}, and BCF relate the concentrations in different phases, and all correlate to the C_{free}. Rearranging the equations, one can find the linear relationship between C_{free}, C_{fiber}, C_{soil}, and C_{biota} (as shown in Fig. 6.9).

Conder et al. [4] investigated the relationship between the C_{soil} and the C_{biota} in amended and unamended soils and found linearity between the C_{soil} and C_{biota} in unamended soil but not in the amended soil. On the other hand, the significant relationships were found between the C_{biota} and the C_{fiber} for both amended and unamended soils. Similar relationship was observed in the accumulation of chlorobenzene and PAHs in PDMS fiber and *E. crypticus* [11] and earthworm [12], respectively. The stability of SPME to biota relationship suggested that SPME was predictive of bioavailability of organic pollutants. In a paper published by Jonker

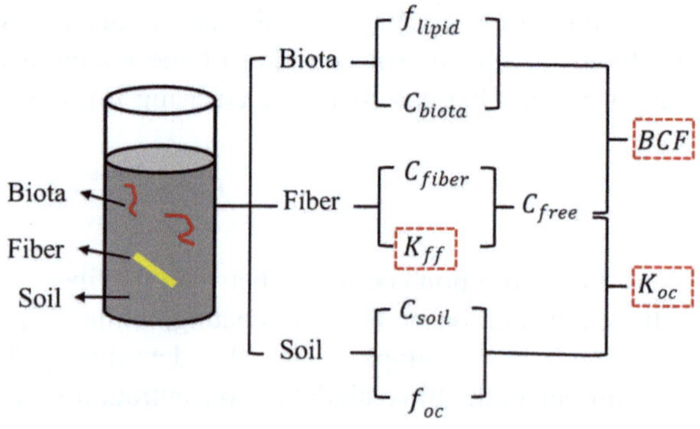

Fig. 6.9 Concentration relationship between the biota, organic matter, SPME fiber, and pore water in soil

et al, they compared the measured concentration of PAHs in earthworm with the ones predicted by SPME measured free concentration and the model published by Nguyen et al. The result was shown in Fig. 6.10, and the C_{biota} predicted by the SPME fiber provided better correlation to the measured C_{biota} than the one predicted by the model. An important advantage of SPME for biomimetic purpose is that only the bioavailable fraction of the toxicants partitions to the fiber is measured. In contrast to exhaustive extraction, biomimetic SPME is more similar to the field situation where uptake by biota does not affect the environmental concentrations.

The SPME has been reported for analysis of a variety of solid samples (see Table 6.1). The most frequently reported one is for sampling the widespread organic pollutant polycyclic aromatic hydrocarbons (PAHs). For PAHs, headspace SPME has been proved to be an efficient method to extract the low-ring PAHs [25], while the high-ring PAHs were difficult to be extracted because of their low volatility and strong adsorption to the particles [19]. However, most high-ring PAHs, such as benzo[a]pyrene and benzo[ghi]perylene, were more toxic and have higher risk of carcinogenesis and mutagenesis than low-ring PAHs [22]. In addition, high-ring PAHs were prone to accumulate in solid matrix due to their lipophilic character. Thus, it was imperative to accurate quantification for high-ring PAHs in the soil samples. DI-SPME has been commonly applied to sample the PAHs in solid samples in combination with other extraction methodologies, such as LLE [5, 9], ultrasonic-assisted [22], and microwave-assisted extraction [19, 21]. For example, with the help of micellar media and ultrasonic, SPME can simultaneously quantify 10 of 16 PAHs in soil sample [22]. SPME coupled with LLE was reported to sample a large range of PAHs with different rings [25].

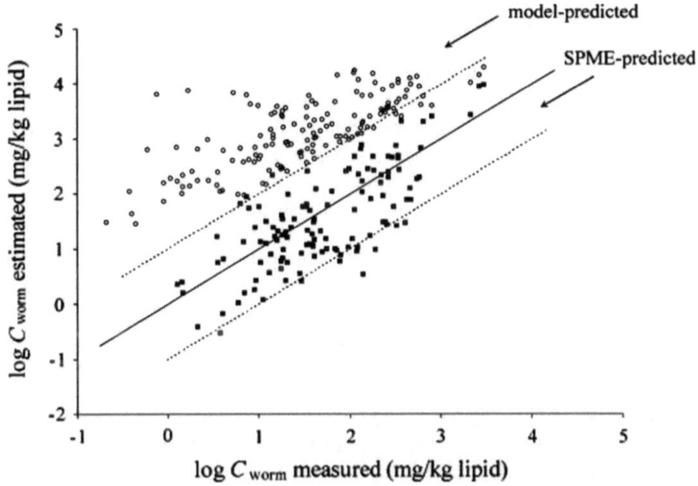

Fig. 6.10 Predicted PAH concentrations in worms versus experimentally determined concentrations (both mg/kg lipid on a logarithmic scale). The model predicted data were calculated from the one-parameter LFER for PAHs as reported by Nguyen. Reprinted with the permission from Ref. [9]. Copyright 2007 American Chemical Society

The SPME method was also widely used for the pesticide residue sampling from environmental samples. However, most of the applications on pesticides were focused on liquid samples, and the soil/sediment sampling was relatively scarce, probably due to some experimental limitations of the technique in the soil application. First of all, the strong interaction among the pesticide molecules and different constituents of the sediments requires other strategies to facilitate the desorption of analytes from the solid matrix. Similar to PAHs, water/organic solvent [10, 30, 79] and microwave [32] or ultrasonic were usually applied to pre-extract the compounds from the soil and sediment before using SPME. Another limitation was the difficulty in calibration due to the complex soil matrix. Some authors investigated the effect of soil organic matter or clay content on the recovery of pesticides present in the soil, and the results indicated that the co-extractants from the analyzed medium complicated the evaporation and diffusion of the pesticides to the SPME fiber during the HS-SPME sampling, and the spiked standard calibration curve cannot be represented for the complex matrix [24, 80].

Acknowledgement We acknowledge financial support from the projects of NNSFC (Grants 21407184, 21225731, 21377172, and 21477166) and the NSF of Guangdong Province (Grant S2013030013474).

References

1. Mayer P, Vaes WHJ, Wijnker F et al (2000) Sensing dissolved sediment porewater concentrations of persistent and bioaccumulative pollutants using disposable solid-phase microextraction fibers. Environ Sci Technol 34:5177–5183. doi:10.1021/es001179g
2. Stringer RD, Burken JG, Elmore AC, Reible DD (2014) Using in situ solid phase microextraction (SPME) for depth profiling in sediments treated with activated carbon. J Soils Sediments. doi:10.1007/s11368-014-0857-9
3. Rico-Rico A, Droge STJ, Hermens JLM (2010) Predicting sediment sorption coefficients for linear alkylbenzenesulfonate congeners from polyacrylate-water partition coefficients at different salinities. Environ Sci Technol 44:941–947. doi:10.1021/es902453s
4. Conder JM, La Point TW (2005) Solid-phase microextraction for predicting the bioavailability of 2,4,6-trinitrotoluene and its primary transformation products in sediment and water. Environ Toxicol Chem 24:1059–1066
5. Bergknut M, Sehlin E, Lundstedt S et al (2007) Comparison of techniques for estimating PAH bioavailability: uptake in *Eisenia fetida*, passive samplers and leaching using various solvents and additives. Environ Pollut 145:154–160. doi:10.1016/j.envpol.2006.03.052
6. Hunter W, Xu Y, Spurlock F, Gan J (2008) Using disposable polydimethylsiloxane fibers to assess the bioavailability of permethrin in sediment. Environ Toxicol Chem 27:568–575. doi:10.1897/07-335
7. Fang H, Chu X, Wang X et al (2010) Using matrix solid-phase microextraction (matrix-SPME) to estimate bioavailability of DDTs in soil to both earthworm and vegetables. Arch Environ Contam Toxicol 58:62–70. doi:10.1007/s00244-009-9329-4
8. Maruya KA, Lao W, Tsukada D, Diehl DW (2015) A passive sampler based on solid phase microextraction (SPME) for sediment-associated organic pollutants: comparing freely-dissolved concentration with bioaccumulation. Chemosphere 137:192–197. doi:10.1016/j.chemosphere.2015.07.042

9. Jonker MTO, Van der Heijden SA, Kreitinger JP, Hawthorne SB (2007) Predicting PAH bioaccumulation and toxicity in earthworms exposed to manufactured gas plant soils with solid-phase microextraction. Environ Sci Technol 41:7472–7478
10. Zhao R, Wang X, Yuan J et al (2006) A novel headspace solid-phase microextraction method for the exact determination of organochlorine pesticides in environmental soil samples. Anal Bioanal Chem 384:1584–1589. doi:10.1007/s00216-006-0341-5
11. Van der Wal L, van Gestel CAM, Hermens JLM (2004) Solid phase microextraction as a tool to predict internal concentrations of soil contaminants in terrestrial organisms after exposure to a laboratory standard soil. Chemosphere 54:561–568. doi:10.1016/j.chemosphere.2003.08.016
12. Van der Wal L, Jager T, Fleuren RRHLJ et al (2004) Solid-phase microextraction to predict bioavailability and accumulation of organic micropollutants in terrestrial organisms after exposure to a field-contaminated soil. Environ Sci Technol 38:4842–4848. doi:10.1021/es035318g
13. Ter Laak TL, Barendregt A, Hermens JLM et al (2006) Freely dissolved pore water concentrations and sorption coefficients of PAHs in spiked, aged, and field-contaminated soils. Environ Sci Technol 40:2184–2190. doi:10.1021/es0524548
14. Hunter W, Yang Y, Reichenberg F et al (2009) Measuring pyrethroids in sediment pore water using matrix-solid phase microextraction. Environ Toxicol Chem 28:36–43. doi:10.1897/08-209.1
15. Bielská L, Šmídová K, Hofman J (2014) Solid phase microextraction of organic pollutants from natural and artificial soils and comparison with bioaccumulation in earthworms. Ecotoxicol Environ Saf 100:44–52. doi:10.1016/j.ecoenv.2013.11.011
16. Chen Y, Geurts M, Sjollema SB et al (2014) Acute toxicity of the cationic surfactant C12-benzalkonium in different bioassays: how test design affects bioavailability and effect concentrations. Environ Toxicol Chem 33:606–615. doi:10.1002/etc.2465
17. Liu H-C, Hwu C-S, Lu C-J (2010) Effect of surfactant on the estimation by solid phase microextraction of bioavailable pyrene in soil samples. World J Microbiol Biotechnol 27:1071–1076. doi:10.1007/s11274-010-0552-y
18. Liu H-C, Hwu C-S, Chu K-C, Lu C-J (2010) Estimation of bioavailability and potential risks of naphthalene in soils with solid phase microextraction. World J Microbiol Biotechnol 26:1311–1316. doi:10.1007/s11274-009-0302-1
19. Pino V, Ayala JH, Afonso AM, González V (2003) Micellar microwave-assisted extraction combined with solid-phase microextraction for the determination of polycyclic aromatic hydrocarbons in a certified marine sediment. Anal Chim Acta 477:81–91. doi:10.1016/S0003-2670(02)01410-1
20. Pino V, Ayala JH, González V, Afonso AM (2007) Determination of the alkyl- and methoxy-phenolic content in wood extractives by micellar solid-phase microextraction and gas chromatography-mass spectrometry. Talanta 73:505–513. doi:10.1016/j.talanta.2007.04.013
21. Guerra-Abreu L, Pino V, Anderson JL, Afonso AM (2008) Coupling the extraction efficiency of imidazolium-based ionic liquid aggregates with solid-phase microextraction-gas chromatography-mass spectrometry. Application to polycyclic aromatic hydrocarbons in a certified reference sediment. J Chromatogr A 1214:23–29. doi:10.1016/j.chroma.2008.10.084
22. Wang Y, Zhang J, Ding Y et al (2009) Quantitative determination of 16 polycyclic aromatic hydrocarbons in soil samples using solid-phase microextraction. J Sep Sci 32:3951–3957. doi:10.1002/jssc.200900420
23. Wei M, Jen J (2003) Determination of chlorophenols in soil samples by microwave-assisted extraction coupled to headspace solid-phase microextraction and gas chromatography-electron-capture detection. J Chromatogr A 1012:111–118
24. Đurović RD, Đorđević TM (2012) Effects of soil composition on solid phase microextraction determination of triazine and organophosphorus pesticides. J Environ Sci Heal Part B 47:851–857. doi:10.1080/03601234.2012.693860

25. Orzechowska G, Kidd R (2011) Analysis of Mars analogue soil samples using solid-phase microextraction, organic solvent extraction and gas chromatography/mass spectrometry. Int J Astrobiol 10:209–219. doi:10.1017/S1473550410000443
26. Rearden P, Harrington PB (2005) Rapid screening of precursor and degradation products of chemical warfare agents in soil by solid-phase microextraction ion mobility spectrometry (SPME–IMS). Anal Chim Acta 545:13–20. doi:10.1016/j.aca.2005.04.035
27. Ezquerro Ó, Ortiz G, Pons B, Tena MT (2004) Determination of benzene, toluene, ethylbenzene and xylenes in soils by multiple headspace solid-phase microextraction. J Chromatogr A 1035:17–22. doi:10.1016/j.chroma.2004.02.030
28. Salgado-Petinal C, Garcia-Chao M, Llompart M et al (2006) Headspace solid-phase microextraction gas chromatography tandem mass spectrometry for the determination of brominated flame retardants in environmental solid samples. Anal Bioanal Chem 385:637–644. doi:10.1007/s00216-006-0440-3
29. Banar M, Ozkan A, Vardar C (2007) Characterization of an urban landfill soil by using physicochemical analysis and solid phase microextraction (SPME)-GC/MS. Environ Monit Assess 127:337–351. doi:10.1007/s10661-006-9284-8
30. Chang SM, Doong RA (2006) Concentration and fate of persistent organochlorine pesticides in estuarine sediments using headspace solid-phase microextraction. Chemosphere 62:1869–1878. doi:10.1016/j.chemosphere.2005.07.023
31. De Souza Silveira CD, Martendal E, Soldi V, Carasek E (2012) Application of solid-phase microextraction and gas chromatography-mass spectrometry for the determination of chlorophenols in leather. J Sep Sci 35:602–607. doi:10.1002/jssc.201100726
32. Carvalho PN, Rodrigues PNR, Alves F et al (2008) An expeditious method for the determination of organochlorine pesticides residues in estuarine sediments using microwave assisted pre-extraction and automated headspace solid-phase microextraction coupled to gas chromatography-mass spectrometry. Talanta 76:1124–1129. doi:10.1016/j.talanta.2008.05.035
33. Herbert P, Silva AL, João MJ et al (2006) Determination of semi-volatile priority pollutants in landfill leachates and sediments using microwave-assisted headspace solid-phase microextraction. Anal Bioanal Chem 386:324–331. doi:10.1007/s00216-006-0632-x
34. Wu S-F, Ding W-H (2010) Fast determination of synthetic polycyclic musks in sewage sludge and sediments by microwave-assisted headspace solid-phase microextraction and gas chromatography-mass spectrometry. J Chromatogr A 1217:2776–2781. doi:10.1016/j.chroma.2010.02.067
35. Zygmunt B, Namiesnik J (2001) Solid-phase microextraction-gas chromatographic determination of volatile monoaromatic hydrocarbons in soil. Fresenius J Anal Chem 370:1096–1099
36. Zhang Z, Pawliszyn J (1995) Quantitative extraction using an internally cooled solid phase microextraction device. Anal Chem 67:34–43. doi:10.1021/ac00097a007
37. Jiang R, Carasek E, Risticevic S et al (2012) Evaluation of a completely automated cold fiber device using compounds with varying volatility and polarity. Anal Chim Acta 742:22–29. doi:10.1016/j.aca.2012.01.010
38. Ghiasvand AR, Hosseinzadeh S, Pawliszyn J (2006) New cold-fiber headspace solid-phase microextraction device for quantitative extraction of polycyclic aromatic hydrocarbons in sediment. J Chromatogr A 1124:35–42. doi:10.1016/j.chroma.2006.04.088
39. Gouliarmou V, Mayer P (2012) Sorptive bioaccessibility extraction (SBE) of soils: combining a mobilization medium with an absorption sink. Environ Sci Technol 46:10682–10689. doi:10.1021/es301515s
40. Guo M, Gong Z, Allinson G et al (2016) Variations in the bioavailability of polycyclic aromatic hydrocarbons in industrial and agricultural soils after bioremediation. Chemosphere 144:1513–1520. doi:10.1016/j.chemosphere.2015.10.027
41. Concha-graña E, Fernández-gonzález V, Grueiro-noche G, Muniategui-lorenzo S (2010) Development of an environmental friendly method for the analysis of organochlorine pesticides in sediments. Chemosphere 79:698–705. doi:10.1016/j.chemosphere.2010.02.052

42. Mäenpää K, Leppänen MT, Reichenberg F et al (2011) Equilibrium sampling of persistent and bioaccumulative compounds in soil and sediment: comparison of two approaches to determine equilibrium partitioning concentrations in lipids. Environ Sci Technol 45:1041–1047. doi:10.1021/es1029969
43. Cui X, Bao L, Gan J (2013) Solid-phase microextraction (SPME) with stable isotope calibration for measuring bioavailability of hydrophobic organic contaminants. Environ Sci Technol 47:9833–9840. doi:10.1021/es4022987
44. Rakowska MI, Kupryianchyk D, Koelmans AA et al (2014) Equilibrium and kinetic modeling of contaminant immobilization by activated carbon amended to sediments in the field. Water Res 67:96–104. doi:10.1016/j.watres.2014.07.046
45. Montes R, Rodríguez I, Cela R (2010) Solid-phase microextraction with simultaneous oxidative sample treatment for the sensitive determination of tetra- to hexa-brominated diphenyl ethers in sediments. J Chromatogr A 1217:14–21. doi:10.1016/j.chroma.2009.11.029
46. Wang W, Delgado-Moreno L, Ye Q, Gan J (2011) Improved measurements of partition coefficients for polybrominated diphenyl ethers. Environ Sci Technol 45:1521–1527. doi:10.1021/es103087a
47. Guez Ä, Vega Moreno G, Sosa Ferrera Z, Santana Rodríguez JJ (2006) Sample extraction method combining micellar extraction-SPME and HPLC for the determination of organochlorine pesticides in agricultural soils. J Agric Food Chem 54:7747–7752
48. Moreno DV, Ferrera ZS, Rodr JS (2006) Microwave assisted micellar extraction coupled with solid phase microextraction for the determination of organochlorine pesticides in soil samples. Anal Chim Acta 571:51–57. doi:10.1016/j.aca.2006.04.046
49. Lu Y, Zhu Y (2013) Determination of chlorobenzenes in textiles by pressurized hot water extraction followed by vortex-assisted liquid-liquid microextraction and gas chromatography-mass spectrometry. J Chromatogr A 1319:27–34. doi:10.1016/j.chroma.2013.10.050
50. Carvalho P, Rodrigues P (2009) Organochlorine pesticides levels in Portuguese coastal areas. Chemosphere 75:595–600. doi:10.1016/j.chemosphere.2009.01.060
51. Carvalho PN, Rodrigues PNR, Alves F et al (2008) An expeditious method for the determination of organochlorine pesticides residues in estuarine sediments using microwave assisted pre-extraction and automated headspace solid-phase microextraction coupled to gas chromatography-mass spectrometry. Talanta 76:1124–1129. doi:10.1016/j.talanta.2008.05.035
52. Montes R, Ramil M, Rodriguez I (2006) Rapid screening of polychlorinated biphenyls in sediments using non-equilibrium solid-phase microextraction and fast gas chromatography with electron-capture detection. J Chromatogr A 1124:43–50. doi:10.1016/j.chroma.2006.04.047
53. Barri T, Bergstr S, Hussen A et al (2006) Extracting syringe for determination of organochlorine pesticides in leachate water and soil-water slurry: a novel technology for environmental analysis. J Chromatogr A 1111:11–20. doi:10.1016/j.chroma.2006.01.097
54. Droge STJ, Hermens JLM (2007) Nonlinear sorption of three alcohol ethoxylates to marine sediment: a combined Langmuir and linear sorption process. Environ Sci Technol 41:3192–3198. doi:10.1021/es062608z
55. Bonansea RI, Amé MV, Wunderlin DA (2013) Determination of priority pesticides in water samples combining SPE and SPME coupled to GC-MS. A case study: Suquía River basin (Argentina). Chemosphere 90:1860–1869. doi:10.1016/j.chemosphere.2012.10.007
56. Monteil-Rivera F, Beaulieu C, Hawari J (2005) Use of solid-phase microextraction/gas chromatography—electron capture detection for the determination of energetic chemicals in marine samples. J Chromatogr A 1066:177–187. doi:10.1016/j.chroma.2005.01.049
57. Santos B, Simonet BM, Ríos A, Valcárcel M (2007) On-line coupling of solid-phase microextraction to commercial CE-MS equipment. Electrophoresis 28:1312–1318. doi:10.1002/elps.200600429

58. Li X, Zeng Z, Xu Y (2006) A solid-phase microextraction fiber coated with diglycidyloxy-calix 4]arene yields very high extraction selectivity and sensitivity during the analysis of chlorobenzenes in soil. Anal Bioanal Chem 384:1428–1437. doi:10.1007/s00216-005-0281-5
59. Liu M, Zeng Z, Fang H (2005) Preparation and application of the sol–gel-derived acrylate/silicone co-polymer coatings for headspace solid-phase microextraction of 2-chloroethyl ethyl sulfide in soil. J Chromatogr A 1076:16–26. doi:10.1016/j.chroma.2005.04.025
60. Mohammadi A, Ameli A, Alizadeh N (2009) Headspace solid-phase microextraction using a dodecylsulfate-doped polypyrrole film coupled to ion mobility spectrometry for the simultaneous determination of atrazine and ametryn in soil and water samples. Talanta 78:1107–1114. doi:10.1016/j.talanta.2009.01.025
61. Farhadi K, Bochani S, Hatami M et al (2014) Gas chromatographic detection of some nitro explosive compounds in soil samples after solid-phase microextraction with carbon ceramic copper nanoparticle fibers. J Sep Sci 37:1578–1584. doi:10.1002/jssc.201400144
62. Amanzadeh H, Yamini Y, Moradi M (2015) Zinc oxide/polypyrrole nanocomposite as a novel solid phase microextraction coating for extraction of aliphatic hydrocarbons from water and soil samples. Anal Chim Acta 884:52–60. doi:10.1016/j.aca.2015.05.018
63. Mayer P, Vaes WHJ, Wijnker F et al (2000) Sensing dissolved sediment porewater concentrations of persistent and bioaccumulative pollutants using disposable solid-phase microextraction fibers. Environ Sci Technol 34:5177–5183. doi:10.1021/es001179g
64. Chen X, Li H, You J (2015) Joint toxicity of sediment-associated permethrin and cadmium to *Chironomus dilutus*: the role of bioavailability and enzymatic activities. Environ Pollut 207:138–144. doi:10.1016/j.envpol.2015.09.012
65. Muijs B, Jonker MTO (2012) Does equilibrium passive sampling reflect actual in situ bioaccumulation of PAHs and petroleum hydrocarbon mixtures in aquatic worms? Environ Sci Technol 46:937–944. doi:10.1021/es202951w
66. Conder JM, La Point TW, Lotufo GR, Steevens JA (2003) Nondestructive, minimal-disturbance, direct-burial solid-phase microextraction fiber technique for measuring TNT in sediment. Environ Sci Technol 37:1625–1632. doi:10.1021/es0260770
67. Baltussen E, Sandra P, David F, Cramers C (1999) Stir bar sorptive extraction (SBSE), a novel extraction technique for aqueous samples: theory and principles. J Microcolumn Sep 11:737–747. doi:10.1002/(SICI)1520-667X(1999)11:10<737::AID-MCS7>3.0.CO;2-4
68. Prieto A, Zuloaga O, Usobiaga A (2008) Use of experimental design in the optimisation of stir bar sorptive extraction followed by thermal desorption for the determination of brominated flame retardants in water. Anal Bioanal Chem 390:739–748. doi:10.1007/s00216-007-1712-2
69. Sánchez-Avila J, Quintana J, Ventura F et al (2010) Stir bar sorptive extraction-thermal desorption-gas chromatography-mass spectrometry: an effective tool for determining persistent organic pollutants and nonylphenol in coastal waters in compliance with existing directives. Mar Pollut Bull 60:103–112. doi:10.1016/j.marpolbul.2009.08.028
70. Ochiai N, Ieda T, Sasamoto K et al (2011) Stir bar sorptive extraction and comprehensive two-dimensional gas chromatography coupled to high-resolution time-of-flight mass spectrometry for ultra-trace analysis of organochlorine pesticides in river water. J Chromatogr A 1218:6851–6860. doi:10.1016/j.chroma.2011.08.027
71. Serôdio P, Nogueira JM (2004) Multi-residue screening of endocrine disrupters chemicals in water samples by stir bar sorptive extraction-liquid desorption-capillary gas chromatography—mass spectrometry detection. Anal Chim Acta 517:21–32. doi:10.1016/j.aca.2004.04.045
72. Chary NS, Herrera S, Gómez MJ, Fernández-Alba AR (2012) Parts per trillion level determination of endocrine-disrupting chlorinated compounds in river water and wastewater effluent by stir-bar-sorptive extraction followed by gas chromatography-triple quadrupole mass spectrometry. Anal Bioanal Chem 404:1993–2006. doi:10.1007/s00216-012-6251-9
73. Llorca-Pórcel J, Martínez-Parreño M, Martínez-Soriano E, Valor I (2009) Analysis of chlorophenols, bisphenol-A, 4-tert-octylphenol and 4-nonylphenols in soil by means of ultrasonic solvent extraction and stir bar sorptive extraction with in situ derivatisation. J Chromatogr A 1216:5955–5961. doi:10.1016/j.chroma.2009.06.043

74. Rodil R, Popp P (2006) Development of pressurized subcritical water extraction combined with stir bar sorptive extraction for the analysis of organochlorine pesticides and chlorobenzenes in soils. J Chromatogr A 1124:82–90. doi:10.1016/j.chroma.2006.05.028
75. Wilcockson JB, Gobas FA (2001) Thin-film solid-phase extraction to measure fugacities of organic chemicals with low volatility in biological samples. Environ Sci Technol 35:1425–1431
76. Reichenberg F, Smedes F, Jönsson J-A, Mayer P (2008) Determining the chemical activity of hydrophobic organic compounds in soil using polymer coated vials. Chem Cent J 2:8. doi:10.1186/1752-153X-2-8
77. Jahnke A, Mayer P, McLachlan MS (2012) Sensitive equilibrium sampling to study polychlorinated biphenyl disposition in baltic sea sediment. Environ Sci Technol 46:10114–10122. doi:10.1021/es302330v
78. Cui X, Mayer P, Gan J (2013) Methods to assess bioavailability of hydrophobic organic contaminants: principles, operations, and limitations. Environ Pollut 172:223–234. doi:10.1016/j.envpol.2012.09.013
79. Zhu F, Ruan W, He M et al (2009) Application of solid-phase microextraction for the determination of organophosphorus pesticides in textiles by gas chromatography with mass spectrometry. Anal Chim Acta 650:202–206. doi:10.1016/j.aca.2009.07.050
80. Zuliani T, Lespes G, Milacic R et al (2006) Influence of the soil matrices on the analytical performance of headspace solid-phase microextraction for organotin analysis by gas chromatography-pulsed flame photometric detection. J Chromatogr A 1132:234–240. doi:10.1016/j.chroma.2006.07.054
81. Drescher SR, Brown SD (2006) Solid phase microextraction-gas chromatographic-mass spectrometric determination of nitrous oxide evolution to measure denitrification in estuarine soils and sediments. J Chromatogr A 1133:300–304. doi:10.1016/j.chroma.2006.08.056
82. De Cesare F, Pantalei S, Zampetti E, Macagnano A (2008) Electronic nose and SPME techniques to monitor phenanthrene biodegradation in soil. Sens Actuat B Chem 131:63–70. doi:10.1016/j.snb.2007.12.012
83. Styrishave B, Mortensen M, Krogh PH et al (2008) Solid-phase microextraction (SPME) as a tool to predict the bioavailability and toxicity of pyrene to the springtail, *Folsomia candida*, under various soil conditions. Environ Sci Technol 42:1332–1336
84. Zuazagoitia D, Millán E, Garcia-Arrona R (2009) Comparison of two extraction methods for the determination of polycyclic aromatic hydrocarbons in surface soils using headspace SPME with GC-FID. J Chromatogr Sci 47:97–102
85. Tölgyessy P, Vrana B, Šilhárová K (2012) An improved method for determination of polychlorinated biphenyls and polybrominated diphenyl ethers in sediment by ultrasonic solvent extraction followed by stir bar sorptive extraction coupled to TD–GC–MS. Chromatographia 76:177–185. doi:10.1007/s10337-012-2364-8
86. Zhang M, Jackson GP, Kruse NA et al (2014) Determination of aroclor 1260 in soil samples by gas chromatography with mass spectrometry with solid-phase microextraction. J Sep Sci 37:2751–2756. doi:10.1002/jssc.201400102
87. Zhuang GC, Lin YS, Elvert M et al (2014) Gas chromatographic analysis of methanol and ethanol in marine sediment pore waters: validation and implementation of three pretreatment techniques. Mar Chem 160:82–90. doi:10.1016/j.marchem.2014.01.011
88. Saraji M, Mehrafza N (2015) Polysiloxane coated steel fibers for solid-phase microextraction of chlorobenzenes. Microchim Acta 182:841–848. doi:10.1007/s00604-014-1395-2
89. Yiantzi E, Kalogerakis N, Psillakis E (2015) Vacuum-assisted headspace solid phase microextraction of polycyclic aromatic hydrocarbons in solid samples. Anal Chim Acta 890:108–116. doi:10.1016/j.aca.2015.05.047

Chapter 7
Application of Solid-Phase Microextraction Combined with Derivatization for Polar Compound Sampling in Environmental Analysis

Lihua Yang and Tiangang Luan

Abstract Solid-phase microextraction (SPME) is a solvent-free sample preparation technique using a fused-silica fiber coated with an appropriate stationary phase. Derivatization is often necessary to determine substances with poor chromatographic behavior, high reactivity, and/or volatility or thermal instability. SPME derivatization technique can be operated under different combined modes and may be influenced by factors such as sample matrix, SPME, derivatization, and desorption condition. Combined with derivatization, HS-SPME, DI-SPME, membrane-protected SPME, and in-tube SPME have been applied to the analysis of a vast amount of chemicals, i.e., aldehydes and acetone, amines, organic acids, phenolic compounds, pesticide, metal and organic metal species, and pharmaceutical and personal care products (PPCPs) in a wide range of environmental matrices, including air, water, soil, and sediment samples. This chapter reviews the mode and method development of the derivatization technique, the derivatization reagents, and reaction mechanism, as well as its applications in environmental analysis.

Keywords Solid-phase microextraction · Derivatization reagents · Reaction mechanism · Polar compound sampling · Environmental analysis

Abbreviations

AMS	Ambient mass spectrometry
APCI	Atmospheric pressure chemical ionization
BSTFA	Bis-(trimethylsilyl)trifluoroacetamide
CW-DVB	Carbowax-divinylbenzene
CW-TR	Carbowax-templated resin
DI	Direct-immersion sampling

L. Yang · T. Luan (✉)
MOE Key Laboratory of Aquatic Product Safety, School of Marine Sciences, Sun Yat-Sen University, Guangzhou 510275, China
e-mail: cesltg@mail.sysu.edu.cn

G. Ouyang and R. Jiang (eds.), *Solid Phase Microextraction*,
DOI 10.1007/978-3-662-53598-1_7

DMS	Dimethylsulfate
DNFB	2,4-Dinitrofluorobenzene
DSTEB	Deuterium-labeled sodium tetraethylborate $NaB(C_2D_5)$
FMOC	9-Fluorenylmethyl chloroformate
GC–ECD	Gas chromatography coupled to electron capture detection
GC-EI-MS	Electron impact-mass spectrometry
GC-ICP-MS	Inductively coupled plasma mass spectrometry
GC-AED	Gas chromatographic separation and atomic emission detection
GC-QqQ MS	Gas chromatography-triple quadrupole mass spectrometry
HS	Headspace sampling
HTFA	1,1,1-Trifluoro-2,4-pentadione (trifluoroacetylacetone)
IT	In tube
LLME	Liquid–liquid microextraction
MSTFA	*n*-Methyl-*n*-(trimethylsilyl)trifluoroacetamide
MTBSTFA/TBDMCS	*n*-Tert-butyl-dimethylsilyl-*n*-methyltrifluoroacetamide) with 1% tert-butyl-dimethylchlorosilane
MTBSTFA	*n*-Methyl-*n*-(tert-butyldimethylsilyl)trifluoroacetamide
NaBEt4	Sodium tetraethylbroate
NaBH4	Sodium tetrahydroborate
NaBPh4	Sodium tetraphenylborate
NaBPr4	Sodium tetrapropylborate
NBD-F	4-Fluoro-7-nitro-2,1,3-benzoxadiazole
NEM	*n*-Ethylmaleimide
OPA-NAC	*o*-Phthalaldialdehyde and *n*-acethyl-*L*-cysteine
PA	Polyacrylate
PDAM	Pyrenyldiazomethane
PDMS/CAR	Polydimethylsiloxane/carboxen
PFBAY	Pentafluorobenzyl aldehyde
PFBBr	Pentafluorobenzyl bromide
PFBHA	1,2,3,4,5-Pentafluorobenzylhydroxylamine
PFBHA	Pentafluorobenzylhydroxylamine
PFPDE	Pentafluorophenyldiazoethane
PFPH	Pentafluorophenylhydrazine
Portable GC-μFID	Portable GC-microflame ionization detection
PPy/PTh	Polypyrrole and polythiophene
PPCPs	Pharmaceutical and personal care products
RP-HPLC-FL	Reversed-phase HPLC–fluorescence
SIBA	*N*-succinimidyl benzoate
SPME	Solid-phase microextraction
TFBza-suc	Tetrafluorobenzoic acid *n*-hydroxysuccinimide ester
TFEH	2,2,2-Trifluoroethylhydrazine

THC	Delta 9-tetrahydrocannabinol
THCCOOH	11-Nor-9-carboxy-delta 9-tetrahydrocannabinol
TMAHS	Tetramethylamonium hydrogen sulfate
TMAOH	Tetramethylammonium hydroxide

7.1 Introduction

Sample preparation is a key procedure in modern chemical analysis. By some estimates, 60–80% of the work activity and operating cost in an analytical laboratory is spent on preparing samples for introduction into an analytical device [1]. Sample preparation can isolate the components of interest from a sample matrix, as well as to purify and concentrate the analytes. The development of environmentally friendly techniques, namely solvent-free sample preparation techniques, began to attract widespread attention recently, and several reviews have been published [2–4].

Among them, solid-phase microextraction (SPME) is an inexpensive, time-efficient, and solvent-free technique, compared to traditional methods such as solid-phase extraction (SPE) and liquid–liquid extraction (LLE) [5]. This technique was firstly introduced by Pawlisyzin et al. in the 1990s [6]. SPME utilizes a small, fused-silica fiber coated with a polymeric coating, to extract the target analytes from the sample matrix. And then, the fiber was transferred into the injector of the gas- or liquid-chromatographic system where they are desorbed by evaporation at high temperature or by dissolution in the mobile phase.

SPME integrates sampling, extraction, concentration, and sample introduction into a single step and preserves all of the advantages of SPE while eliminating the disadvantages of plugging and solvent use [4]. SPME has been used routinely in combination with gas chromatography (GC) and gas chromatography–mass spectrometry (GC–MS) and successfully applied to a wide variety of compounds, especially for the extraction of volatile and semi-volatile organic compounds from complex sample matrices [7]. SPME was also introduced for direct coupling with high-performance liquid chromatography (HPLC) and liquid chromatography–mass spectrometry (LC–MS) in order to analyze weakly volatile or thermally labile compounds not amenable to GC or GC–MS [7].

The main challenge in organic analysis is polar compounds, which are difficult to extract from complex matrices and difficult to separate on the chromatographic column. Chemical derivatization is frequently used to address this challenge [8]. Analyte derivatization is often necessary because of the ionic/polar nature of the target compounds. Derivatization before, during, or after the SPME procedure can be used to enhance the detection power by introducing select functional groups into the analytes. The use of derivatization in the solid phase or in solution in analytical methods based on SPME has made possible low detection limits (LODs) to determine substances with poor chromatographic behavior, high reactivity, and/or volatility or thermal instability [9]. Analytical derivatizations of various classes of compounds [10] and for different types of extraction techniques [9, 11] have been reviewed.

Combined with derivatization, SPME has been applied to the analysis of a vast amount of chemicals, i.e., aldehydes and acetone, amines, organic acids, phenolic compounds, pesticide, metal and organic metal species, and pharmaceutical and personal care products (PPCPs). Related review articles about the application of SPME in environmental analysis can refer to references [12–14]. This chapter will focus on the sample preparation technique of SPME combined with derivatization, including two main aspects: the introduction and method development of the technique, and their applications in environment analysis such as air, water, soil, and sediment.

7.2 Derivatization Mode

Derivatization converts polar analytes into less polar analogues, therefore increasing their coating/water or coating/gas partition coefficients and improving SPME efficiency and method sensitivity. Derivatization changes analytes with low volatilities into more volatile derivatives, thus improving their GC separations, detection, and quantitation [15]. Derivatization combined with SPME was mainly performed in the following ways summarized in Fig. 7.1 [1, 9, 12, 15, 16]: direct derivatization in the sample matrix (in-vial/in situ derivatization), simultaneous or post-derivatization and extraction in the SPME fiber coating (on-fiber in situ or post-derivatization), and derivatization in the GC or LC injector port (in-port derivatization).

7.2.1 In-Vial Derivatization

In-vial derivatization is similar to a well-established approach used in solvent extraction. Derivatizing agent is first added to the sample matrix and reacts for a

Fig. 7.1 SPME derivatization approaches

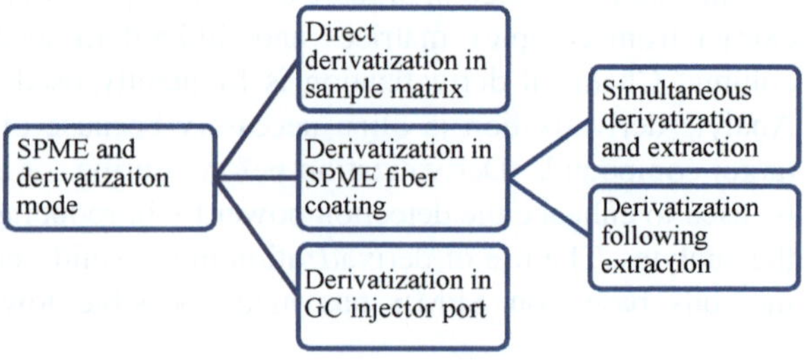

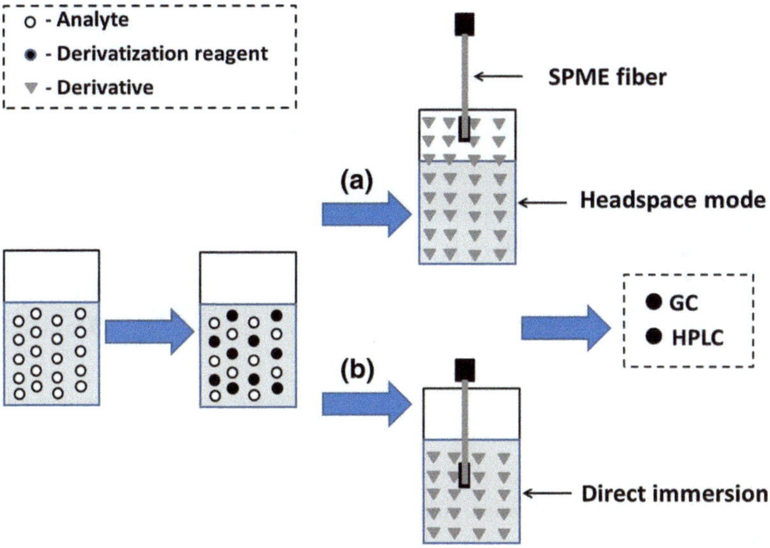

Fig. 7.2 In-vial derivatization. **a** Sampling by headspace mode and **b** sampling by direct immersion. Adapted from Ref. [9], Copyright 2004, with permission from Elsevier

definite time, and then, the target derivatives are extracted by headspace-mode (HS) or direct-immersion (DI) SPME and introduced into the analytical instrument (Fig. 7.2) [8]. There is no change in the performance of SPME. Derivatization increases the extraction yield because of higher volatility and lipophilicity of the derivatives. The reagent must be sufficiently stable to hydrolysis since, as a rule, SPME is performed from aqueous samples. For example, this approach has been applied to extract and separate phenolic compounds and their acetate derivatives from aqueous samples by firstly conducting in situ derivatization by adding dimethyl sulfate or acetic anhydride to the sample matrix [16–19]. Similar in situ derivatization technique can be applied to acetic acids and their alkylated derivatives by adding sulfuric acid/methanol, sulfuric acid/ethanol, dimethyl sulfate, and alkyl halides [20–23], to carbonyl compounds by adding fluorogenic reagents [24–28], and to organometallic compounds by adding sodium tetraethylborate ($NaBEt_4$) [29–33].

7.2.2 *On-Fiber In Situ Derivatization*

The procedure of this technique is first to dope the fiber with a derivatization reagent of low volatility and subsequently expose it to the sample (shown in Fig. 7.3). With the derivatizing reagent in the fiber stationary phase, the SPME device can act as a concentrator, isolator, and reactor to effectively isolate and

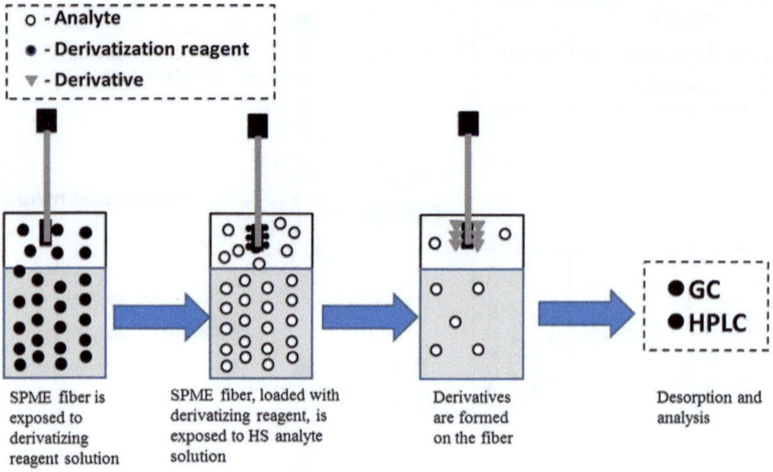

Fig. 7.3 On-fiber in-situ derivatization. Adapted from Ref. [9], Copyright 2004, with permission from Elsevier

derivatize target analytes in the sample preparation process [9]. This approach was first reported by Pan et al. [34] who determined short-chain fatty acids using 1-pyrenyldiazomethane (PDAM) as derivatization reagent. The application also includes on-fiber in situ derivatization with pentafluorophenylhydrazine with HS SPME for GC analysis for aldehydes [24–28], on-fiber in situ derivatization with dibutylamine(DBA) for analysis of isocyanates [35], on-fiber in situ derivatization with MTBSTFA/TDMCS for analysis of alkylphenols [36], and on-fiber in situ derivatization with ethyl chloroformate (ECF) for analysis of anti-inflammatory drugs and steroids [37].

7.2.3 On-Fiber Post-Derivatization

For on-fiber derivatization following extraction, the fiber is exposed after SPME to the vapor phase above the derivatization reagent in a separate vessel (illustrated in Fig. 7.4). Reagents used for this purpose have high volatility, such as silylation reagents including *n*-methyl-*n*-(trimethylsilyl)trifluoroacetamide (MSTFA) [38], *n*-methyl-*n*-(tert-butyldimethylsilyl) trifluoroacetamide (MTBSTFA) [39], hexamethyldisilazane [40], *n*-methylbis(trifluoroacetamide) [41], pentafluorobenzoyl chloride [42], or trifluoroacetic anhydride [43]. After DI SPME with appropriate fiber, on-fiber post-derivatization with silylation reagents have mainly focused on the analysis of aliphatic amines, phenolic compounds, acid pesticides,

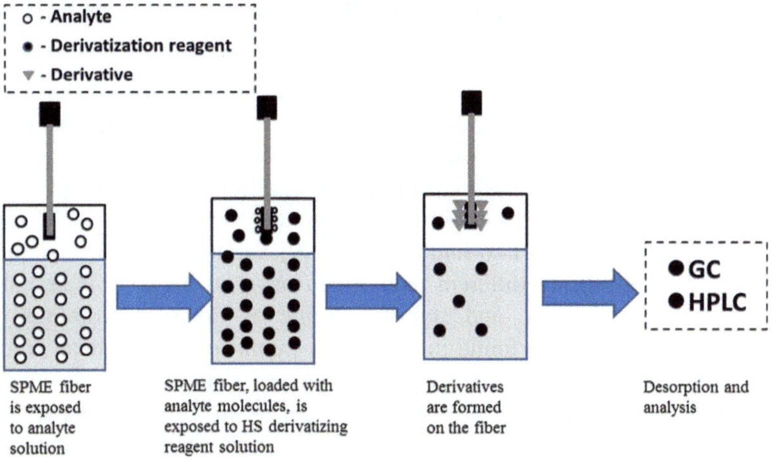

Fig. 7.4 On-fiber post-derivatization. Adapted from Ref. [9], Copyright 2004, with permission from Elsevier

anti-inflammatory drugs, clenbuterol, parabens, steroids, PAHs, salicylate- and benzophenone-type UV filters, and triclosan and possible related compounds [36, 44–61].

7.2.4 In-Port Derivatization

Derivatization in the GC or LC injector port is an approach analogous to post-derivatization, but it is performed at high temperatures in GC injection port or by elution by organic solution in LC injection port. For example, long-chain fatty acids along with the ion-pair reagents (TMA-OH and TMA-HSO_4) were added to the sample and isolated by SPME fiber. After the desired extraction time was reached, the fiber was transferred into the hot GC injector port. During volatilization, tetramethylammonium hydroxide and tetramethylammonium hydrogen sulfate were employed to convert long-chain fatty acids into their volatile methyl esters via in-injector derivatization analytes [15]. Another example is the direct ion-pair extraction of alkylbenzensulfonates in the presence of tetrabutylammonium ions on a PDMS fiber. The ion pairs are transformed into the corresponding sulfonated butyl esters in the injection port of the gas chromatograph at 300 °C [62]. In the same way, carboxylic acids are methylated by coextracted phenyltrimethylammonium ions in the injection port [63]. Bizkarguenaga et al. [64] reviewed online in-port derivatizations where the derivatization reaction is simultaneously carried out with the analysis step by injecting the sample/reagent mixture directly into the hot GC inlet after a sorptive extraction step.

7.3 Derivatization Reaction and Reagents

The optimization of derivatization conditions includes selection of derivatization agent, derivatization reaction mode, derivatization time, and temperature [1]. Commonly used derivatization reactions include silylation, alkylation, acetylation, and fluorescence/ultraviolet derivatization. The reaction mechanism of analytical derivatization is summarized as a subset of functional group analysis [5, 65]. The review published by Stashenko and Martinez [9] summarizes the application of derivatization in SPME to different analytes in order to improve their volatility, chromatographic properties, and stability. Some of the most commonly used derivatization agents can be found in the review published by Quintana et al. [66]. Stalikas and Fiamegos [11] have recently reviewed derivatization techniques in SPME, but they expanded their review to other microextraction techniques, such as liquid-phase microextraction or membranes. In addition, selective derivatization to analogues containing high detector response groups will result in enhancement in sensitivity and selectivity of detection.

7.3.1 Silylation

Silylation is the introduction of a (usually) trialkylsilyl group, typically trimethylsilyl (–$SiMe_3$, Me=CH_3) to a molecule to replace a proton (Fig. 7.5). A general procedure is to deprotonate the substrate with a suitable strong base and then allow it to react with a silyl chloride (e.g., trimethylsilyl chloride). Trimethylsilyl (TMS), bis(trimethylsilyl) acetamide (BTSA), bis(trimethylsilyl) trifluoroacetamide (BTMSTFA), *N*-methyl-*N*-(tert-butyldimethylsilyl) trifluoroacetamide (MTBSTFA), and pentafluorophenyldimethylsilyl (PFPDMS) reagents have been used as typical trimethylsilylating agents (Table 7.1). Most silylation reagents and their derivatives are sensitive to moisture, so the silylation reactions are most effective under anhydrous reaction conditions. But silylating reagents containing sterically crowded groups, such as tert-butyldimethylsilyl (TBDMS), are more stable to hydrolysis and present enhanced reactivity.

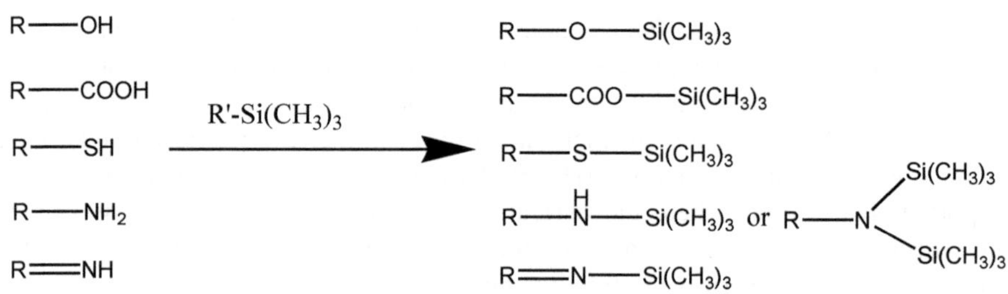

Fig. 7.5 Mechanism of derivatization by trimethylsilylating agents

Table 7.1 Information about the commonly used derivatization reagents and targeted compounds

Categories	Abberivation	Chemical name	CAS No.	Targeted compounds	References
Silylation	BSTFA	Bis(trimethylsilyl)trifluoroacetamide	25561-30-2	Halophenols; alkylphenols; steroids	[44–51]
	MTBSTFA	*n*-(ter-Butyldimethylsilyl)-*n*-methyltrifluoroacetamide	77377-52-7	Alkylphenols; BPA; polar chlorinated herbicides; anti-inflammatory drugs; parabens; salicylate- and benzophenone-type UV filters; triclosan and possible related compounds; steroids; PAHs; pesticides; aldrin (ALD); heptachlor (HEPT); and 2,4′-DDD	[36, 52–58]
	MSTFA	*n*-Methyl-*n*-(trimethylsilyl)trifluoroacetamide	24589-78-4	Salicylate- and benzophenonetype UV filters; cannabinoids	[60, 61]
	HMDS	Hexamethyldisilazane	999-97-3	β-Agonist clenbuterol	[59]

(continued)

Table 7.1 (continued)

Categories	Abberivation	Chemical name	CAS No.	Targeted compounds	References
Alkylation	–	Benzyl bromide	100-39-0	Acetic acids; acidic pesticides	[67, 68]
	ECF	Ethyl chloroformate	541-41-3	Acidic pesticides; anti-inflammatory drugs; steroids; homocysteine (Hcy), cysteine (Cys); and methionine (Met)	[37, 69, 70]
	MCF	Methyl chloroformate	79-22-1	Acidic pesticides	[69]
	HCF	Hexyl chloroformate	6092-54	Acetic acids; anatoxin-a	[67, 71]
	BuCF	Butyl chloroformate	592-34-7	Acidic pesticides; homocysteine (Hcy), cysteine (Cys); and methionine (Met)	[70, 72]
	iBCF	Iso-butyl chloroformate	543-27-1	Amphetamine-type stimulants (ATSs); homocysteine (Hcy); cysteine (Cys); and methionine (Met)	[70, 73]
	PCF	Propyl chloroformate	109-61-5	Homocysteine (Hcy); cysteine (Cys); and methionine (Met); and perfluorocarboxylic acids (PFCAs); hydrazine	[70, 74, 75]
	iPCF	Isopropyl chloroformate	108-23-6	Homocysteine (Hcy); cysteine (Cys); and methionine (Met)	[70]
	2,6-DMP	2,6-Dimethylphenol	576-26-1	Bromate	[76]
	–	*n*-Butanol	71-36-3	Perfluorocarboxylic acids (PFCAs)	[77]
	C_2H_5I	Iodoethane/sodium hydride	75-03-6	Phenylurea herbicides (PUHs)	[78]
	CH_3I	Iodomethane	74-88-4	Phenoxy acid herbicide	[79]

(continued)

Table 7.1 (continued)

Categories	Abberivation	Chemical name	CAS No.	Targeted compounds	References
	CH_2N_2	Diazomethane(/diethylether)	334-88-3	Phthalic monoesters (MPs)	[80, 81]
	DMS	Dimethyl sulfate/NaOH	77-78-1	Acetic acids; halophenol; alkylphenol; anti-inflammatory drugs	[17, 23, 82, 83]
	–	Sulfuric acid/methanol	–	Haloacetic acids	[20, 22]
	–	Sulfuric acid/ethanol	–	Haloacetic acids; fluoroacetic acid	[21, 84]
Alkylboration	$NaBEt_4$	Sodium tetraethylborate	15523-24-7	Tin species; mercury species; lead species; and acaricide fenbutatin oxide	[29–31, 33, 126–138]
	$NaBH_4$	Sodium tetrahydroborate	16940-66-2	Tin species	[87, 139]
	$NaBPr_4$	Sodium tetra(*n*-propyl)borate	45067-99-0	Tin species; mercury species; lead species	[86, 88, 129]
	$NaBPh_4$	Sodium tetraphenylborate, $BaB(C_6H_5)_4$	143-66-8	Mercury species; organomanganese	[33, 140, 141]
	KBH_4	Potassium tetrahydroborate	13762-51-1	Tin species; mercury species	[142–144]
	DSTEB	Deuterium sodium tetraethylborate, $NaB(C_2D_5)_4$	–	Lead species	[89]
	PBA	Phenylboronic acid	–	Mercury species	[85]
Aceylation	AA	Acetic anhydride	108-24-7	Halophenols; alkylphenols; selective serotonin reuptake inhibitors (SSRIs); parabens; triclosan and possible related compounds; and alkylmines	[18, 19, 90–97]

(continued)

Table 7.1 (continued)

Categories	Abberivation	Chemical name	CAS No.	Targeted compounds	References
Fluorescence/ultraviolet derivatization	FMOC	9-Fluorenylmethyl chloroformate	28920-43-6	Aliphatic amines; amphetamine-type stimulants (ATSs); and alkylmines	[101–103]
	–	OPA-NAC/FMOC	–	Aliphatic amines	[102]
	PFPH	Pentafluorophenyl hydrazine	828-73-9	Aldehydes	[105, 106]
	TFEH	2,2,2-Trifluoroethyl hydrazine	5042-30-8	Aldehydes; acetone	[107, 108]
	PFBHA·HCl	*o*-(2,3,4,5,6-Pentafluorobenzyl)hydroxylamine hydrochloride	57981-02-9	Aldehydes; acetone	[24, 25, 27, 106, 111, 114–119]
	DNBC	3,5-Dinitrobenzoyl chloride	99-33-2	4,4'-Methylenebis (2-chloroaniline)	[99]
	PFBCF	2,3,4,5,6-Pentafluorobenzyl chloroformate	53526-74-2	Alkylmines	[145]
	PFBBr	Pentafluorobenzyl bromide	1765-40-8	Phenoxy acid herbicides	[68]
	1-NC	1-Naphthoyl chloride	879-18-5	Alcohol ethoxylate (Brij 56) and 1-hexadecanol	[100]
	PFBAY	Pentafluorobenzaldehyde	653-37-2	Aliphatic amines	[120]
	HTFA	1,1,1-Trifluoro-2,4-pentadione (trifluoroacetyl acetone)	–	Chromium	[146, 147]
	H(FHD)	1,1,1,2,2,6,6,7,7,7-Decafluoro-3,5-heptanedione	–	Beryllium and lanthanide	[148]
	NBD-F	4-Fluoro-7-nitro-2,1,3-benzoxadiazole	29270-56-2	Anatoxin-a	[112, 113]
	DNPH	2,4-Dinitrophenylhydrazine	119-26-6	Aldehydes	[109, 110]
Diazotation/iodination	–	Hydroiodic acid/sodium nitrite	–	Aromatic amines	[121, 122]

(continued)

Table 7.1 (continued)

Categories	Abberivation	Chemical name	CAS No.	Targeted compounds	References
Other reagents	NEM	*n*-Ethylmaleimide	128-53-0	Thiol compounds	[123, 149]
	BTF	Boron trifluoride	7637-07-2	Perfluorocarboxylic acids (PFCAs)	[150]
	–	Carbon disulfide	75-15-0	Aromatic primary amines	[151]
	TFBza-suc	Tetrafluorobenzoic acid *n*-hydroxysuccinimide ester	122894-73-9	Aliphatic amines	[104]
	–	1,3-Propanedithiol	109-80-8	Lewisite oxide	[152]
	–	*n*-Methylhydrazine	60-34-4	Malonaldehyde	[124]
	SIBA	*n*-Succinimidyl benzoate	60444-78-2	Aliphatic amines	[125]

– Not indicated

After DI SPME with appropriate fibers, phenolic compounds, drugs, herbicides, and PAHs could all be derivatized by on-fiber in situ or post-mode silylation [36, 44–61]. Silyl derivatives are generally less polar and more volatile and more thermally stable than their precursor organic compound, making the derivatives suitable for analysis by GC–EI–MS. Silylation reactions generally proceed very rapidly (within 5 min) with pyridine being the most frequently used solvent. Disadvantages include the need to operate under anhydrous conditions, to heat the sample, and to inject a reactive mixture onto the column.

7.3.2 Alkylation

Alkylation is the replacement of active hydrogen in R–COOH, R–OH, R–SH, and R–NH_2 with an alkyl group or, sometimes aryl group in a process referred to esterification (Fig. 7.6). In structural formula, the symbol R is used to designate a generic (unspecified) alkyl group. Alkyls form homologous series with the general formula of C_nH_{2n+1}, including methyl ($CH_3\cdot$), ethyl ($C_2H_5\cdot$), propyl ($C_3H_7\cdot$), butyl ($C_4H_9\cdot$), and pentyl ($C_5H_{11}\cdot$), hexyl ($C_6H_{13}\cdot$). Alkyl groups also contain one ring having the formula C_nH_{2n-1}, e.g. cyclopropyl and cyclohexyl.

Alky halide such as benzyl bromide, ethyl chloroformate, methyl chloroformate, hexyl chloroformate, butyl chloroformate, iso-butyl chloroformate, propyl chloroformate, and iso-propyl chloroformate were widely used by in vial derivatization with acid pesticide, amine acids, anatoxin-a, hydrazine, perfluorocarboxylic acids, and so on (Table 7.1) [37, 67–75]. It has been reported that *n*-hexyl chloroformate reacts with carboxylic acids to give *n*-hexyl esters via the mixed anhydride. The derivatives were extracted by HS or DI SPME and were subsequently analyzed by GC.

Moreover, in vial alkylation with 2,6-dimethylphenol, *N*-butanol, iodoethane (/sodium hydride), and diazomethane (/diethylether) have also been reported with phenylurea and phenoxy acid herbicides, phthalic monoesters, bromate, and PFCAs [76–81]. Esterification of acetic acids was carried out by reacting with dimethyl sulfate/NaOH, sulfuric acid/methanol, sulfuric acid/ethanol with in vial

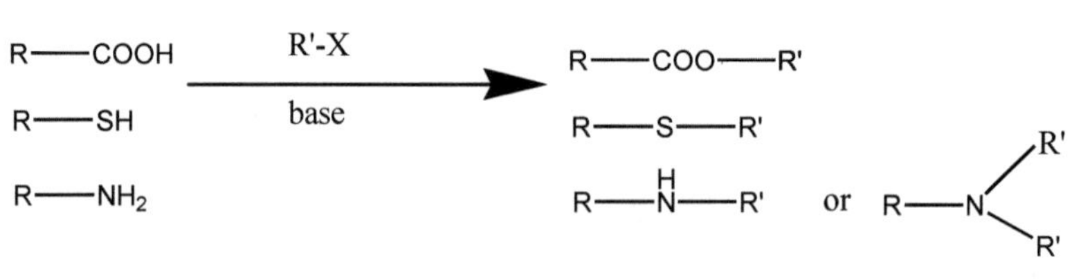

Fig. 7.6 Mechanism of derivatization by alkylating agents

derivatization, and then, the products were extracted by HS SPME mode for GC analysis of acetic acids, halophenols, and alkylphenols. And in some cases, acid-actalyzed derivatization of acetic acids was carried out after evaporation of the water sample to dryness and the esters were extracted by HS SPME [17, 20–23, 82–84].

7.3.3 Alkylborate Derivatization

GC is the most frequently employed technique for analysis of organometallic compounds, due to its excellent separation efficiency and the availability of many suitable detectors. Because organometallic compounds differ significantly in volatility, stability, and polarity, a suitable derivatization technique is essential to make them amenable for GC separation. A variety of derivatizing agents including hydride generation, ethylation, propylation, or Grignard alkylation have been employed to coordinate metal ions, thus increasing the volatility of the species and promoting more analytes to the sample headspace [33].

Using SPME, derivatization in aqueous medium is preferable and tetralkylborates such as sodium tetraethylborate ($NaBEt_4$) and sodium tetrapropylborate ($NaBPr_4$) shown in Table 7.1, which are fairly soluble in water and are most commonly used nowadays for this purpose (Fig. 7.7). However, the use of the tetraalkylborates has been seen as the most suitable method when extractions from water samples are contemplated as this reagent can tolerate aqueous conditions. Moreover, for efficient derivatization of inorganic mercury and its organic compounds, reagents such as phenylboronic acid (PBA) have been used [85].Typical reactions that have been proposed for tin, lead, and mercury with sodium tetraethylborate are ascribed as follows [33].

A number of metal species have been reported to be successfully derivatized by tetraethylborate and include the following: the mono-, di-, and trichloride forms of methyl-, ethyl-, butyl-, phenyl-, and cyclohexyl-tin compounds; poly-substituted methyl-, and ethyl-lead compounds; and methyl- and ethyl-mercury chlorides [32, 33, 86–89].

$$3NaB(Et)_4 + RSn^{3+} \longrightarrow (Et)_3SnR + 3Et_3B + 3Na^+$$

$$4NaB(Et)_4 + 2Pb^{2+} \longrightarrow Et_4Pb + 4Et_3B + Pb + 4Na^+$$

$$NaB(Et)_4 + RHg^+ \longrightarrow EtHgR + Et_3B + Na^+$$

Fig. 7.7 Mechanism of derivatization by alkylborate reagents, Et = C_2H_5

7.3.4 Acetylation

Acetylation refers to the process of introducing an acetyl functional group for an active hydrogen atom into a compound. A reaction involving the replacement of the hydrogen atom of a hydroxyl group with an acetyl group (CH_3CO) yields a specific ester, the acetate. Acetyl is a functional group with chemical formula $COCH_3$ and sometimes represented by the symbol Ac (Fig. 7.8). The main aceylating reagent acetic anhydride (AA) was widely used in halophenol, alkylphenol, alkylamines, and PPCPs such as parabens, triclosan, and drugs [18, 19, 90–97]. The procedure is often with in vial derivatization and HS or DI SPME sampling, and then for GC–MS analysis.

Furthermore, the reaction medium is acidic that damages the GC column, which cannot be used for acid-sensitive compounds. To solve this problem, basic compounds, such as sodium carbonate, NaOH, or potassium hydrogen carbonate, can be used as an acceptor for acidic by-products and also as a catalyst of the derivatization reaction (Table 7.1) [5].

7.3.5 Fluorescence/Ultraviolet Derivatization

As some compounds, such as aliphatic amines and aldehydes, lack UV-absorption chromophores, or fluorescent signal in the UV/FLD spectrum, their detection by HPLC is not easy. To overcome this limitation, the aliphatic amines are derivatized or tagged with some strongly absorbing chromophore [98]. Except the following traditional methods, some newly fluorescence/ultraviolet reagents, such as HTFA for chromium, H(FHD) for beryllium and lanthanide, DNBC for 4,4′-methylenebis (2-chloroaniline) [99], 1-NC for alcohol ethoxylate (Brij 56), and 1-hexadecanol [100], have been developed with in vial derivatization combined with DI or HS SPME and HPLC or LC analysis.

FMOC and PFBCF: The most commonly used derivatization reagents are a wide range of chloroformates, among which FMOC is used in HPLC applications, which is used widely for the analysis of aliphatic amines, alkylmines, and

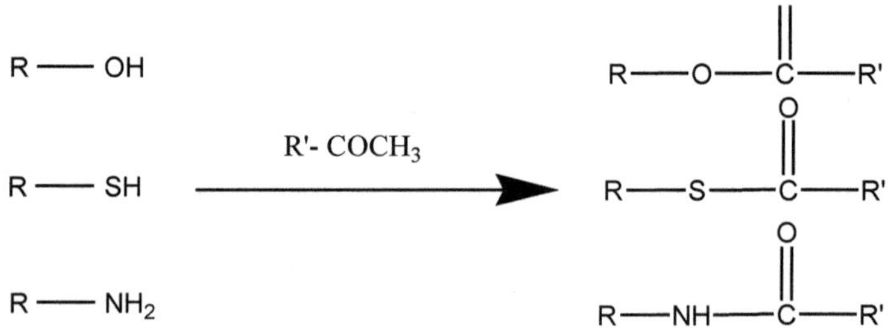

Fig. 7.8 Mechanism of derivatization by acetylating agents

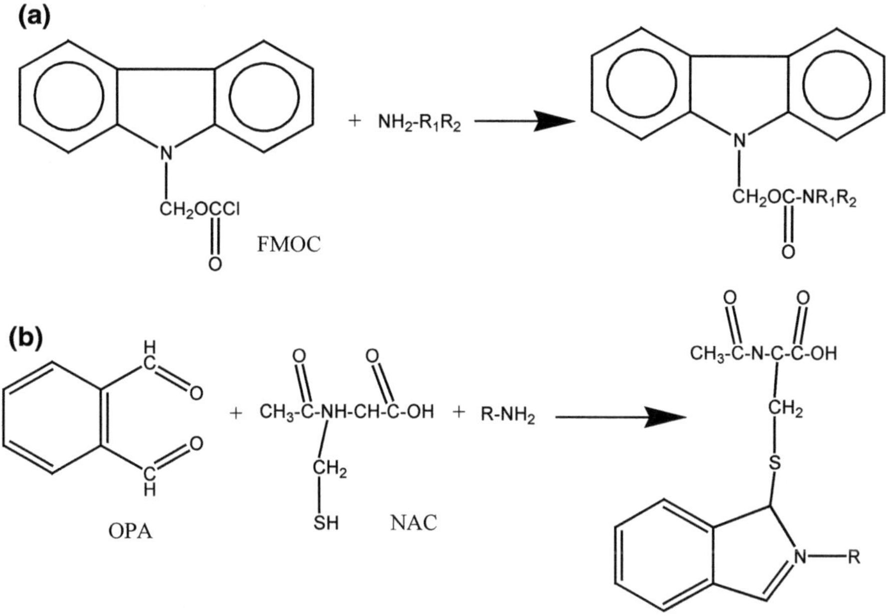

Fig. 7.9 Reaction schemes of the derivatization of amines **a** with FMOC and **b** with OPA-NAC

amphetamines [101–103]. Moreover, Cai et al. [104] proposed a sequential derivatization with *o*-phthaldialdehyde and *N*-acetylcysteine (OPA–NAC) and then with FMOC (Fig. 7.9). OPA–NAC is only reactive toward primary amines; in the first stage, PA is transformed into PA–OPA–NAC. In the second stage, FMOC only reacts with DMA. Since the PA–OPA–NAC and DMA–FMOC derivatives present very different features, they can be satisfactorily resolved under typical reversed phase conditions. On the basis of the results, a new method is developed for the selective determination of DMA in water.

PFPH, TFEH, and DNPH: As to the assay of aldehydes and acetone, many assays are based on reactions with derivatization reagents that usually involve 5,5-dimethyl-1,3-cyclohexanedione, *N*-methylbenzothiazolon-(2)-hydrazone, pentafluorophenylhydrazine (PFPH), and 2,4-dinitrophenylhydrazine (2,4-DNPH). Otherwise, 2,2,2-trifluoroethylhydrazine (TFEH), a very volatile hydrazine, reacts also with carbonyl compounds to produce volatile hydrazones [105–110]. Thus, it is possible to analyze the hydrazone with HS-SPMEGC-MS. The most commonly reported derivatizing agent for the determination of acetone in both freshwater and seawater is 2,4-dinitrophenylhydrazine (DNPH). The reaction of acetone with DNPH creates a hydrazone, which can be analyzed using LC with detection by ultraviolet–visible (UV/Vis) absorbance.

PFBHA, PFBAY, PFBBr, and NBD-F: For GC determination of aldehydes and acetone, derivatization has also been accomplished using in vial derivatization with *O*-(2,3,4,5,6-pentafluorobenzyl)-hydroxylamine (PFBHA) hydrochloride and

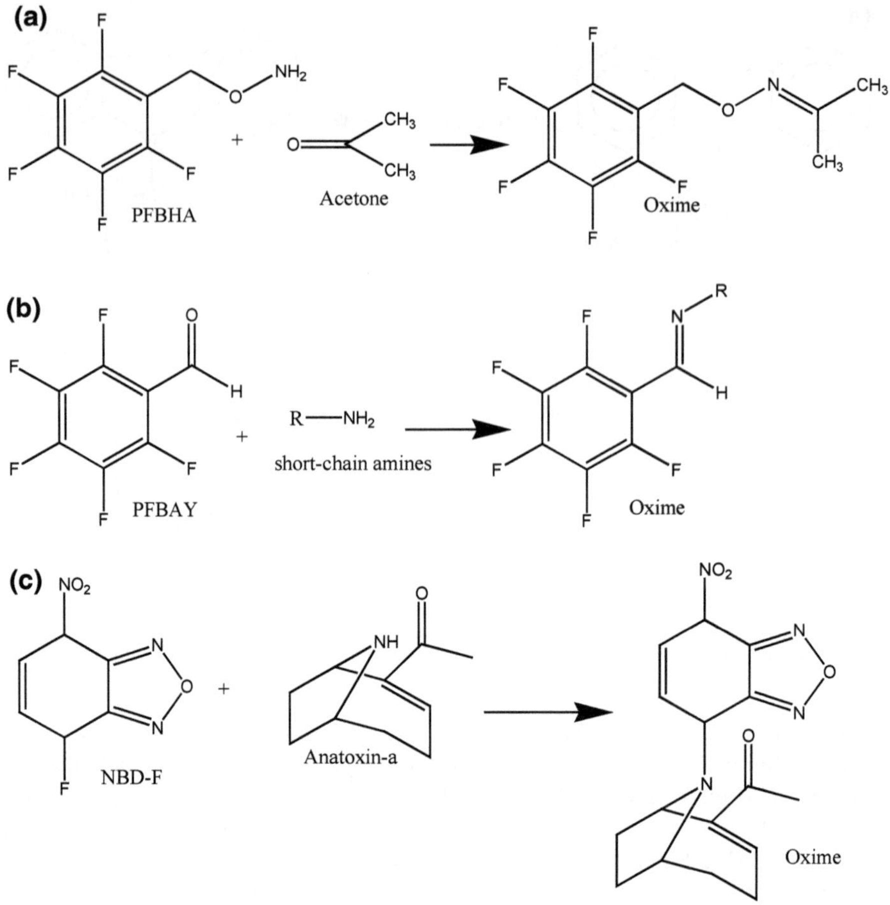

Fig. 7.10 Derivatization reaction of **a** acetone with PFBHA, **b** short-chain amines with PFBAY, **c** anatoxin-a with NBD-F

HS SPME extraction (Fig. 7.10a) [111]. As to anatoxin-a, after DI SPME, the fibers were exposed for on-fiber derivation by NBD-F and then for HPLC analysis (Fig. 7.10b) [112, 113].

Moreover, free amines are difficult to extract from sample matrices, cannot be chromatographed well, and often give tailing peak, due to their high polarity. Therefore, it is often necessary to convert free amines into suitable derivatives prior to analysis. Among the existing derivatizing agents, pentafluorobenzaldehyde (PFBAY) is widely used for short-chain aliphatic amines, and the derivatizing reaction can proceed rapidly in aqueous solution at room temperature and provides very good yield. After derivatization, the formed derivatives required further extraction and concentration prior to GC analysis [24, 25, 27, 68, 106, 111–120]. Aliphatic amines derivatization with PFBAY to corresponding pentafluorobenzylimines (Fig. 7.10c) are compatible with GC. The reaction can be performed under very mild conditions, and it provided relatively high sensitivity and good chromatographic resolution [120].

7.3.6 Other Derivatization Method

Diazotation/iodination for aromatic amides: As to the analysis of aromatic amides, the analytes were in vial derivatized by diazotation and subsequent iodination prior to the HS SPME extraction as the following procedure [121, 122]. Sample of water was acidified with hydroiodic acid and then mixed with sodium nitrite for a reaction time of 20 min; amidosulfonic acid was added to destroy the surplus of nitrite; and the surplus of iodine was destroyed with the addition of a saturated aqueous solution of sodium as shown in Fig. 7.11.

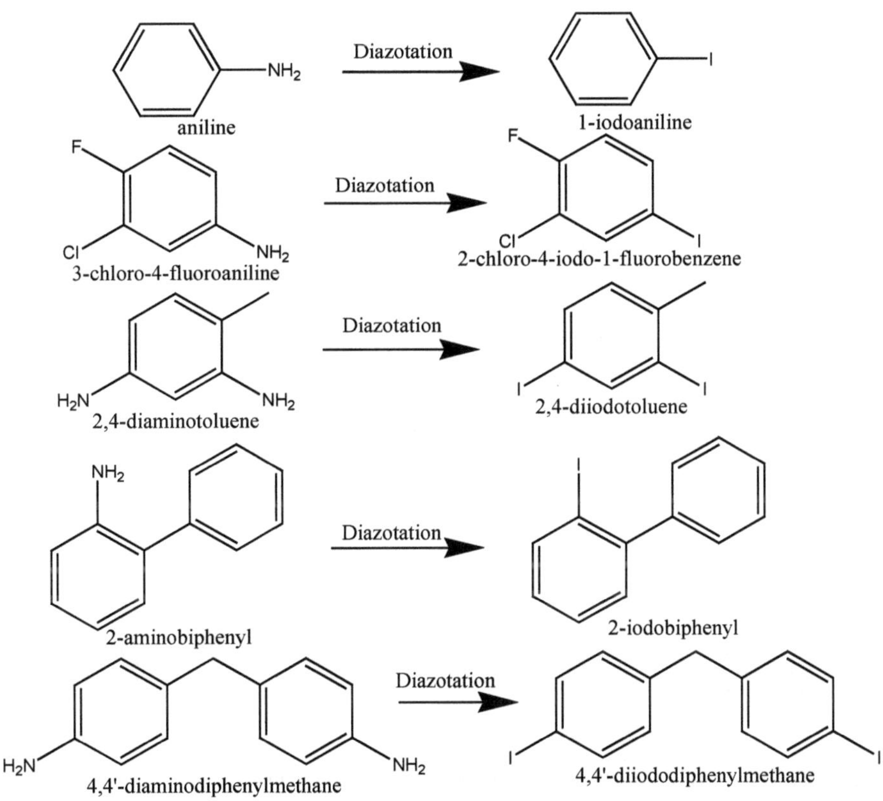

Fig. 7.11 Chemical compounds with name, chemical structure and their iodinated derivatives

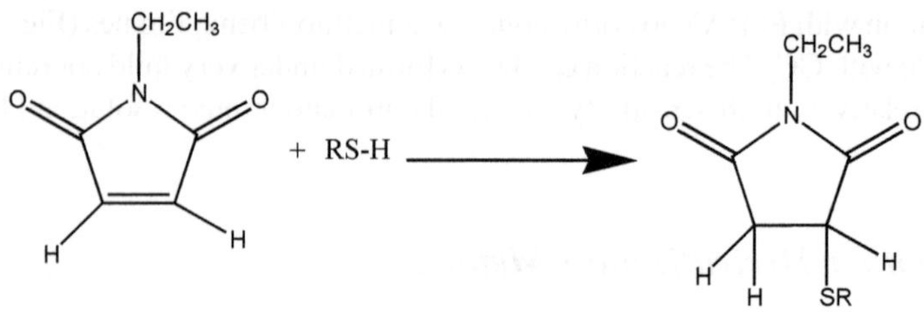

Fig. 7.12 Schematic of the derivatization reaction of alkanethiols in aqueous medium and their *N*-ethylmaleimide derivatives

N-ethylmaleimide for thiol compounds: *N*-ethylmaleimide (NEM) is chosen for the derivatization of thiol compounds because it behaves as an electrophilic molecule that forms covalent bonds with free SH groups (Fig. 7.12). It is an efficient reagent for thiols in aqueous solution yielding stable derivatives, and it is generally considered to be highly specific for thiol groups [123].

Tetrafluorobenzoic acid N-hydroxysuccinimide ester (TFBza-suc) for aliphatic amines: The reaction of TFBza-suc with the aliphatic amines is summarized in Fig. 7.13. TFBza-suc reacts specifically with aliphatic amines—it reacts readily with aliphatic primary and secondary amines under mild conditions and does not react with aromatic amines, hydroxyl, thiol, and phenol groups. Primary and secondary aliphatic amines can be easily converted into their TFBza-suc-amine derivatives by the following procedures described above. The TFBza-suc-amine derivatives are easy to separate by GC and detectable by FID and MS.

N-methylhydrazine for aldehydes: This method involves the derivatization of MA with *N*-methylhydrazine into a stable 6.1-methylpyrazole (1-MP) under mild conditions (Fig. 7.14); subsequently, 1-MP is determined by a gas chromatograph (GC) equipped with a nitrogen–phosphorus detector (NPD) [124].

N-succinimidyl benzoate for aliphatic amines: A new derivatizing reagent, *N*-succinimidyl benzoate (SIBA), was synthesized, characterized, and used to determine methylamine, dimethylamine, ethylamine, *n*-propylamine, *n*-butylamine, *n*-pentylamine, and *n*-hexylamine. SIBA is specific to aliphatic primary and

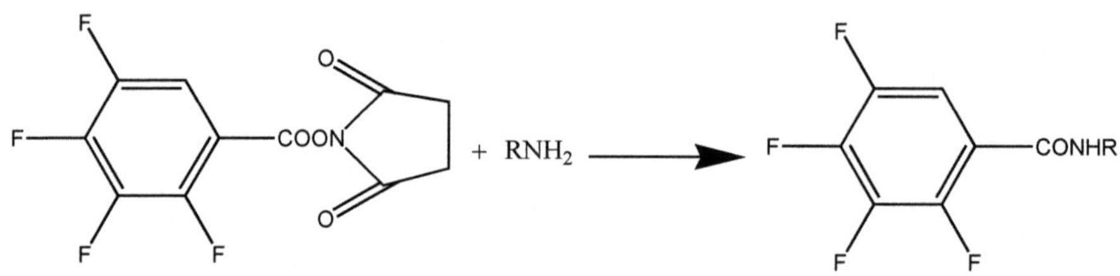

Fig. 7.13 Derivatization reaction of TFBza-suc with aliphatic amines

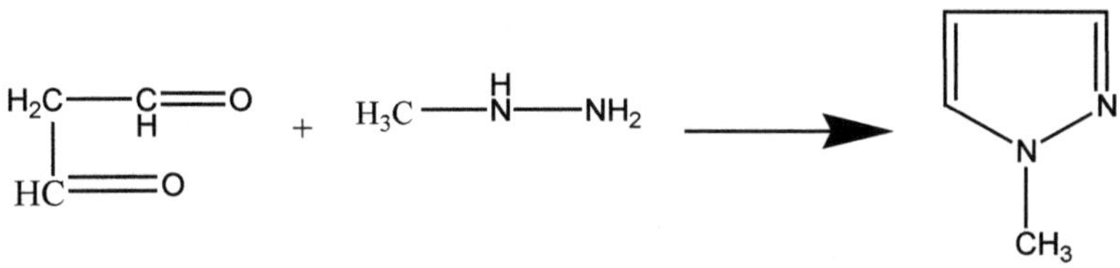

Fig. 7.14 1-Methylpyrazole formation from malonaldehyde and *N*-methylhydrazine

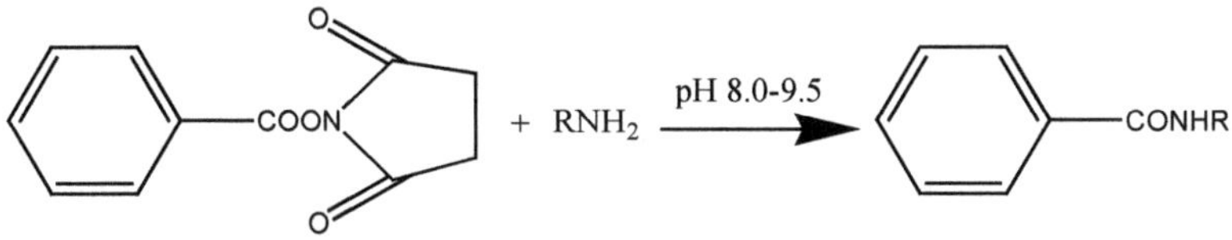

Fig. 7.15 Reaction of *N*-succinimidyl benzoate with aliphatic amines

secondary amines under mild conditions and cannot react with aromatic amines, hydroxyl, thiol, and phenol group. The reaction of SIBA with aliphatic amines is demonstrated in Fig. 7.15 [125].

7.4 Application in Environmental Analysis

Combined with derivatization, SPME have been used for analysis of different kinds of pollutants in environmental analysis such as air, aerosols, water, soil, and sediment samples. Based on the previously reviewed manuscript by Ouyang and Pawliszyn [12], recent development of this technique including new fibers and reagents, application in new analytes and complex matrixes, have been covered in this section.

7.4.1 Air and Aerosols

Indoor air quality has become a predominant topic in terms of public policy. Combination of derivatization with SPME has been utilized for air and particle matters sampling in the field [116, 117, 153] or in the laboratory work [102, 109]. Commercial SPME fibers such as PDMS, PDMS/DVB, CW/DVB, and CW/TPR have been widely used for most applications [102, 109, 114, 116, 117, 153], and also, novel nanoporous polymeric (polypyrrole and polythiophene, PPy/PTh) fiber coating were used [99].

During the "dwellings campaign," 20 kinds of compounds were identified as recurrent pollutants, such as formaldehyde, acetaldehyde, styrene, and benzene, toluene, ethylbenzene, and xylenes (BTEX). Of these pollutants, formaldehyde is of particular concern as it was recognized as "carcinogenic to humans" (class 1) by the International Agency for Research on Cancer, and it is one of the OQAI's ten priority substances [106]. As to carbonyl compounds such as aldehydes and acetone, target analytes are extracted by the commercial SPME fiber by either DI or HS mode [106, 109, 114–117], and combined by in situ derivatization of either in vial or on fiber with *o*-(2,3,4,5,6-pentafluorobenzyl)hydroxylamine hydrochloride (PFBHA) [78–82] or 2,4-dinitrophenylhydrazine/acetylacetone [109] and then for GC analysis. Moreover, online analysis of carbonyl compounds with derivatization by DNPH in aqueous extracts of atmospheric particulate matters by in-tube SPME coupled to capillary LC was reported [110, 154].

Alkylamines play an important role in atmospheric chemistry and are of concern for human health. Determining them from the vapor phase is challenging owing to their high polarity and volatility, water solubility, low concentrations, and poor chromatographic properties [145]. Parshintsev et al. [145] proposed on-fiber derivatization SPME to increase the sensitivity and selectivity for the determination of alkylamines in air samples. SPME fibers coated in headspace with PFBCF (10 min) or FMOC chloride were exposed to the sample, after which the derivatized alkylamines were thermally desorbed in the GC injection port and analyzed by GC–MS [145]. Chafer-Pericas et al. [167] described a new approach for the determination of trimethylamine (TMA) in air by LC. The assay is based on the employment of a SPME fiber for sampling and for derivatization of the analyte with the fluorogenic reagent FMOC. The fiber, with a Carbowax-templated resin of 50 μm coating, was first immersed into a solution of the reagent. Once loaded with the reagent, the fiber was immersed into the water samples or exposed to the air samples in order to extract and to derivatize the analyte. Finally, the fiber was placed into a HPLC-SPME interface to desorb and transfer the TMA-FMOC derivative to the LC equipment [102].

Phenol is widely employed as disinfectant and antiseptic in medicinal preparations such as mouthwash and sore throat lozenges and in biotechnology-derived products such as vaccines and antiserum. It is also a constituent of phenolic resins and used in the manufacturing of nylon and other synthetic fibers. Due to its high vapor pressure, this compound can be released to the air and may pollute the occupational environment. A method based on SPME followed by on-fiber derivatization and GC–MS for determination of phenol in air was developed. Three different types of SPME fibers, polar and non-polar PDMS, and PEG were synthesized using solgel technology, and their feasibility to the sampling of phenol was investigated [90].

In order to evaluate the outdoor atmospheric contamination by non-agricultural pesticide, an analytical methodology using accelerated solvent extraction (ASE) and a silylation procedure coupled to SPME and GC/MS was developed for the determination of 31 pesticides of different chemical classes (urea, phenoxy acids, pyrethrenoids, etc.). Pesticides were simultaneously sampled on glass fiber filters and on XAD-2 resin traps by using a low volume sampler (Partisol) for 1 week. Traps were extracted by ASE with acetonitrile and concentrated to 1 mL by using a rotary evaporator. 500 μL of the extract was dissolved in 19.5 mL of 1.5% NaCl acidified water (pH = 3) and SPME extracted by PA fiber for 55 min at 50 °C. Since most of the studied pesticides are polar or thermo-labile, a derivatization step by injection of 2 mu L of MtBSTFA just before SPME desorption was done. MtBSTFA was chosen since it delivers very specific ions on electronic impact (m/z = M-57) [52].

Haloacetic acids (HAAs) are always as chlorination disinfection by-products (CDBPs) present in chlorine disinfected waters and are a concern for human health. A SPME method was used for the analysis of nine HAAs in air (aerosols) from indoor swimming pools (ISPs). The analysis is characterized by derivatization of HAAs to their methyl esters with dimethyl sulfate, HS-SPME with a CAR-PDMS fiber, and GC/ECD.

MOCA (4,4′-methylenebis(2-chloroaniline)) in the polymer industry was used as a curing agent for urethane and epoxy resins. Classic techniques for air sampling of MOCA were enhanced by incorporating a in vial derivatization step (3,5-dinitrobenzoyl chloride solution in toluene), thus increasing the limit of detection and limit of quantification. Sampling of MOCA from water solution was performed HS SPME using novel nanoporous polymeric (polypyrrole and polythiophene) fiber. Samples were analyzed by HPLC coupled with a UV detector [99] (Table 7.2).

7.4.2 Water

Environmental water samples include drinking water, seawater, rainwater, river water, and wastewater. Commercial SPME fibers including PA, PDMS, CAR-PDMS, PDMS-DVB, and CW-DVB, as well as fibers with new coating materials, such as NiTi-ZrO_2 [18], oxidized multiwalled carbon nanotubes [155], polyaniline (PANI) [156–158] have been used for analysis of environmental pollutants in aqueous samples.

Various kinds of derivatization reagents are used for different aims. In vial derivatization, acetic anhydride [18, 91, 159], and dimethyl sulfate/NaOH [17, 82] were used to derivatizate phenolic compounds; sodium tetrahydroborate ($NaBH_4$), $NaBPr_4$, and $NaBEt_4$ were used to derivatizate organometallic compounds [29, 31, 86, 87, 89, 127–129, 138–140, 160]. Silylating reagents such as MTBSTFA, BSTFA, and MSTFA were widely used for different kinds of chemicals in on-fiber post-derivatization [39, 45, 53, 54, 56, 161–163]. Pentafluorophenylhydrazine (PFPH) and PFBHA were widely used for on-fiber in situ derivatization of aldehyde compounds [27, 105, 164, 165]. Tetrabutylammonium hydrogensulfate as an ion-pairing reagent was used for in-port derivatization of perfluorocarboxylic acids [166] and linear alkylbenzensulfonates (LASs) [62] in GC. In HPLC-FID or LC–MS analysis, 9-fluorenylmethyl chloroformate (FMOC) was used for methylamine (MA), dimethylamine, and trimethylamine (TMA) [103, 167, 168]; 4-fluoro-7-nitro-2,1,3-benzoxadiazole (NBD-F) was used for anatoxin-a [112, 113];

Table 7.2 Application of SPME and derivatization in air and particulate PM10 and PM2.5 analysis

Chemicals	SPME method	SPME fiber	Derivatization method	Derivatization regents	Detection	References
Aldehydes and acetones	HS	PDMS	On fiber (in situ)	2,4-Dinitrophenylhydrazine/acetylacetone	GC–MS	[109]
	HS; DI	PDMS/DVB; PDMS; CW/DVB	In vial; on fiber (in situ)	PFBHA	GC–MS, GC–FID	[106, 114–117]
	IT	–	In vial; on fiber (in situ)	DNPH	LC-DAD; CapLC-MS	[110, 154]
Alkylamines	HS	PDMS/DVB	On fiber (in situ)	PFBCF; FMOC	GC–MS	[145]
	DI	CW/TPR	On fiber (in situ)	FMOC	LC	[102]
Phenol	DI	PDMS/PEG	On fiber (post)	Acetic anhydride	GC–MS	[90]
Pesticides	DI	PA	In vial	MtBSTFA	GC–MS	[52]
Haloacetic acids	HS	CAR-PDMS	In vial	Dimethyl sulfate (DMS)	GC–ECD	[83]
4,4′-methylenebis (2-chloroaniline)	HS	PPy/PTh	In vial	3,5-Dinitrobenzoyl chloride	HPLC-UV	[99]

– Not indicated

and photochemically induced fluorimetry derivatization was used for phenylurea herbicides [169] and pyrethroids [170].

HS-SPME technique is widely used for extracting various kinds of chemicals. Among them, in-vial (in situ) derivatization is frequently used for haloacetic acids (HAAs) [21, 23] and acetic acid [67], aldehyde compounds [25, 124, 165], phenoxy acid herbicides [68], phenolic pollutants [17, 18, 82, 91, 159], aliphatic amines [120, 125], acaricide fenbutatin oxide (FBTO) [131], acetone [111], selective serotonin reuptake inhibitors (SSRIs) such as venlafaxine, fluvoxamine, fluoxetine, citalopram, and sertraline [97], and organometallic compounds such as organotin [29, 31, 86, 87, 127, 128, 138, 139, 160], organolead [31, 86, 89, 160], and organomercury [31, 128, 129, 140, 160]. On-fiber in situ derivatization is used for analyzing chemicals such as aldehyde compounds [27, 105, 164], carbonyls compounds [28], and thiol compounds [171]. Moreover, on-fiber post-derivatization [172], as well as column derivatization [46], has also been used to the analysis of phenolicEDCs.

DI-SPME has also been widely applied to the analysis of different classes of contaminants: In-vial derivatization is used for homocysteine (Hcy), cysteine (Cys), and methionine (Met) [70], phenylurea herbicides (PUHs) [78], polar aromatic amines [122], degradation products of chemical warfare agents [58], methylamine (MA) [168], alkanethiols and dihydrogen sulfide [123], phenolic acids [69], anatoxin-a [71], acidic pesticides [72], phenoxy acid herbicides [68], methyl mercury and Hg(II) [141], and chromium [146, 147]. On-fiber post-derivatization is widely used for anatoxin-a [112, 113], *p*-hydroxybenzoic acid (parabens) [56], phenolic compounds [45, 53, 161, 162], β-agonist clenbuterol [40], anti-inflammatory drugs [39], dimethylamine [103], salicylate- and benzophenone-type UV filters [163], phenoxy acid herbicides, and dicamba [54]. On-fiber in situ derivatization is used for compounds isocyanates [35] and trimethylamine (TMA) [102]. In-port derivatization techniques in GC are used for analyzing perfluorocarboxylic acids [166] and linear alkylbenzensulfonates (LASs) [62]. Moreover, post-column photochemically induced fluorimetry derivatization and fluorescence detection by HPLC are developed for the determination of phenylurea herbicides [169], pyrethroids [170], an alcohol ethoxylate (Brij 56), and 1-hexadecanol [100].

Moreover, the combination of different SPME and derivatization techniques was also used to determine metabolism and degradation products of chemical warfare agents [58], volatile and semi-volatile compounds [89], octanol [81], methylmercury [173], hydroxyketone [174, 175], and PAHs [176]. SPME analysis of aqueous samples has occasionally been assisted by use of microwaves [177]. In response to the treatment of massive number of samples, fast and automated analytical methods were recently developed for determination of fenbutatin oxide (FBTO) [131], thiol compounds [171], phenolic compounds [162], PAHs [176], and organometallic compounds [137, 160] (Table 7.3).

Table 7.3 Application of SPME and derivatization in water analysis

Chemicals		SPME method	SPME fiber	Derivatization method	Derivatization regents	Detection	References
Acetic acid	Haloacetic acids (HAAs)	HS	PDMS; CAR/PDMS	In vial	Sulfuric acid/methanol; sulfuric acid/ethanol; dimethyl sulfate (DMS)	GC–ECD; GC–MS	[20–23]
	Fluoroacetic acid (FAA)	HS	PDMS; PA	In vial	Sulfuric acid/ethanol	GC–FID; GC–MS	[84]
	Acetic acid	DI; HS	CW/DVB; PA	In vial	*n*-Hexyl chloroformate; benzyl bromide	GC–FID	[67]
Carbonyl compounds	Aldehydes	HS	PDMS/DVB	In vial	PFBHA	GC–ECD; GC–MS	[24–28]
		HS	PDMS/DVB	On fiber (in situ)	PFPH	GC–ECD	[105]
		HS	PDMS/DVB; CAR/PDMS	In vial	TFEH	GC–MS	[107, 108]
	Acetone	HS	PDMS; PDMS/DVB; CAR/PDMS	In vial	PFBHA	GC–MS	[111]
		HS	PDMS/DVB	In vial	TFEH	GC–MS	[107]
	Isocyanates		PDMS	On fiber (in situ)	Dibutylamine (DBA)	GC–FID; GC–MS	[35]
Amines	Aliphatic amines	HS	Dibenzo-18-crown-6	In vial	TFBza-suc	GC–FID	[104]
		DI	CW-TR	On fiber (post)	OPA-NAC/FMOC; FMOC	LC	[103]
		DI	PDMS; PDMS/DVB; DVB/CAR/PDMS; CW/TPR	On fiber (post)	FMOC	HPLC-FLD; HPLC-DAD	[168]
		HS	PDMS/DVB; PA; CAR/PDMS; CW/DVB; DVB/CAR/PDMS	In vial	PFBAY	PortableGC-μFID	[120]
		HS	PDMS	In vial	SIBA	GC–FID; GC–MS	[125]
	Aromatic amines	HS	PDMS/DVB	In vial	Diazotation	GC/MS	[121, 122]

(continued)

Table 7.3 (continued)

Chemicals		SPME method	SPME fiber	Derivatization method	Derivatization regents	Detection	References
Metal species	Tin species	HS	PDMS; PDMS/CAR; PDMS/DVB	In vial	$NaBEt_4$	GC–MS; GC–MS/MS;GC-ICP–TOF–MS	[29–33]
		HS	PDMS; PA; PDMS/DVB	In vial	$NaBH_4$	GC–FPD	[87, 139]
		HS	PDMS	In vial	KBH_4	GC–AFS	[142].
		HS; DI	PDMS	In vial	NaBPr4	GC–MIP–AES; GC–AED	[86, 88, 128]
	Mercury speciation	HS	PDMS	In vial	NaBPr4	GC–ICP–MS	[129]
		HS	PDMS	In vial	NaBPh4	GC–ECD	[140]
		HS	PDMS; PDMS/DVB; CAR/PDMS	In vial	NaBEt4	GC–ECD; GC–MS; GC-MIP–AES; GC–ICP–TOF–MS	[31–33, 128, 129, 178]
		–	poly(ethylene glycol)	In vial	Phenylboronic acid	GC–FID	[85]
	Lead species	HS		in vial	DSTEB	GC/MS; GC/FID	[89]
		HS	PDMS; PDMS/DVB; CAR/PDMS	In vial	NaBEt4	GC–MS; GC-ICP–TOF–MS; GC–AED	[31–33, 88]
		HS	PDMS	In vial	NaBPr4	GC–AED	[86, 88]
	Chromium	HS	PDMS	In vial	HTFA	GC–ECD; GC–EI-MS; GC-ICP–MS; GC–FPD	[146, 147]
	Organomanganese	HS	PDMS	In vial	$NaBPh_4$	GC–AED	[88]
	Beryllium and lanthanide	DI	PA (polydimethylsiloxane)	On fiber	H(FHD)	GC–MS	[148]

(continued)

Table 7.3 (continued)

Chemicals		SPME method	SPME fiber	Derivatization method	Derivatization regents	Detection	References
Phenolic compounds	Halophenols (chlorophenol, trichlorobromophenols)	HS	NiTi alloy coated with ZrO2	HS	Sodium carbonate/acetic anhydride		[18]
		HS	CAR/PDMS/DVB; PDMS	In vial	Dimethyl Sulfate/NaOH	GC–MS	[17]
		HS	PA	Column silylation	BSTFA	GC–MS	[46]
		HS	CAR/PDMS; PDMS	In vial	acetic anhydride/NaOH	GC–MS	[91, 94]
		IT	–	In vial	–	GC–FID	[92]
		DI; HS	PA; CAR/PDMS; PDMS; PDMS/DVB; DVB/CAR/PDMS	In vial	acetic anhydride	GC–MS	[96]
		DI	PA	On fiber (post)	MTBSTFA	GC–MS	[53]
		DI	PA	On fiber (post)	BSTFA	GC–MS	[44]
	Alkylphenols (APs)	DI	Etched stainless steel wire/ionic liquid; PA	HS	BSTFA	GC–MS; HPLC-UV	[48, 49, 179]
		HS	CW/DVB	In vial	Dimethyl sulfate/NaOH	GC–MS	[82]
		HS	PA	Column silylation	BSTFA	GC–MS	[46]
		IT	–	In vial	–	GC–FID	[92]
		HS	PA; PDMS/DVB	On fiber (in situ)	MTBSTFA/TBDMCS	GC–MS	[36]
		Microwave-assisted HS	PDMS/DVB	In vial	Acetic anhydride/NaOH	GC–MS	[19]
	Bisphenol A (BPA)	DI	PA	On fiber (post)	MTBSTFA	GC–MS	[47]

(continued)

Table 7.3 (continued)

Chemicals		SPME method	SPME fiber	Derivatization method	Derivatization regents	Detection	References
Herbicides	Acaricide fenbutatin oxide (FBTO)	HS	PDMS	in vial	$NaBEt_4$	GC–MS	[131]
	Phenylurea herbicides (PUHs)	–	PA	In vial	Iodoethane and sodium hydride	GC–MS	[78]
	Acidic pesticides	Online	PDMS	In vial	Butylchloroformate (BuCF),	GC–MS	[72]
		DI	PA	On fiber (post)	MTBSTFA	GC–MS	[54]
		DI	PA; PDMS; CAR/PDMS; PDMS/DVB	In vial	Ethyl and methyl chloroformate	GC–FID	[69]
			PDMS/DVB	In vial	Benzyl bromide	GC–MS	[68]
	Polar chlorinated herbicides	DI	DMS	HS	MTBSTFA	GC–MS	[55]
	Phthalic monoesters (MPs)	DI	PDMS/DVB; PA; PDMS; CW/DVB	On fiber	Diazomethane	GC–MS	[80]
	Aldrin (ALD), heptachlor (HEPT),2,4'-DDD	DI	PA	On fiber (post)	MTBSTFA	GC–MS	[47]

(continued)

Table 7.3 (continued)

Chemicals		SPME method	SPME fiber	Derivatization method	Derivatization regents	Detection	References
PPCPs	Anti-inflammatory drugs	HS	PDMS	In vial	Dimethyl sulfate	GC–MS	[180, 181]
		DI	PA	On fiber (post)	MTBSTFA	GC–MS	[47, 57]
		DI	PDMS/DVB; PA; PDMS/CAR; PDMS;	On fiber (in situ)	ECF	GC–MS	[37]
	Selective serotonin reuptake inhibitors (SSRIs)	DI	PDMS/DVB	In vial	Potassium hydrogen carbonate and acetic anhydride	GC–MS	[97]
	Amphetamine type stimulants (ATSs)	DI	PDMS/DVB	In vial	Iso-butyl chloroformate (iBCF)	GC–MS	[73]
		DI	CW/TR	In vial; On fiber (post)	FMOC	LC	[101]
	β-agonist clenbuterol		PA	On fiber (post)	Hexamethyldisilazane	GC–FID, GC–MS	[59]
	Cannabinoids (THC, THCCOOH)	DI	DVB/CAR/PDMS	On fiber	MSTFA	GC	[61]
	Parabens	DI	PA; CAR/PDMS; PDMS; PDMS/DVB; DVB/CAR/PDMS	On fiber (post)	MTBSTFA	GC–MS/MS	[56]
		DI; HS	PA; CAR/PDMS; PDMS; PDMS/DVB; DVB/CAR/PDMS	In vial	Acetic anhydride	GC–MS	[96]
	Salicylate- and benzophenone-type UV filters	DI	PDMS-DVB	HS	MSTFA	GC–MS/MS	[60]
		DI	PA	On fiber (post)	MTBSTFA	GC–MS	[47]
	Triclosan and possible related compounds	DI	PA	On fiber (post)	MTBSTFA	GC–MS	[53]
		DI; HS	PA; CAR/PDMS; PDMS; PDMS/DVB; DVB/CAR/PDMS	In vial	Acetic anhydride	GC–MS	[96]
	Steroids	DI	PA	HS	BSTFA	GC–MS	[48, 49]
		DI	PDMS/DVB; PA; PDMS/CAR; PDMS;	On fiber (in situ)	ECF	GC–MS	[37]

(continued)

Table 7.3 (continued)

Chemicals		SPME method	SPME fiber	Derivatization method	Derivatization regents	Detection	References
		DI	PA	on fiber (post)	MTBSTFA	GC–MS	[47]
	Alcohol ethoxylate (Brij 56) and 1-hexadecanol		PDMS/DVB	Pre-column, SPME-HPLC online derivatization	1-naphthoyl chloride (4-(dimethylamino) pyridine (DMAP) as catalyst)	HPLC–FID	[100]
Other compounds	Anatoxin-a	DI	PDMS	In vial	Hexylchloroformate,	HPLC-FLD	[71]
		DI	PDMS; PDMS/DVB; PA; CW/TPR	on fiber (post)	NBD-F	HPLC; HPLC–FLD	[112, 113]
	4,4′-methylenebis (2-chloroaniline)	HS	PPy/PTh	In vial	3,5-dinitrobenzoyl chloride	HPLC-UV	[99]
	Homocysteine (Hcy), cysteine (Cys), and methionine (Met)	DI	PDMS, PA	In vial	Ethyl, propyl, isopropyl, butyl, isobutyl and alkyl chloroformate	GC–MS	[70]
	Thiol compounds	DI	PDMS, PDMS/DVB, CW/DVB, DVB/CAR/PDMS	In vial	*n*-ethylmaleimide	GC–MS	[123]
		HS	PDMS/Carboxen	On fiber	*n*-ethylmaleimide(NEM)	GC–MS	[149]
	Perfluorocarboxylic acids (PFCAs)	DI	PDMS	In-port	Boron trifluoride	GC–NCI–MS	[166]
		HS	CAR/PDMS	In vial	Propylchloroformate/propanol	GC–QqQ–MS	[74]
	Bromate	HS	CAR/PDMS	In vial	2,6-DMP	GC–MS	[76]
	Hydrazine	DI	PDMS/DVB	In vial	Propyl chloroformate	GC–QqQ–MS	[75]
	PAHs	DI	PA	On fiber (post)	MTBSTFA	GC–MS	[51]

– Not indicated

7.4.3 Soil and Sediment

Pollutants discharged into estuarine and coastal marine waters are usually adsorbed onto particle surfaces and incorporated into sediments. Hence, the sediments represent an important source of contamination to both overlying water column and benthic organism with undoubtedly ecotoxicological implications. Only a few papers have been published describing methods for determining environmental pollutants from soil and sediment samples, and in most applications, analytes have been firstly extracted by other methods such as pressurized fluid extraction (PFE), supercritical fluid extraction (SEE), microwave-assisted extraction (MSE), ultrasonic extraction (USE), or acid digestion, and then by in vial derivatization and HS-SPME technique [132, 135, 150, 152]. Commercial fiber PDMS are frequently used for most occasions, followed by PDMS-DVB and CAR-PDMS.

Perfluorocarboxylic acids (PFCAs) have one end with polar groups attached and another end with nonpolar groups in which all the aliphatic hydrogens present in the fatty acids are replaced by fluorides that lessen water surface tension more than hydrocarbon-based surfactants, so they are more powerful wetting agents. Since the 1940s, PFCA compounds are used as surfactants and surface protectors in various kinds of products including paints, lubricants, PTFE synthesis, polishers, food packaging, and fire-fighting foams. These highly fluorinated compounds have unusual features with high thermal and chemical stabilities because of the high energy of the bonds between carbon and fluorine. A procedure for the determination of perfluorocarboxylic acids (i.e., PFC(7-10)A) in sediment by pressurized fluid extraction (PFE), derivatization, HS-SPME, and GC–MS determination in the negative ion chemical ionization mode was developed [150]. HS-SPME in situ supercritical fluid extraction (SEE) was investigated for the determination of trace amounts of perfluorocarboxylic acids (PECAs) in sediments. Quantitation was performed by using gas chromatography coupled to negative chemical ionization-tandem mass spectrometry (GC–NCI–MS/MS) [77].

Mercury is considered a highly toxic element because of its accumulative and persistent character in the environment. Inorganic and organic species of mercury are generated in different industrial activities, mainly pharmaceutical, paper, electrochemical, and plaguicide industries. Jitaru et al. [32] reported the development of an analytical approach for speciation analysis of mercury at ultra-trace levels on the basis of SPME and multicapillary GC hyphenated to inductively coupled plasma time-of-flight mass spectrometry. HS SPME with a CAR/PDMS fiber is used for the extraction/preconcentration of mercury species after derivatization with sodium tetraethylborate and subsequent volatilization. Isothermal separation of methylmercury (MeHg), inorganic mercury (Hg^{2+}), and propylmercury (PrHg) used as internal standard is achieved within a chromatographic run below 45 s without the introduction of spectral skew [134]. Furthermore, a solvent-free SPME method has been developed to determine methyl mercury and Hg^{2+} in sediment samples from TTC area (Mumbai, India) using GC–MS. The analytical method consists of phenylation with $Na[B(C_6H_5)_4]$, simultaneous SPME of the derivatives, followed by a final GC–MS analysis [141].

Organotin compounds have been introduced into the environment mainly through their use as insecticides, fungicides, bactericides, wood preservatives, plastic stabilizers, and biocides in antifouling paints for boats and ships. Tin is considered an essential element. However, organotin compounds, due to their lipid solubility, may penetrate into tissues and the central nervous system with potentially greater toxicity to organisms. Ten superficial sediments from river estuaries of Gipuzkoa (North Spain) were collected and analyzed for nine metals (including tin and Sn) and butyltin compounds (monobutyltin, MBT; dibutyltin, DBT; tributyltin, TBT). Total metal concentration in the fine fraction (<63 mum) of the sediment was determined in aqua regia–hydrofluoric acid extracts by atomic absorption spectrometry. The butyltin species (BTs) were firstly extracted from the sediments with hydrochloric acid–methanol mixture. After derivatization with sodium tetraethyl borate, the organotin compounds were extracted by SPME in HS mode. A fiber coated with 100 mum poly(dimethylsiloxane) was used for SPME. The organotin species were analyzed by GC–FID [135]. A unique approach was developed to improve the precision of quantification of tributyltin (TBT) in sediments by SPME using isotope dilution GC/MS [133]. A method for the determination of organotin compounds (monobutyl = MBT, dibutyl = DBT, and tributyltin = TBT) in marine sediments by HS-SPME has been developed. The analytical procedure involved (1) extraction of TBT, DBT, and MBT from sediments with HCl and methanol mixture, (2) in situ derivatization with sodium tetraethylborate, and (3) HS-SPME extraction using a fiber coated with poly(dimethylsiloxane). The derivatized organotin compounds were desorbed into the splitless injector and simultaneously analyzed by GC–MS [132]. An automated method for the simultaneous determination of six important organotin compounds, namely monobutyltin (MBT), dibutyltin (DBT), tributyltin (TBT), monophenyltin (MPhT), diphenyltin (DPhT), and triphenyltin (TPhT) in water and sediment samples, is described. The method is based on derivatization with sodium tetraethylborate followed by automated HS-SPME combined with GC–MS under retention time-locked (RTL) conditions [137]. A rapid and solvent-free procedure for the simultaneous determination of mono-, di-, and tributyltin in sediments by HS-SPME is described. The butyltin chlorides present in sediments were converted into their hydride forms by potassium tetrahydroborate prior to extraction. A quartz fiber coated with 100 μm film of polydimethylsiloxane (PDMS) was introduced into the headspace over the matrix to extract butyltin hydrides. After extraction, the fiber was directly inserted into the injector of the gas chromatography (GC) for desorption, separation, and quantification. A laboratory-modified flame photometric detector using quartz surface-induced tin emission (QSIL-FPD) was used throughout the experiment [143]. A procedure for determination of tetraethyltin (TeET) and tetrabutyltin (TeBT) in water by SPME using the headspace approach has been developed. The method has been adapted for the simultaneous determination of mono-, di-, and tributyltin species (MBT, DBT, and TBT) after derivatization with sodium tetraethylborate in water and sediment samples [138].

Table 7.4 Application of SPME and derivatization in soil and sediment analysis

Chemicals	Pretreatment	SPME method	SPME fiber	Derivatization method	Derivatization regents	Detection	References
Perfluorocarboxylic acids (PECAs)	PFE	HS	PDMS	In vial	Boron trifluoride	GC–NCI–MS	[150]
	SEE	HS	PDMS	–	*n*-butanol	GC–NCI–MS/MS	[77]
Mercury and mercury species	Digestion	HS	PDMS/DVB	In vial	NaBPh4	GC–MS	[141]
	Digestion	HS	PDMS	In vial	$NaBEt_4$	GC–MS	[134]
	Digestion	HS	–	In vial	KBH_4	GC–AAS	[144]
Tin and butyltin species	Digestion; MSE; USE	HS	PDMS	In vial	$NaBEt_4$, KBH_4	GC–FID, GC–MS	[131–133, 135, 138, 182]
Lewisite oxide	USE	–	PDMS	–	1,3-propanedithiol	GC–MS	[152]

– not indicated

Lewisite (syn. 2-chlorovinyldichloroarsine or L, CAS registry No. [541-25-3]) is an organoarsenical chemical blister agent originally developed during World War I and produced worldwide ever since by various agencies, including the US Department of Defense. A rapid, sensitive, and convenient method is described for determining Lewisite oxide in soil. Samples are initially fortified with phenylarsine oxide (surrogate), and then, both species are extracted using ascorbic acid solutions containing 1,3-propanedithiol (derivatizing reagent). The corresponding filtered supernatant is sampled using a SPME fiber. Collected analytes are thermally desorbed in a heated gas chromatographic inlet, separated using fused-silica capillary columns ("primary" and "confirmatory"), and detected with either a mass spectrometric (selected ion monitoring mode) or flame photometric (sulfur-selective mode) detector [152] (Table 7.4).

7.5 Summary and Future Trends

A large number of applications of SPME combined with derivatization to environmental analysis have recently been reported in the literatures. The target analytes include aldehydes and acetone, amines, organic acids, phenolic compounds, pesticide, metal and organic metal species, and PPCPs. Combined with derivatization, HS-SPME, DI-SPME, membrane-protected SPME, and in-tube SPME have been used for analysis of a wide range of environmental matrices, including air, water, soil, and sediment samples.

Moreover, the new specific derivatization reagents, such as chirality compounds, should be synthesized and developed. New techniques need to be developed for the existing reagents, and the sampling preparation time should be decreased. The built-up database of the derivatization products is recommended. Although the use of derivatization reactions improves the extractability of some analytes, the development of new coatings is expected to enhance the applicability of SPME to the determination of polar species [66].

The ambient ionization techniques based on atmospheric pressure chemical ionization (APCI) mechanism recently were developed and implemented into analytical practices. Coupling SPME to ambient mass spectrometry (AMS) is an effective strategy to improve the sensitivity and suppress the matrix effects for direct analysis of complex samples [183–185]. Thus, coupling SPME, the derivatization with AMS gives promising prospects in environmental applications.

References

1. Risticevic S, Lord H, Gorecki T et al (2010) Protocol for solid-phase microextraction method development. Nat Protoc 5:122–139
2. Nerin C, Salafranca J, Aznar M et al (2009) Critical review on recent developments in solventless techniques for extraction of analytes. Anal Bioanal Chem 393:809–833

3. Boland AB, Chai M, Luo YZ et al (1994) New solvent-free sample preparation techniques based on fiber and polymer technologies. Environ Sci Technol 28:569A–574A
4. Zhang ZY, Yang MJ, Pawliszyn J (1994) Solid-phase microextraction.A solvent-free alternative for sample preparation. Anal Chem 66:844A–853A
5. Farajzadeh MA, Nouri N, Khorram P (2014) Derivatization and microextraction methods for determination of organic compounds by gas chromatography. TrendsAnal Chem 55:14–23
6. Arthur CJ, Pawliszyn J (1990) Solid phase microextraction with thermal desorption using fused silica optical fibers. Anal Chem 62:2145–2148
7. Kataoka H, Lord HL, Pawliszyn J (2000) Applications of solid-phase microextraction in food analysis. J Chromatogr A 880:35–62
8. Lord H, Pawliszyn J (2000) Evolution of solid-phase microextraction technology. J Chromatogr A 885:153–193
9. Stashenko SE, Martinez JR (2004) Derivatization and solid-phase microextraction. Trends Anal Chem 23:553–561
10. Rosenfeld JM (2003) Derivatization in the current practice of analytical chemistry. Trends Anal Chem 22:785–798
11. Stalikas CD, Fiamegos YC (2008) Microextraction combined with derivatization. Trends Anal Chem 27:533–542
12. Ouyang G, Pawliszyn J (2006) SPME in environmental analysis. Anal Bioanal Chem 386:1059–1073
13. Penalver A, Pocurull E, Borrull F et al (1999) Trends in solid-phase microextraction for determining organic pollutants in environmental samples. Trends Anal Chem 18:557–568
14. Risticevic S, Niri VH, Vuckovic D et al (2009) Recent developments in solid-phase microextraction. Anal Bioanal Chem 393:781–795
15. Pan L, Pawliszyn J (1997) Derivatization/ solid-phase microextraction: new approach to polar analytes. Anal Chem 69:196–205
16. Buchholz KD, Pawliszyn J (1994) Optimization of solid-phase microextraction conditions for determination of phenols. Anal Chem 66:160–167
17. Diaz A, Ventura F, Galceran MT (2006) Analysis of odorous trichlorobromophenols in water by in-sample derivatization/solid-phase microextraction GC/MS. Anal Biotech Chem 386:293–298
18. Budziak D, Martendal E, Carasek E (2007) Application of NiTi alloy coated with ZrO_2 as a new fiber for solid-phase microextraction for determination of halophenols in water samples. Anal Chim Acta 598:254–260
19. Wu YP, Wang YC, Ding WH (2012) Rapid determination of alkylphenols in aqueous samples by in situ acetylation and microwave-assisted headspace solid-phase microextraction coupled with gas chromatography-mass spectrometry. J Sep Sci 35:2122–2130
20. Hammami B, Driss MR (2013) Development of dry derivatization and headspace solid-phase microextraction technique for the GC-ECD determination of haloacetic acids in tap water. J Anal Chem 68:671–679
21. Sarrion MN, Santos FJ, Galceran MT (1999) Solid-phase microextraction coupled with gas chromatography-ion trap mass spectrometry for the analysis of haloacetic acids in water. J Chromatogr A 859:159–171
22. Wu FW, Gabryelski W, Froese K (2002) Improved gas chromatography methods for micro-volume analysis of haloacetic acids in water and biological matrices. Analyst 127:1318–1323
23. Sarrion MN, Santos FJ, Galceran MT (2000) In situ derivatization/solid phase microextraction for the determination of haloacetic acids in water. Anal Chem 72:4865–4873
24. Beranek J, Kubatova A (2008) Evaluation of solid-phase microextraction methods for determination of trace concentration aldehydes in aqueous solution. J Chromatogr A 1209:44–54
25. Cancho B, Ventura F, Galceran MT (2002) Determination of aldehydes in drinking water using pentafluorobenzylhydroxylamine derivatization and solid-phase microextraction. J Chromatogr A 943:1–13

26. Martos PA, Pawliszyn J (1998) Sampling and determination of formaldehyde using solid-phase microextraction with on-fiber derivatization. Anal Chem 70:2311–2320
27. Tsai SW, Chang CM (2003) Analysis of aldehydes in water by solid-phase microextraction with on-fiber derivatization. J Chromatogr A 1015(143–150):28
28. Schmarr HG, Potouridis T, Ganss S et al (2008) Analysis of carbonyl compounds via headspace solid-phase microextraction with on-fiber derivatization and gas chromatographic-ion trap tandem mass spectrometric determination of their *O*-(2,3,4,5,6-pentafluorobenzyl)oxime derivatives. Anal Chim Acta 617:119–131
29. Chou CC, Lee MR (2005) Determination of organotin compounds in water by headspace solid phase microextraction with gas chromatography-mass spectrometry. J Chromatogr A 1064:1–8
30. Cui ZY, Ge N, Cao YZ et al (2013) Speciation of organotin compounds in seawater by headspace solid phase microextraction and gas chromatography-tandem mass after ethylation. Chin J Anal Chem 41:1887–1892
31. Centineo G, Gonzalez EB, Sanz-Medel A (2004) Multielemental speciation analysis of organometallic compounds of mercury, lead and tin in natural water samples by headspace-solid phase microextraction followed by gas chromatography-mass spectrometry. J Chromatogr A 1034:191–197
32. Jitaru P, Infante HG, Adams FC (2004) Simultaneous multi-elemental speciation analysis of organometallic compounds by solid-phase microextraction and multicapillary gas chromatography hyphenated to inductively coupled plasma-time-of-flight-mass spectrometry. J Anal Atom Spectrom 19:867–875
33. Parkinson DR, Bruheim I, Christ I et al (2004) Full automation of derivatization-solid-phase microextraction-gas chromatography-mass spectrometry with a dual-arm system for the determination of organometallic compounds in aqueous samples. J Chromatogr A 1025:77–84
34. Pan L, Adams M, Pawliszyn J (1995) Determination of fatty acids using solid-phase microextraction. Anal Chem 67:4396–4403
35. Batlle R, Colmsjo A, Nilsson U (2001) Development of a personal isocyanate sampler based on DBA derivatization on solid-phase microextraction fibers. Fresenius J Anal Chem 371:514–518
36. Pan YP, Tsai SW (2008) Solid phase microextraction procedure for the determination of alkylphenols in water by on-fiber derivatization with *N*-tert-butyl-dimethylsilyl-*N*-methyltrifluoroacetamide. Anal Chim Acta 624:247–252
37. Gomes PCFD, Barletta JY, Nazario CED et al (2011) Optimization of in situ derivatization SPME by experimental design for GC-MS multi-residue analysis of pharmaceutical drugs in wastewater. J Sep Sci 34:436–445
38. Musshoff F, Junker HP, Lachenmeier DW et al (2002) Fully automated determination of cannabinoids in hair samples using headspace solid-phase microextraction and gas chromatography-mass spectrometry. J Anal Toxicol 26:554–560
39. Rodriguez I, Carpinteiro J, Quintana JB et al (2004) Solid-phase microextraction with on-fiber derivatization for the analysis of anti-inflammatory drugs in water samples. J Chromatogr A 1024:1–8
40. Engelmann MD, Hinz D, Wenclawiak BW (2003) Solid-phase micro extraction (SPME) and headspace derivatization of clenbuterol followed by GC-FID and GC-SIMMS quantification. Anal Bioanal Chem 375:460–464
41. Musshoff F, Junker HP, Lachenmeier DW et al (2002) Fully automated determination of amphetamines and synthetic designer drugs in hair samples using headspace solid-phase microextraction and gas chromatography-mass spectrometry. J Chromatogr Sci 40:359–364
42. Koster EH, Bruins CH, de Jong GJ (2002) On-fiber derivatization for direct immersion solid-phase microextraction. Part II. Acylation of amphetamine with pentafluorobenzoyl chloride for urine analysis. Analyst 127:598–602

43. Jurado C, Gimenez MP, Soriano T et al (2000) Rapid analysis of amphetamine, methamphetamine, MDA, and MDMA in urine using solid-phase microextraction, direct on-fiber derivatization, and analysis by GC-MS. J Anal Toxicol 24:11–16
44. Wang XW, Chen RH, Luan TG et al (2012) Full automatic determination of chlorophenols in water using solid-phase microextraction/on-fiber derivatization and gas chromatography-mass spectrometry. J Sep Sci 35:1017–1026
45. Chang CM, Chou CC, Lee MR (2005) Determining leaching of bisphenol A from plastic containers by solid-phase microextraction and gas chromatography-mass spectrometry. Anal Chim Acta 539:41–47
46. Helaleh MIH, Fujii S, Korenaga T (2001) Column silylation method for determining endocrine disruptors from environmental water samples by solid phase micro-extraction. Talanta 54:1039–1047
47. Basaglia G, Pietrogrande MC (2012) Optimization of a SPME/GC/MS method for the simultaneous determination of pharmaceuticals and personal care products in waters. Chromatographia 75:361–370
48. Yang LH, Lan CY, Liu HT et al (2006) Full automation of solid-phase microextraction/on-fiber derivatization for simultaneous determination of endocrine-disrupting chemicals and steroid hormones by gas chromatography-mass spectrometry. Anal Biotech Chem 386:391–397
49. Yang LH, Luan TG, Lan CY (2006) Solid-phase microextraction with on-fiber silylation for simultaneous determinations of endocrine disrupting chemicals and steroid hormones by gas chromatography-mass spectrometry. J Chromatogr A 1104:23–32
50. Karaisz KG, Snow NH (2001) The use of solid-phase microextraction/gas chromatography-mass spectrometry for the determination of degradation products of volatile and semivolatile compounds. J Microcol Sep 13:1–7
51. Luan TG, Yu KSH, Zhong Y et al (2006) Study of metabolites from the degradation of polycyclic aromatic hydrocarbons (PAHs) by bacterial consortium enriched from mangrove sediments. Chemosphere 65:2289–2296
52. Raeppel C, Fabritius M, Nief M et al (2014) Coupling ASE, sylilation and SPME GC/MS for the analysis of current-used pesticides in atmosphere. Talanta 121:24–29
53. Canosa P, Rodriguez I, Rubi E et al (2005) Optimization of solid-phase microextraction conditions for the determination of triclosan and possible related compounds in water samples. J Chromatogr A 1072:107–115
54. Rodriguez I, Rubi E, Gonzalez R et al (2005) On-fibre silylation following solid-phase microextraction for the determination of acidic herbicides in water samples by gas chromatography. Anal Chim Acta 537:259–266
55. Prieto A, Rodil R, Quintana JB et al (2014) Evaluation of polyethersulfone performance for the microextraction of polar chlorinated herbicides from environmental water samples. Talanta 122:264–271
56. Canosa P, Rodriguez I, Rubi E et al (2006) Optimisation of a solid-phase microextraction method for the determination of parabens in water samples at the low ng per litre level. J Chromatogr A 1124:3–10
57. Rodriguez I, Carpinteiro J, Quintana JB et al (2004) Solid-phase microextraction with on-fiber derivatization for the analysis of anti-inflammatory drugs in water samples. J Chromatogr A 1024:1–8
58. Sng MT, Ng WF (1999) In-situ derivatisation of degradation products of chemical warfare agents in water by solid-phase microextraction and gas chromatographic mass spectrometric analysis. J Chromatogr A 832:173–182
59. Engelmann MD, Hinz D, Wenclawiak BW (2003) Solid-phase micro extraction (SPME) and headspace derivatization of clenbuterol followed by GC-FID and GC-SIMMS quantification. Anal Biotech Chem 375:460–464
60. Negreira N, Rodriguez I, Ramil M et al (2009) Sensitive determination of salicylate and benzophenone type UV filters in water samples using solid-phase microextraction, derivatization and gas chromatography tandem mass spectrometry. Anal Chim Acta 638:36–44

61. Racamonde I, Villaverde-de-Saa E, Rodil R et al (2012) Determination of Delta 9-tetrahydrocannabinol and 11-nor-9-carboxy-Delta 9-tetrahydrocannabinol in water samples by solid-phase microextraction with on-fiber derivatization and gas chromatography-mass spectrometry. J Chromatogr A 1245:167–174
62. Alzaga R, Pena A, Ortiz L et al (2003) Determination of linear alkylbenzensulfonates in aqueous matrices by ion-pair solid-phase microextraction-in-port derivatization-gas chromatography-mass spectrometry. J Chromatogr A 999:51–60
63. Liu Y, Cho SR, Danielson ND (2002) Solid-phase microextraction and on-line methylation gas chromatography for aliphatic carboxylic acids. Anal Bioanal Chem 373:64–69
64. Bizkarguenaga E, Iparragirre A, Navarro P et al (2013) In-port derivatization after sorptive extractions. J Chromatogr A 1296:36–46
65. Rosenfeld JM (1999) Solid-phase analytical derivatization: enhancement of sensitivity and selectivity of analysis. J Chromatogr A 843:19–27
66. Quintana JB, Rodriguez I (2006) Strategies for the microextraction of polar organic contaminants in water samples. Anal Bioanal Chem 384:1447–1461
67. Wittmann G, Van Langenhove H, Dewulf J (2000) Determination of acetic acid in aqueous samples, by water-phase derivatisation, solid-phase microextraction and gas chromatography. J Chromatogr A 874:225–234
68. Nilsson T, Baglio D, Galdo-Miguez I et al (1998) Derivatisation/solid-phase microextraction followed by gas chromatography mass spectrometry for the analysis of phenoxy acid herbicides in aqueous samples. J Chromatogr A 826:211–216
69. Citova I, Sladkovsky R, Solich P (2006) Analysis of phenolic acids as chloroformate derivatives using solid phase microextraction-gas chromatography. Anal Chim Acta 573:231–241
70. Myung SW, Kim M, Min HK et al (1999) Determination of homocysteine and its related compounds by solid-phase microextraction-gas chromatography-mass spectrometry. J Chromatogr B 727:1–8
71. Rodriguez V, Yonamine M, Pinto E (2006) Determination of anatoxin-a in environmental water samples by solid-phase microextraction and gas chromatography-mass spectrometry. J Sep Sci 29:2085–2090
72. Henriksen T, Svensmark B, Lindhardt B et al (2001) Analysis of acidic pesticides using in situ derivatization with alkylchloroformate and solid-phase microextraction (SPME) for GC-MS. Chemosphere 44:1531–1539
73. Racamonde I, Rodil R, Quintana JB et al (2013) In-sample derivatization-solid-phase microextraction of amphetamines and ecstasy related stimulants from water and urine. Anal Chim Acta 770:75–84
74. Monteleone M, Naccarato A, Sindona G et al (2012) A rapid and sensitive assay of perfluorocarboxylic acids in aqueous matrices by headspace solid phase microextraction-gas chromatography-triple quadrupole mass spectrometry. J Chromatogr A 1251:160–168
75. Gionfriddo E, Naccarato A, Sindona G et al (2014) Determination of hydrazine in drinking water: development and multivariate optimization of a rapid and simple solid phase microextraction-gas chromatography-triple quadrupole mass spectrometry protocol. Anal Chim Acta 835:37–45
76. Lim HH, Shin HS (2012) Sensitive and robotic determination of bromate in sea water and drinking deep-sea water by headspace solid-phase micro extraction and gas chromatography-mass spectrometry. Anal Chim Acta 741:32–37
77. Liu WL, Hwang BH, Li ZG et al (2011) Headspace solid phase microextraction in-situ supercritical fluid extraction coupled to gas chromatography-tandem mass spectrometry for simultaneous determination of perfluorocarboxylic acids in sediments. J Chromatogr A 1218:7857–7863
78. Gerecke AC, Tixier C, Bartels T et al (2001) Determination of phenylurea herbicides in natural waters at concentrations below 1 ng 1(-1) using solid-phase extraction, derivatization, and solid-phase microextraction gas chromatography mass spectrometry. J Chromatogr A 930:9–19

79. Lin HL, Ma XX, Li QL et al (2013) Determination of phenoxy-acid herbicides in water with solid phase microextraction fiber coated with single-walled carbon nanotubes. Chin J Anal Chem 41:1758–1761
80. Alzaga R, Pena A, Bayona JM (2003) Determination of phthalic monoesters in aqueous and urine samples by solid-phase microextraction-diazomethane on-fibre derivatization-gas chromatography-mass spectrometry. J Sep Sci 26:87–96
81. Guillard C, Theron P, Pichat P et al (2002) Evaluation of 1-octanol degradation by photocatalysis and ultrasound using SPME. Water Res 36:4263–4272
82. Diaz A, Ventura F, Galceran MT (2002) Simultaneous determination of estrogenic short ethoxy chain nonylphenols and their acidic metabolites in water by an in-sample derivatization/solid-phase microextraction method. Anal Chem 74:3869–3876
83. Sa CSA, Boaventura RAR, Pereira IB (2012) Analysis of haloacetic acids in water and air (aerosols) from indoor swimming pools using HS-SPME/GC/ECD. J Environ Sci Health A47:176–183
84. Koryagina NL, Savelieva EI, Khlebnikova NS et al (2006) Determination of fluoroacetic acid in water and biological samples by GC-FID and GC-MS in combination with solid-phase microextraction. Anal Biotech Chem 386:1395–1400
85. Sarafraz-Yazdi A, Fatehyan E, Amiri A (2014) Determination of mercury in real water samples using in situ derivatization followed by sol-gel-solid-phase microextraction with gas chromatography-flame ionization detection. J Chromatogr Sci 52:81–87
86. Crnoja M, Haberhauer-Troyer C, Rosenberg E et al (2001) Determination of Sn- and Pb-organic compounds by solid-phase microextraction-gas chromatography-atomic emission detection (SPME-GC-AED) after in situ propylation with sodium tetrapropylborate. J Anal Atom Spectrom 16:1160–1166
87. Jiang GB, Liu JY, Yang KW (2000) Speciation analysis of butyltin compounds in Chinese seawater by capillary gas chromatography with flame photometric detection using in-situ hydride derivatization followed by headspace solid-phase microextraction. Anal Chim Acta 421:67–74
88. Penalver R, Campillo N, Hernandez-Cordoba M (2011) Comparison of two derivatization reagents for the simultaneous determination of organolead and organomanganese compounds using solid-phase microextraction followed by gas chromatography with atomic emission detection. Talanta 87:268–275
89. Yu XM, Pawliszyn J (2000) Speciation of alkyllead and inorganic lead by derivatization with deuterium-labeled sodium tetraethylborate and SPME-GC/MS. Anal Chem 72:1788–1792
90. Es-haghi A, Baghernejad M, Bagheri H (2012) In situ solid-phase microextraction and post on-fiber derivatization combined with gas chromatography-mass spectrometry for determination of phenol in occupational air. Anal Chim Acta 742:17–21
91. Llompart M, Lourido M, Landin P et al (2002) Optimization of a derivatization-solid-phase microextraction method for the analysis of thirty phenolic pollutants in water samples. J Chromatogr A 963:137–148
92. Olejniczak J, Staniewski J (2007) Enrichment of phenols from water with in-situ derivatization by in-tube solid phase microextraction-solvent desorption prior to off-line gas chromatographic determination with large-volume injection. Anal Chim Acta 588:64–72
93. Morales R, Ortiz MC, Sarabia LA (2012) Optimization of headspace experimental factors to determine chlorophenols in water by means of headspace solid-phase microextraction and gas chromatography coupled with mass spectrometry and parallel factor analysis. Anal Chim Acta 754:20–30
94. Morales R, Sarabia LA, Sanchez MS et al (2013) Experimental design for the optimization of the derivatization reaction in determining chlorophenols and chloroanisoles by headspace-solid-phase microextraction-gas chromatography/mass spectrometry. J Chromatogr A 1296:179–195

95. Abbasghorbani M, Attaran A, Payehghadr M (2013) Solvent-assisted dispersive micro-SPE by using aminopropyl-functionalized magnetite nanoparticle followed by GC-PID for quantification of parabens in aqueous matrices. J Sep Sci 36:311–319
96. Regueiro J, Becerril E, Garcia-Jares C et al (2009) Trace analysis of parabens, triclosan and related chlorophenols in water by headspace solid-phase microextraction with in situ derivatization and gas chromatography-tandem mass spectrometry. J Chromatogr A 1216:4693–4702
97. Lamas JP, Salgado-Petinal C, Garcia-Jares C et al (2004) Solid-phase microextraction-gas chromatography-mass spectrometry for the analysis of selective serotonin reuptake inhibitors in environmental water. J Chromatogr A 1046:241–247
98. Szulejko JE, Kim KH (2014) A review of sampling and pretreatment techniques for the collection of airborne amines. Trends Anal Chem 57:118–134
99. Buszewski B, Olszowy P, Szultka M et al (2012) New approaches to extraction techniques in determination of 4,4'-methylenebis(2-chloroaniline) in air and water solutions. Talanta 93:117–121
100. Aranda R, Burk RC (1998) Determination of a non-ionic surfactant by solid-phase microextraction coupled with high-performance liquid chromatography and on-line derivatization. J Chromatogr A 829:401–406
101. Chafer-Pericas C, Campins-Falco P, Herraez-Hernandez R (2004) Application of solid-phase microextraction combined with derivatization to the determination of amphetamines by liquid chromatography. Anal Biochem 333:328–335
102. Chafer-Pericas C, Campins-Falco P, Herraez-Hernandez R (2006) Comparative study of the determination of trimethylamine in water and air by combining liquid chromatography and solid-phase microextraction with on-fiber derivatization. Talanta 69:716–723
103. Chafer-Pericas C, Herraez-Hernandez R, Campis-Falco P (2005) A new selective method for dimethylamine in water analysis by liquid chromatography using solid-phase microextraction and two-stage derivatization with *o*-phthalaldialdehyde and 9-fluorenylmethyl chloroformate. Talanta 66:1139–1145
104. Cai LS, Zhao YY, Gong SL et al (2003) Use of a novel sol-gel dibenxo-18-crown-6 solid-phase microextraction fiber and a new derivatizing reagent for determination of aliphatic amines in lake water and human urine. Chromatographia 58:615–621
105. Stashenko EE, Puertas MA, Salgar W et al (2000) Solid-phase microextraction with on-fibre derivatisation applied to the analysis of volatile carbonyl compounds. J Chromatogr A 886:175–182
106. Bourdin D, Desauziers V (2014) Development of SPME on-fiber derivatization for the sampling of formaldehyde and other carbonyl compounds in indoor air. Anal Biotech Chem 406:317–328
107. Kim HJ, Shin HS (2011) Simple and automatic determination of aldehydes and acetone in water by headspace solid-phase microextraction and gas chromatography-mass spectrometry. J Sep Sci 34:693–699
108. Lim HH, Shin HS (2012) Simple determination of acrolein in surface and drinking water by headspace SPME GC-MS. Chromatographia 75:943–948
109. Bolta SV, Zupancic-Kralj L, Marsel J (1998) Gas chromatographic determination of formaldehyde in air using solid-phase microextraction sampling. Chromatographia 48:95–100
110. Prieto-Blanco MC, Lopez-Mahia P, Campins-Falco P (2011) On-line analysis of carbonyl compounds with derivatization in aqueous extracts of atmospheric particulate PM10 by in-tube solid-phase microextraction coupled to capillary liquid chromatography. J Chromatogr A 1218:4834–4839
111. Hudson ED, Okuda K, Ariya PA (2007) Determination of acetone in seawater using derivatization solid-phase microextraction. Anal Biotech Chem 388:1275–1282
112. Namera A, So A, Pawliszyn J (2002) Analysis of anatoxin-a in aqueous samples by solid-phase microextraction coupled to high-performance liquid chromatography with fluorescence detection and on-fiber derivatization. J Chromatogr A 963:295–302

113. Rellan S, Gago-Martinez A (2007) Improved conditions for the application of solid phase microextraction prior to HPLC-FLD analysis of anatoxin-a. J Sep Sci 30:2522–2528
114. Pieraccini G, Bartolucci G, Pacenti M et al (2002) Gas chromatographic determination of glutaraldehyde in the workplace atmosphere after derivatization with *O*-(2,3,4,5,6-pentafluorobenzyl)hydroxylamine on a solid-phase microextraction fibre. J Chromatogr A 955:117–124
115. Koziel JA, Noah J, Pawliszyn J (2001) Field sampling and determination of formaldehyde in indoor air with solid-phase microextraction and on-fiber derivatization. Environ Sci Technol 35:1481–1486
116. Pacolay BD, Ham JE, Slaven JE et al (2008) Feasibility of detection and quantification of gas-phase carbonyls in indoor environments using PFBHA derivatization and solid-phase microextraction (SPME). J Environ Monitor 10:853–860
117. Pacolay BD, Ham JE, Wells JR (2006) Use of solid-phase microextraction to detect and quantify gas-phase dicarbonyls in indoor environments. J Chromatogr A 1131:275–280
118. Trenholm RA, Rosario-Ortiz FL, Snyder SA (2008) Analysis of formaldehyde formation in wastewater using on-fiber derivatization-solid-phase microextraction-gas chromatography-mass spectrometry. J Chromatogr A 1210:25–29
119. Alvarez EG, Viidanoja J, Munoz A et al (2007) Experimental confirmation of the dicarbonyl route in the photo-oxidation of toluene and benzene. Environ Sci Technol 41:8362–8369
120. Li HQ, Deng CH, Zhang XM (2008) Fast field analysis of short-chain aliphatic amines in water using solid-phase microextraction and a portable gas chrornatograph. J Sep Sci 31:3225–3230
121. Rubio L, Sanllorente S, Sarabia LA et al (2014) Optimization of a headspace solid-phase microextraction and gas chromatography/mass spectrometry procedure for the determination of aromatic amines in water and in polyamide spoons. Chemometrics IntelLabSys 133:121–135
122. Zimmermann T, Ensinger WJ, Schmidt TC (2004) In situ derivatization/solid-phase microextraction: determination of polar aromatic amines. Anal Chem 76:1028–1038
123. Salgado-Petinal C, Alzaga R, Garcia-Jares C et al (2005) Low part per trillion determination of reactive alkanethiols in wastewater by in situ derivatization-solid-phase microextraction followed by GC/MS. Anal Chem 77:6012–6018
124. Fujioka K, Shibamoto T (2005) Improved malonaldehyde assay using headspace solid-phase microextraction and its application to the measurement of the antioxidant activity of phytochemicals. J Agri Food Chem 53:4708–4713
125. Zhao YY, Cai LS, Jing ZZ et al (2003) Determination of aliphatic amines using *N*-succinimidyl benzoate as a new derivatization reagent in gas chromatography combined with solid-phase microextraction. J Chromatogr A 1021:175–181
126. Mothes S, Wennrich R (2000) Coupling of SPME and GC-AED for the determination of organometallic compounds. Mikrochim Acta 135:91–95
127. Vercauteren J, De Meester A, De Smaele T et al (2000) Headspace solid-phase microextraction-capillary gas chromatography-ICP mass spectrometry for the determination of the organotin pesticide fentin in environmental samples. J Anal Atom Spectrom 15:651–656
128. Botana JC, Rodriguez RR, Diaz AMC et al (2002) Fast and simultaneous determination of tin and mercury species using SPME, multicapillary gas chromatography and MIP-AES detection. J Anal Atom Spectrom 17:904–907
129. Bravo-Sanchez LR, Encinar JR, Martinez JIF et al (2004) Mercury speciation analysis in sea water by solid phase microextraction-gas chromatography-inductively coupled plasma mass spectrometry using ethyl and propyl derivatization. Matrix effects evaluation. Spectrochimica Acta B 59:59–66
130. Kovacs MH, Ristoiu D, Voica C et al (2013) Optimization of organometallic compounds extraction from aqueous samples in order to improve their gas chromatography-mass spectrometry analysis performance. Romanian J Phys 58:204–210

131. Devos C, Moens L, Sandra P (2005) Determination of the acaricide fenbutatin oxide in water samples by automated headspace-SPME-GC/MS. J Sep Sci 28:665–668
132. Cardellicchio N, Giandomenico S, Decataldo A et al (2001) Speciation of butyltin compounds in marine sediments with headspace solid phase microextraction and gas chromatography-mass spectrometry. Fresenius J Anal Chem 369:510–515
133. Bancon-Montigny C, Maxwell P, Yang L et al (2002) Improvement of measurement precision of SPME-GC/MS determination of tributyltin using isotope dilution calibration. Anal Chem 74:5606–5613
134. Jitaru P, Adams FC (2004) Speciation analysis of mercury by solid-phase microextraction and multicapillary gas chromatography hyphenated to inductively coupled plasma-time-of-flight-mass spectrometry. J Chromatogr A 1055:197–207
135. Arambarri I, Garcia R, Millan E (2003) Assessment of tin and butyltin species in estuarine superficial sediments from Gipuzkoa, Spain. Chemosphere 51:643–649
136. Beichert A, Padberg S, Wenclawiak BW (2000) Selective determination of alkylmercury compounds in solid matrices after subcritical water extraction, followed by solid-phase microextraction and GC-MS. Appl OrganometChem 14:493–498
137. Devos C, Vliegen M, Willaert B et al (2005) Automated headspace-solid-phase micro extraction—retention time locked-isotope dilution gas chromatography–mass spectrometry for the analysis of organotin compounds in water and sediment samples. J Chromatogr A 1079:408–414
138. Millan E, Pawliszyn J (2000) Determination of butyltin species in water and sediment by solid-phase microextraction-gas chromatography-flame ionization detection. J Chromatogr A 873:63–71
139. Jiang GB, Liu JY (2000) Determination of butyltin compounds in aqueous samples by gas chromatography with flame photometric detector and headspace solid-phase microextraction after in-situ hydride derivatization. AnalSci 16:585–588
140. Carro AM, Neira I, Rodil R et al (2002) Speciation of mercury compounds by gas chromatography with atomic emission detection. Simultaneous optimization of a headspace solid-phase microextraction and derivatization procedure by use of chemometric techniques. Chromatographia 56:733–738
141. Mishra S, Tripathi RM, Bhalke S et al (2005) Determination of methylmercury and mercury (II) in a marine ecosystem using solid-phase microextraction gas chromatography-mass spectrometry. Anal Chim Acta 551:192–198
142. Shi JB, Jiang GB (2011) Application of gas chromatography-atomic fluorescence spectrometry hyphenated system for speciation of butyltin compounds in water samples. Spectrosc Lett 44:393–398
143. Liu JY, Jiang GB, Zhou QF et al (2001) Headspace solid-phase microextraction of butyltin species in sediments and their gas chromatographic determination. J Sep Sci 24:459–464
144. He B, Jiang GB (1999) Analysis of organomercuric species in soils from orchards and wheat fields by capillary gas chromatography on-line coupled with atomic absorption spectrometry after in situ hydride generation and headspace solid phase microextraction. Fresenius J Anal Chem 365:615–618
145. Parshintsev J, Ronkko T, Helin A et al (2015) Determination of atmospheric amines by on-fiber derivatization solid-phase microextraction with 2,3,4,5,6-pentafluorobenzyl chloroformate and 9-fluorenylmethoxycarbonyl chloride. J Chromatogr A 1376:46–52
146. Abranko L, Yang L, Sturgeon RE et al (2004) Solid phase microextraction for the determination of chromium in sea-water. J Anal Atom Spectrom 19:1098–1103
147. Ding TH, Lin HH, Whang CW (2005) Determination of chromium(III) in water by solid-phase microextraction with a polyimide-coated fiber and gas chromatography-flame photometric detection. J Chromatogr A 1062:49–55
148. Harvey SD, Lucke RB, Douglas M (2012) Rapid separation of beryllium and lanthanide derivatives by capillary gas chromatography. J Sep Sci 35:2750–2755

149. Frerot E, Bagnoud A, Cicchetti E (2014) Quantification of hydrogen sulfide and methanethiol and the study of their scavenging by biocides of the isothiazolone family. Chempluschem 79:77–82
150. Alzaga R, Salgado-Petinal C, Jover E et al (2005) Development of a procedure for the determination of perfluorocarboxylic acids in sediments by pressurised fluid extraction, headspace solid-phase microextraction followed by gas chromatographic-mass spectrometric determination. J Chromatogr A 1083:1–6
151. Jain A, Reddy-Noone K, Pillai AKKV et al (2013) Conversion to isothiocyanates via dithiocarbamates for the determination of aromatic primary amines by headspace-solid phase microextraction and gas chromatography. Anal Chim Acta 801:48–58
152. Tomkins BA, Sega GA, Ho CH (2001) Determination of Lewisite oxide in soil using solid-phase microextraction followed by gas chromatography with flame photometric or mass spectrometric detection. J Chromatogr A 909:13–28
153. Koziel JA, Noah J, Pawliszyn J (2001) Field sampling and determination of formaldehyde in indoor air with solid-phase microextraction and on-fiber derivatization. Environ Sci Technol 35:1481–1486
154. Prieto-Blanco MC, Moliner-Martinez Y, Lopez-Mahia R et al (2013) Determination of carbonyl compounds in particulate matter PM2.5 by in-tube solid-phase microextraction coupled to capillary liquid chromatography/mass spectrometry. Talanta 115:876–880
155. Liu XY, Ji YS, Zhang YH et al (2007) Oxidized multiwalled carbon nanotubes as a novel solid-phase microextraction fiber for determination of phenols in aqueous samples. J Chromatogr A 1165:10–17
156. Bagheri H, Mir A, Babanezhad E (2005) An electropolymerized aniline-based fiber coating for solid phase microextraction of phenols from water. Anal Chim Acta 532:89–95
157. Ghassempour A, Najafi NM, Mehdinia A et al (2005) Analysis of anatoxin-a using polyaniline as a sorbent in solid-phase microextraction coupled to gas chromatography-mass spectrometry. J Chromatogr A 1078:120–127
158. Huang MJ, Tai C, Zhou QF et al (2004) Preparation of polyaniline coating on a stainless-steel wire using electroplating and its application to the determination of six aromatic amines using headspace solid-phase microextraction. J Chromatogr A 1048:257–262
159. Regueiro J, Becerril E, Garcia-Jares C et al (2009) Trace analysis of parabens, triclosan and related chlorophenols in water by headspace solid-phase microextraction with in situ derivatization and gas chromatography-tandem mass spectrometry. J Chromatogr A 1216:4693–4702
160. Parkinson DR, Bruheim I, Christ I et al (2004) Full automation of derivatization–solid-phase microextraction-gas chromatography-mass spectrometry with a dual-arm system for the determination of organometallic compounds in aqueous samples. J Chromatogr A 1025:77–84
161. Yang L, Luan T, Lan C (2006) Solid-phase microextraction with on-fiber silylation for simultaneous determinations of endocrine disrupting chemicals and steroid hormones by gas chromatography-mass spectrometry. J Chromatogr A 1104:23–32
162. Deng CH, Li N, Ji J et al (2006) Development of water-phase derivatization followed by solid-phase microextraction and gas chromatography/mass spectrometry for fast determination of valproic acid in human plasma. Rapid Commun Mass Spectrom 20:1281–1287
163. Negreira N, Rodriguez I, Ramil M et al (2009) Sensitive determination of salicylate and benzophenone type UV filters in water samples using solid-phase microextraction, derivatization and gas chromatography tandem mass spectrometry. Anal Chim Acta 638:36–44
164. Trenholm RA, Rosario-Ortiz FL, Snyder SA (2008) Analysis of formaldehyde formation in wastewater using on-fiber derivatization-solid-phase microextraction-gas chromatography-mass spectrometry. J Chromatogr A 1210:25–29

165. Beranek J, Kubatova A (2008) Evaluation of solid-phase microextraction methods for determination of trace concentration aldehydes in aqueous solution. J Chromatogr A 1209:44–54
166. Alzaga R, Bayona JM (2004) Determination of perfluorocarboxylic acids in aqueous matrices by ion-pair solid-phase microextraction-in-port derivatization-gas chromatography-negative ion chemical ionization mass spectrometry. J Chromatogr A 1042:155–162
167. Chafer-Pericas C, Campins-Falco P, Herraez-Hernandez R (2006) Application of solid-phase microextraction combined with derivatization to the enantiomeric determination of amphetamines. J Pharm Biomed Anal 40:1209–1217
168. Herraez-Hernandez R, Chafer-Pericas C, Campins-Falco P (2004) Analysis of methylamine by solid-phase microextraction and HPLC after on-fibre derivatization with 9-fluorenylmethyl chloroformate. Anal Chim Acta 513:425–433
169. Mughari AR, Vazquez PP, Galera MM (2007) Analysis of phenylurea and propanil herbicides by solid-phase microextraction and liquid chromatography combined with post-column photochemically induced fluorimetry derivatization and fluorescence detection. Anal Chim Acta 593:157–163
170. Vazquez PP, Mughari AR, Galera MM (2008) Application of solid-phase microextraction for determination of pyrethroids in groundwater using liquid chromatography with post-column photochemically induced fluorimetry derivatization and fluorescence detection. J Chromatogr A 1188:61–68
171. Wang Q, Chong JM, Pawliszyn J (2006) Determination of thiol compounds by automated headspace solid-phase microextraction with in-fiber derivatization. Flavour Frag J 21:385–394
172. Pan YP, Tsai SW (2008) Solid phase microextraction procedure for the determination of alkylphenols in water by on-fiber derivatization with *N*-tert-butyl-dimethylsilyl-*N*-methyltrifluoroacetamide. Anal Chim Acta 624:247–252
173. Abranko L, Jokai Z, Fodor P (2005) Investigation of the species-specific degradation behaviour of methylmercury and ethylmercury under microwave irradiation. Anal Biotech Chem 383:448–453
174. Baker J, Arey J, Atkinson R (2005) Rate constants for the reactions of OH radicals with a series of 1,4-hydroxyketones. J Photochem Photobio A-Chem 176:143–148
175. Baker J, Arey J, Atkinson R (2005) Formation and reaction of hydroxycarbonyls from the reaction of OH radicals with 1,3-butadiene and isoprene. Environ Sci Technol 39:4091–4099
176. Luan T, Fang S, Zhong Y et al (2007) Determination of hydroxy metabolites of polycyclic aromatic hydrocarbons by fully automated solid-phase microextraction derivatization and gas chromatography-mass spectrometry. J Chromatogr A 1173:37–43
177. Yan CT, Jen JF (2004) Determination of aniline in water by microwave-assisted headspace solid-phase microextraction and gas chromatography. Chromatographia 59:517–520
178. Kralova Z, Vavrova M, Sucman E et al (2014) Optimization of the hs-spme method coupled with gc/mu ecd for determination of methylmercury in water. Fresenius Environ Bull 23:3362–3365
179. Cui MY, Qiu JX, Li ZH et al (2015) An etched stainless steel wire/ionic liquid-solid phase microextraction technique for the determination of alkylphenols in river water. Talanta 132:564–571
180. Araujo L, Wild J, Villa N et al (2008) Determination of anti-inflammatory drugs in water samples, by in situ derivatization, solid phase microextraction and gas chromatography–mass spectrometry. Talanta 75:111–115
181. Huang SM, Zhu F, Jiang RF et al (2015) Determination of eight pharmaceuticals in an aqueous sample using automated derivatization solid-phase microextraction combined with gas chromatography–mass spectrometry. Talanta 136:198–209
182. Liu JT, Hara K, Kashimura S et al (2001) New method of derivatization and headspace solid-phase microextraction for gas chromatographic–mass spectrometric analysis of amphetamines in hair. J Chromatogr B 758:95–101

183. Gomez-Rios GA, Pawliszyn J (2014) Development of coated blade spray ionization mass spectrometry for the quantitation of target analytes present in complex matrices. Angew Chem Int Edit 53:14503–14507
184. Deng JW, Yang YY, Wang XW et al (2014) Strategies for coupling solid-phase microextraction with mass spectrometry. Trends Anal Chem 55:55–67
185. Deng JW, Yang YY, Fang L et al (2014) Coupling solid-phase microextraction with ambient mass spectrometry using surface coated wooden-tip probe for rapid analysis of ultra trace perfluorinated compounds in complex samples. Anal Chem 86:11159–11166

Chapter 8
Application of Solid Phase Microextraction in Food Analysis—Flavor and Off-Flavor Sampling

Henryk Jeleń, Małgorzata Majcher and Anna Gracka

Abstract The chapter focuses on the use of solid-phase microextraction (SPME) in the field of flavor analysis. Due to its robustness, versatility, and sensitivity, SPME is the dominating technique used for the extraction of flavor and volatile compounds in food nowadays. New developments in SPME technologies, including mainly fiber coatings and also cold fiber SPME, sampling in vacuum conditions, are discussed in this chapter. Specificity of food matrix in terms of extraction techniques, and specificity of food volatiles as well as the influence of sample preparation on extraction efficiency are described. The special emphasis is put on the quantitative aspects of SPME in the analysis of food flavors and off-flavors. Different extraction and calibration approaches are demonstrated. SPME is also compared to other extraction methods used in flavor and off-flavor analyses such as simultaneous distillation extraction (SDE), solvent-assisted flavor evaporation (SAFE), liquid–liquid extraction, purge and trap, and hydrodistillation. The usefulness of SPME in profiling of volatile compounds profiles is also outlined.

Keywords Food flavors · Off-flavors · Quantitative SPME · Calibration · Fiber coatings · Cold fiber SPME

8.1 Developments for Increased Applicability of SPME for Food Flavors Analysis

Solid-phase microextraction (SPME) was developed 25 years ago, as a robust technique, allowing the extraction of volatile compounds from various matrices using syringe-like device with fiber in needle coated with sorbent. The construction of the device, its simplicity and versatility, together with analytical performance made it probably the fastest developing microextraction technique in various fields,

H. Jeleń (✉) · M. Majcher · A. Gracka
Faculty of Food Science and Nutrition, Poznań University of Life Sciences,
WojskaPolskiego 31, 60-624 Poznań, Poland
e-mail: henrykj@up.poznan.pl

G. Ouyang and R. Jiang (eds.), *Solid Phase Microextraction*,
DOI 10.1007/978-3-662-53598-1_8

ranging from analytical chemistry to biochemistry, food science or environmental analysis. As there have been numerous (approx. 200) reviews written on various aspects of SPME throughout a period of 25 years, including many reviews concerning applications in food science (53) and flavor (17), this chapter will focus on the developments and use of SPME in flavor analysis in a period of last 10 years (2005–2015). In this period, review articles discussing this technique can be found in the field of extraction methods for food volatiles [1], wine analysis [2], quality of foods [3], flavor compounds [4], off-flavors and taints [5, 6], and natural products aroma [7].

Among SPME main features that resulted in broad application fields are selectivity (based on K values), good sensitivity (total sample transfer to GC), solvent-less technology, integration of sampling and sample preparation in one step, non-exhaustive extraction, possibilities of automation [8] and also a good reproducibility in both automated and manual operation mode. These features contributed to wide-ranging use of SPME in food and especially flavor chemistry. When SPME word is searched in Web of Science (as topic), it reveals 7639 documents since 1990, including 5602 in a period of last 10 years. When the term "flavor" is searched within SPME results, it reveals 1292 documents, including 1017 in the period 2005–2015. Figure 8.1 shows the increase in the number of publications in food and flavor compared to all SPME papers published since 1990. There is a higher increase of flavor papers compared to all SPME papers when 5-year periods are compared. SPME became in last 25 years one of the most often used techniques for sample preparation, both in flavor/volatile compounds screening and profiling and in quantitative analysis.

The increasing applicability of SPME in flavor research is a result of the following factors:

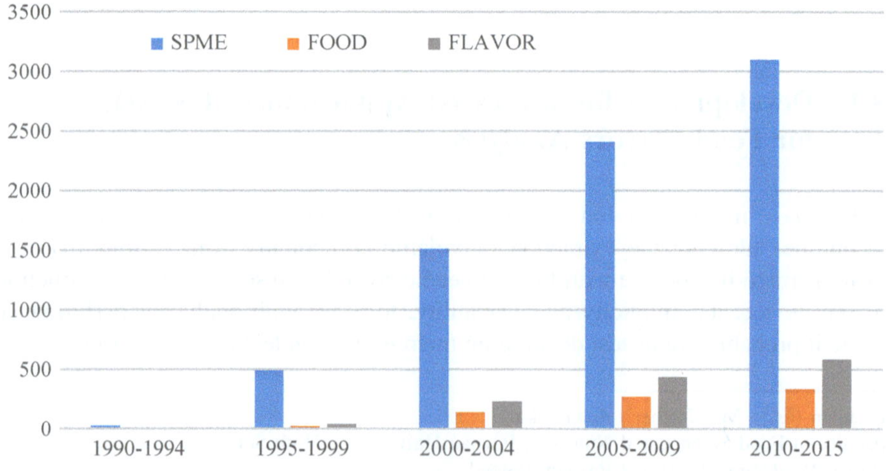

Fig. 8.1 Number of documents in Web of Science published on SPME as a topic and on food and flavor terms within SPME publications (as of 01.04.2015)

- Developments in SPME technology, especially new fiber coatings providing high selectivity and sensitivity;
- Its use as a robust and easy to automate tool for qualitative analysis especially volatiles profiling in complex food samples;
- Suitability of SPME for quantitative trace analysis, especially in the field of targeted analysis of food odorants and off-odorants;
- Suitability of SPME for sampling from a variety of matrices which food products provide.

The success of SPME, not only in flavor research, has been attributed among other factors to developments and commercial availability of various extraction phases. The proper choice of extracting phase for a particular group of analytes, especially in trace target analysis, is a first and often a crucial step in method development. After the initial introduction of polydimethylsiloxane (PDMS) and polyacrylate (PA) absorbents, which represented nonpolar and polar extraction phases, more fiber coatings, including mixed-phase sorbents, have been introduced commercially. Recent trends in extraction phases the development address selectivity, biocompatibility, and sensitivity of SPME fibers. The most often used absorption-type phase—PDMS, apart from being used in SPME, is also used in thin-film extraction (TFE), SBSE (introduced in 1999), in-tube SPME, and equilibrium sorptive enrichment (ESE). For a sampling system based on partitioning properties, the extraction time profile at specific temperature is determined by the volume of PDMS phase, its surface-to-volume ratio, sample agitation conditions, partition coefficient of the analyte of interest, boundary layer conditions, as well as the analyte's diffusion coefficient within the absorbent [9]. One of the disadvantages of PDMS in direct sampling, especially of biological fluids and some foods, is the non-specific absorption of proteins to the hydrophobic surface of the polymer. PDMS is also used in SPME fiber coatings using polymers, such as Carboxene or divinylbenzene. PDMS is also modified for use in SBSE by adding other sorbents, such as Carbopack B, Tenax TR or used with solvents such as acetone, ethyl acetate stored inside a PDMS tube [10]. PDMS is still probably the best phase to start method optimization for flavor analysis with, especially for the analysis of nonpolar compounds, such as terpenes. Its advantages are good sensitivity, durability of the coating, and for majority of compounds the best linear range among SPME fibers. Polyacrylate (PA) fibers are composed of partially cross-linked acrylic acid monomers, and recommended for use mainly for polar compounds. They are more resistant to organic solvents than PDMS fibers; however, they tend to swell in water and polar solvents [11]. The application of PA for polar compounds in food is decreasing in popularity in favor to adsorption type, polymer-based fibers, such as Carboxene/Divinylbenzene/PDMS or Carboxene/PDMS. Compounds are adsorbed in the macro- (>500 Å), meso- (20–500 Å), and micropores (<20 Å) of DVB or Carboxene particles, that are "glued" to the fiber using PDMS. One of the most extensively used fiber in flavor analysis is Carboxene/Divinylbenzene/PDMS, which is the only fiber nowadays produced in 2-cm length to increase the volume of sorbent—designed especially for off-flavors analysis. For the analysis of polar compounds as an alternative to PA fiber, Carbowax fiber was introduced, in which polyethylene glycol (PEG) is mixed with DVB or template resin. When different fibers were examined for the type and number of volatiles detected in milk, wine, fruit/vegetable products, the highest number of compounds was detected using DVB/CAR/PDMS fiber regardless of the compound class. It dominated in the extraction of alcohols, esters, aldehydes, ketones, acids, and also terpenes, furans, and phenols. Depending on the class of food volatiles, the next-generation fiber in terms of efficiency of flavor compounds extraction are PDMS and CAR/PDMS fibers, the first one being more effective in the extraction of terpenes and esters [11].

Novel coatings and their preparation technologies are rapidly developing areas of research on SPME fundamentals, which results in better selectivity and sensitivity of methods in various application areas including flavor analysis. The coating procedure has a significant effect on the stability of the fiber, both mechanical and thermal [12]. Among coating procedures are dipping, solgel technology, chemical grafting, electrochemical methods, electrospinning, and liquid-phase deposition. ZrO_2 deposited electrolytically on a NiTi alloy (NiTi-ZrO_2) as a SPME fiber was successfully employed to determine haloanisoles in red wine samples at ng/L level and showed durability to withstand more than 700 extractions [13]. TiNi (nitinol)-oxidized rods were also used to extract volatiles and semivolatiles from citrus fruit [14]. Nanomaterials as sorbents have large loading capacity, high extraction efficiency, and high extraction rate. Carbon nanotubes (CNTs) are the most widely used nanomaterials in SPME. In a recent review paper on carbon-based sorbents [15], use of carbon nanotubes in SPME was summarized. However, among volatile compounds for which CNTs were used as SPME coatings were mainly environmental contaminants, pesticides, and drugs detected in water mainly, not in the field of flavor analysis. Molecularly imprinted solid-phase microextraction (MISPME) utilizes molecularly imprinted polymers (MIPs) advantages—predetermined recognition ability, stability, potential application to a wide range of target molecules, and high selectivity. From the field of food science, its applications in the detection of Sudan dyes, chlorogenic acid, pesticides, ochratoxin A were reported [16]; however, its use in food volatiles/flavors analysis is restricted to the determination of acetaldehyde migrating from PET bottles to soft drinks [17]. Construction of SPME fibers with different coatings incorporated into one fiber enables the extraction of compounds of different physical properties in so-called space-resolved SPME fibers. It was demonstrated on extraction of TCA and DBP from wines, where graphene (G) was used for TCA extraction, whereas graphene oxide (GO) was used for DBP using solgel coating, and the extraction was performed by immersing the 1-cm fiber coated with GO in liquid matrix, whereas the other 1-cm fiber coated with G was adsorbing compounds from the headspace [18].

Enhancement of extraction efficiency in SPME can be achieved in two different approaches: (1) by increasing the magnitude of K_{fs} (distribution coefficient between the coating and sample matrix) and (2) by increasing the volume or active surface area of extraction phase. Coating type has effect on the both of factors. Although the manipulation with extraction phase volume is limited in SPME fibers, introduced mainly by varying coating thickness or length of the coating, it found out application in other extraction technologies: For 100-μm PDMS fiber, the volume of extraction phase is approx. 0.5 μL, whereas SBSE volume is 50–250 times larger than that of fiber SPME, and RDSE (rotating disk sorbent extraction) has even more volume than SBSE [19]. To overcome the disadvantages of high volume of the extracting phase, as in SBSE, which results in long time required to reach equilibrium increasing the surface area (thin-film microextraction, thin-film SPME) can be applied, though at present stage wide use of this approach suffers from the lack of automation.

Internally cooled or cold fiber SPME (CF-SPME) was introduced in automated form in 2006, to influence partition coefficients of analytes. It (cooled PDMS) has been demonstrated to produce better results in terms of number of compounds detected than traditional fibers (PDMS, CAR/PDMS, DVB/CAR/PDMS) in acerola, guava, tamarind, cashew, yellow passion fruit pulps [20]. It also proved to provide several times lower LODs when chloroanisoles were analyzed with this technique (for 2,4,6-TCA in cork, it was 0.25 ng/g vs. 1.2 ng/g when CF-HS-SPME was compared to HS-SPME) [21]. For compounds characterized with low vapor pressure, which are hard to transfer from aqueous solution to headspace, their extraction can be enhanced using gas purging [22].

Sampling under vacuum conditions is another modification of SPME technique, in which sample is introduced into pre-evacuated large volume sampling chamber [23] or smaller (22 mL) modified sampling vial [24] prior to HS-SPME sampling and was elaborated in detail by this author, though before that Brunton used reduced pressure in HS-SPME sampling of raw turkey meat homogenates [25]. Reduced pressure increases extraction rates compared to HS-SPME under atmospheric pressure due to the enhancement of evaporation rates in the presence of air evacuated headspace. The advantages of vacuum SPME sampling are especially noticeable for semi- and low-volatile compounds.

The originally developed fiber SPME device has been used in the majority of applications. Though reproducible when used in manual form, it has been automated, especially with the XYZ type of autosamplers. Automation of SPME not only allows a large number of sample sets to be analyzed, but also automatic fiber exchange and on fiber derivatization processes. Also automation of SPME resulted in the introduction of different variations in solid-phase microextraction, such as in-tube extraction or solid-phase dynamic extraction (SPDE), which is characterized by increase of the extraction phase volume, robustness, and larger surface-to-volume ratio than SPME [26]. In-fiber derivatization can be also automated using SPME, similarly as fiber cooling. Automation of sampling using SPME allows reproducible, various derivatization procedures involving in-port alkylation or in-port silylation, to name the most often used, following SPME [27]. Flavor compounds analysis by SPME involves mainly the derivatization of aldehydes using PFBOA; however, automation offers robust analysis of other derivatizable aroma compounds.

Automation of SPME is also a prerequisite in reproducible analyses of large sample lots in metabolomics (foodomics or flavoromics) studies. SPME offers a fast and reproducible sampling opportunity to obtain complex profiles of food volatile compounds that can be used for various purposes:

- Authenticity testing of foods based on volatiles fingerprinting and multivariate statistical analysis;
- Monitoring the changes in flavor and volatile fraction on foods induced by technological processes, storage of products, detrimental changes caused by chemical, biochemical, and microbial reactions.

New perspectives have been opened by using SPME with comprehensive gas chromatography-mass spectrometry (GCxGC-(ToF)MS), so large datasets of compounds could be used for chemometric procedures in mentioned above research.

Specificity of use of SPME for food flavors analysis is related to several factors:

- Various chemicals character and stability of food flavor compounds. They involve compounds belonging to different chemical classes, of different polarities and volatilities. Very often flavor compounds are unstable and can react with other compounds or matrix. Food odorants can also undergo interaction if prolonged sampling at elevated temperatures is applied.
- Very different concentrations of important aroma compounds (key odorants) occurring in particular product, ranging from ng/L up to mg/L.
- The nature of food matrix determines extraction mode. As direct extraction is used only in very clean matrices, majority of food products allow only for headspace extraction (HS-SPME), in which fiber is exposed to headspace above aqueous or solid sample. It provides accelerated extraction rates for volatile compounds, but the main reason for its use is to protect the fiber from matrix interferences and high molecular weight compounds, that might interfere especially in the desorption process.
- Nature of the matrix, mainly its non-homogeneity and often chemical instability and physical changes induced by temperature (i.e., protein denaturation), affects quantitation, and a choice of proper quantitation method is the most challenging area in flavor analysis using SPME.

8.2 SPME Quantitative Analysis for Food Flavors

8.2.1 Matrix Preparation Considerations

Food flavors and off-flavors quantitative analysis is generally performed to achieve two different goals. The first one is to determine the amount of specific compound (s) responsible for the characteristic odor (off-odor), or being a marker for undergoing chemical, biochemical, or biological changes in food during processing and storage (i.e., hexanal for lipid oxidation monitoring). The other one is to determine concentrations of compounds that are essential to particular flavor (key odorants) of a product, and it combines gas chromatography–olfactometry with gas chromatography–mass spectrometry to determine these compounds and then quantify them. For target analysis of specific flavors and off-flavors markers, due to their often very low odor thresholds, extraction methods need to provide high analyte enrichment, and both exhaustive extraction methods (LLE, P&T, SDE), as well as methods based on partition (SPME, SPE, SBSE) are used.

For the second goal, sampling and sample preparation is one of the most challenging tasks to perform because the composition of obtained extract should reflect the composition of flavor compounds in a real food sample. Therefore, exhaustive extraction methods, usually performed at vacuum dominate. Extraction methods should allow for isolation of all compounds that contribute to the flavor of the food product and not alter the profile of characteristic volatiles by neglecting key odorants, damaging labile compounds, or forming artifacts. This task is especially difficult to perform when trace levels of odor-active compounds (ppt, ppq) or compounds with strong affinity to food products need to be extracted and quantified.

SPME has proved to be precise and sensitive flavor compounds extraction method with low detection limits [4] and can be recommended for quantitative analysis. There are, however, some reported data on difficulties with performing SPME, especially from heterogeneous food samples. Proper preparation of food matrices prior to extraction in majority of cases improves method precision, linear range, and sensitivity.

Reviewing SPME methods described in scientific articles published during last 10 years, there is a tendency observed to handle SPME method with minimized sample preparation steps: in more than 40% of studies, there were no special sample preparation steps before the extraction of volatile with SPME and in additional 20% only grinding was performed (Fig. 8.2).

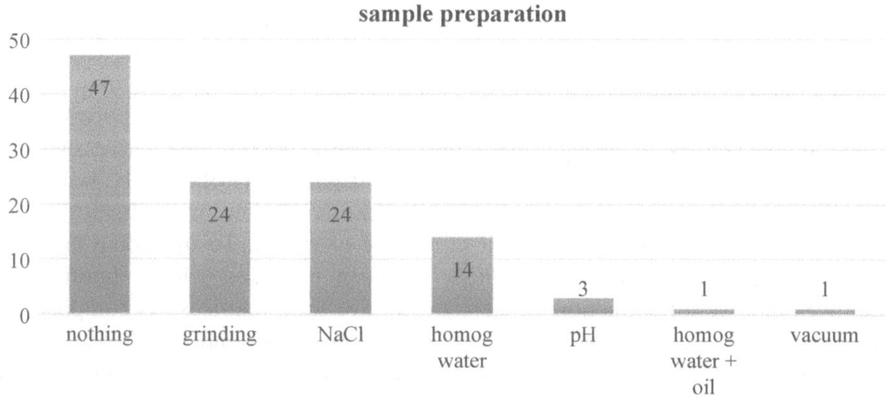

Fig. 8.2 Sample preparation for the extraction of flavor compounds by SPME technique from food products. Data from the Web of Science, years 2005–2015

Vast majority of food is heterogeneous, comprising of diverse components such as proteins, carbohydrates, lipids minerals, or vitamins. Depending on the amount of water and macroconstituents (proteins, fat, and carbohydrates), their form and texture, as well as rheological properties vary a lot. This diversity in chemical and physical properties of foods has to be taken into consideration when performing extraction of analytes. Food can often be regarded as heterogeneous sample containing porous solid particles in which the analyte is adsorbed into the porous matrix. Constituents will be released in several steps involving desorption from the surface, diffusion through organic part of the matrix, solvated by extraction phase and diffuse to reach the bulk of extraction phase, and the kinetics of this process can be described using mass transfer in porous media [28]. Modification of the matrix by sample grinding, adding salt, or changing the pH or temperature is a usual procedure to increase the extraction efficiency.

One of the most popular tactics to deal with food matrix when performing SPME for flavor analysis is sample grinding (Fig. 8.2). Increasing surface area as a result of grinding favors and accelerates the extraction rates, due to increased surface ratio and improved diffusion of analyte from solid material. Changes in the matrix composition by grinding can be furthermore facilitated by the addition of water or NaCl solution. Water is an effective additive to enhance the release of analytes from the matrix, influencing their partition coefficients, and is often used to improve extraction [29]. Salting out by addition of NaCl to liquid matrix improves analytes' migration into the headspace by decreasing solubility of especially polar compounds, thus improving sensitivity for these compounds. Salting out is a common procedure for static headspace analysis, and as in food analysis headspace SPME is usually performed; this approach increases analyte concentration in gas phase. For some flavor compounds, change in the ionic strength may either decrease the aqueous solubility or enhance the extraction (more polar compounds), or for some volatiles the effect is opposite. Therefore, the effect of salt addition on SPME extraction efficiency should be verified experimentally as one of the obligatory steps to properly optimize extraction parameters, as usually flavor compounds in particular food comprises of compounds of various polarities.

The other parameter important to optimize is extraction temperature, especially in sampling food matrices. Raising the temperature of the extraction may at one point increase the extraction rate but on the other hand decrease the fiber/air distribution constant. By increasing the temperature of extraction, the native analytes dissociate from the matrix and move into the headspace more effectively; however, the decrease in the coating/sample distribution coefficient may result in the diminution in the equilibrium amount of analyte extracted. To overcome this setback, the cold fiber extraction method has been proposed which significantly improves the overall extraction efficiency [30]. For food matrices, its composition greatly influences extraction parameters, extraction temperature being one of the most matrix-dependent parameter. Too high temperature of extraction can result in such matrix reactions as lipid oxidation, protein denaturation, viscosity changes, to name only few.

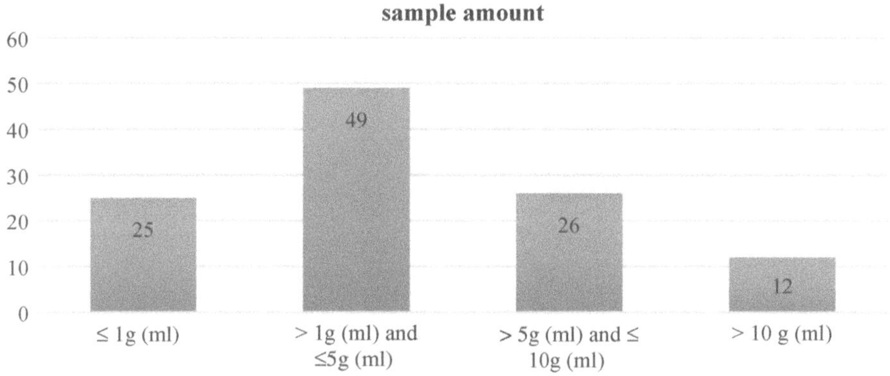

Fig. 8.3 Sample amount prepared for the extraction of flavor compounds by SPME technique from food products. Data from the Web of Science, years 2005–2015

As majority of flavor analysis from foods by SPME is performed from the headspace of the sample, it is important to consider the effect of the headspace volume in the vial, which practically depends on sample amount in a selected vial. Between food flavor analysts, there is a large diversity in the choice of the sample amount and in the setup of headspace volume/sample amount ratio when preparing the samples. As presented in Fig. 8.3, the most common sample amount (over 43%) prepared for flavor analysis is between 1 and 5 g for solid samples or mL for liquids. Additionally in more than 50% of studies, the ratio between headspace volume and sample amount (phase ratio) was set between 0.2 and 0.5. The effect of sample volume on quantitation, time to reach equilibrium, and method precision in SPME analysis is evident and may be neglected only in rare cases, where the sample volume is much larger than the fiber capacity. Usually, testing the effect of sample amount on extraction efficiency is often limited in laboratories by the size of available sample vials, where typically 1.5–20-mL vials are used, especially when automated SPME is operated. The one of the advantages of SPME, as compared to conventional sample preparation techniques, is the possibility to extract analytes from very small samples. The results of reviewing research papers on food flavor analysis, presented on the Fig. 8.3 are showing that in more than 22% of studies (years 2005–2015) extremely small sample amounts, <1 g or mL, were used for SPME extraction.

8.2.2 *Quantitative and Semiquantitative Analyses*

The application of SPME technique in food flavor for quantitative analysis, despite its potential, has not been so widely used. In between 120 research papers found in the Web of Science database only, 38 (31%) were using full quantitative methods such as external calibration (EC), standard addition (SA), multiple headspace

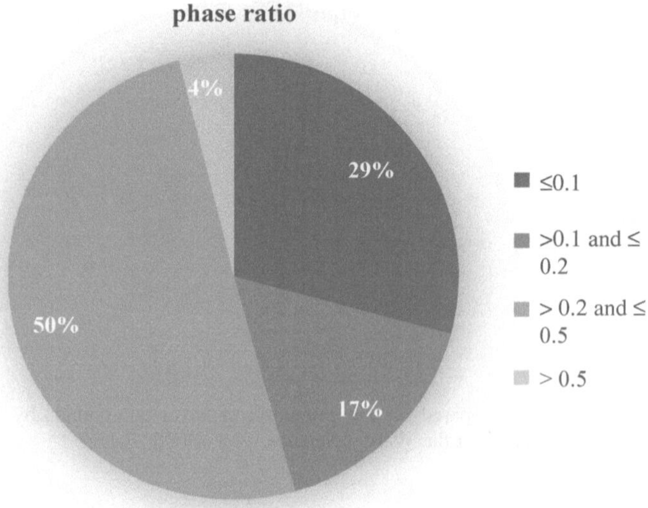

Fig. 8.4 Phase ratio (headspace volume/sample amount) of food samples prepared for extraction of flavor compounds by SPME technique. Data from the Web of Science, years 2005–2015

extraction (MHE), or stable isotopes dilution analysis (SIDA). More than 69% of researchers uses SPME as a semi-quantitative method either with presentation of peak area (PA), percent composition (%), or addition of an internal standard (IS) (Fig. 8.4). The low-level exploitation of full quantitation of flavor compounds (in profiling food flavor) is not only a case using SPME but also with many other extraction methods, which is probably mainly due to the matrix complexity of most food products which causes problems with linearity of calibration curves. This is in particular a problematic situation when trace levels of volatiles have to be extracted and quantified from food main constituents such as lipids, proteins, and carbohydrates which interact continually and affect the retention of volatiles within the food and hence their levels in the gaseous phase.

The importance of proper quantification of flavor compounds is necessary when key odorants of a certain food product need to be identified. With the so-called "sensomic" approach developed and described by Schieberle and his coworkers, only those compounds present in food product in the concentration higher than their odor threshold (OT) reflect the true aroma of a product. In their recent paper [31], they have shown that only about 230 aroma compounds from about 10,000 volatiles occurring in food products are odor-active and define the volatile stimulus space of most of foods and beverages. It means that only by 3 to circa 40 compounds (key odorants), the aroma of a given food product can be characterized. However, to find out their role in flavor, a comprehensive and proper quantitation has to be performed. Taking into consideration that some of the aroma compounds have their OT lower than ng/kg level (i.e., 2-methyl-3-furanthiol: 0.03 ng/kg water or

(Z)-1,5-octadien-3-one: 0.3 ng/kg water), their quantitation to rival the sensitivity of human nose is a challenging task.

SPME method allows for proper quantitation of flavor compounds with different calibration procedures; however, some principles have to be followed with each action.

External calibration (EC) is based on the comparison of instrument responses from the sample to the responses from the target compounds in the calibration standard which are run separately from the actual samples in a prepared solution. The principal advantages of external standard calibration are that it is simple to perform and that calibration curve can serve a multitude of samples. Unfortunately, they do not correct for losses that may occur during the preparation of the sample, such as extraction, centrifugation, evaporation, and moreover they do not compensate for the interaction of volatiles with other food constituents, often named as "matrix effect." If the matrix has a real effect on the release of volatiles, from which sample can be already noticed during typical optimization of SPME extraction conditions, which at one point requires comparison of extraction time profile in a pure matrix with that of real sample. There are examples in the literature of using external calibration with SPME technique for more complex matrices, where the matrix effect has been overcame by preparing the calibration curve in the matrix of the same composition. Of course this can be achieved when the matrix is well defined or there is blank matrix available without target components. Lojzova and coworkers [32] have quantified alkylpyrazines in fried potato chips by preparing external calibration curve in a matrix-matched sample of pre-fried potato chips with the verified absence of target compounds, reaching linearity of regression coefficient higher than 0.995. Similar situation has been performed by Vichi group when analyzing volatile phenols in virgin olive oils using an external calibration prepared in deodorized sunflower oil [33] or by using reconstituted whey protein isolates [34]. Refined, deodorized oil can be used for the matrix preparation in the analysis of volatile compounds in virgin oils (olive oil or other cold pressed oils) [35].

A standard addition (SA) method involves adding known amounts of analyte standard to one or more aliquots of the processed sample solution, compensating for the presence of analyte in the matrix. In cases where external calibration cannot be performed due to a complex matrix (such as coffee), standard addition allows reliable quantitation [36]. Unfortunately for each sample, there must be a separate calibration curve prepared which makes quantitation time-consuming. When using SPME with SA, calibration sample should be prepared as homogenic as possible to assure a complete release of native analytes from the sample, which may not be unequivocal in heterogenic samples. Moreover, SA method is rather recommended with the application of absorption-type fibers such as PDMS or PA where we can assume that the response between original analyte concentration and spiked concentration is linear in the concentration range used. When using porous polymers coatings, e.g., PDMS/DVB Carbowax/DVB, which extract analytes by adsorption rather than absorption, the linear range is narrower due to the limited number of sorption sites on the surface. It is important, therefore, to check the linearity of the response using standard solutions. Nevertheless, the SA approach should be recommended for precise quantitation of volatiles from complex food matrices. During last decade, it has been applied to several food products for the quantitation of key odorants in olive oil [35], wine [37], and fish oil emulsion [38].

Multiple headspace extraction (MHE) technique was proposed by Suzuki in 1970 and is based on performing several extractions from a single vial. It is a type of a quantitative approach introduced by Kolb and Ettre [39] in which the total peak area of an analyte can be calculated directly in the matrix, as a sum of the areas of each individual extraction (exhaustive extraction). Although this quantitative technique has evident advantage as it is not affected by the matrix composition, in the last decade (2005–2015) with SPME approach, it has been used for food flavor analysis only in 6 research papers, Fig. 8.5. One of the reasons may be the fact that this MHE technique is useful when extensive quantity is removed at each extraction which is true only if the partition coefficient between the matrix and analyte is small. MHE may be also considered as labor-intensive and time-consuming quantitation technique; however, it has been successfully introduced for both solid and liquid food products such as roasted coffee, mushrooms, or wine [40–42].

Stable isotope dilution assay (SIDA) uses stable isotopes twin molecule of the flavor compound, mostly ^{13}C or ^{2}H, as a most suitable internal standard. This

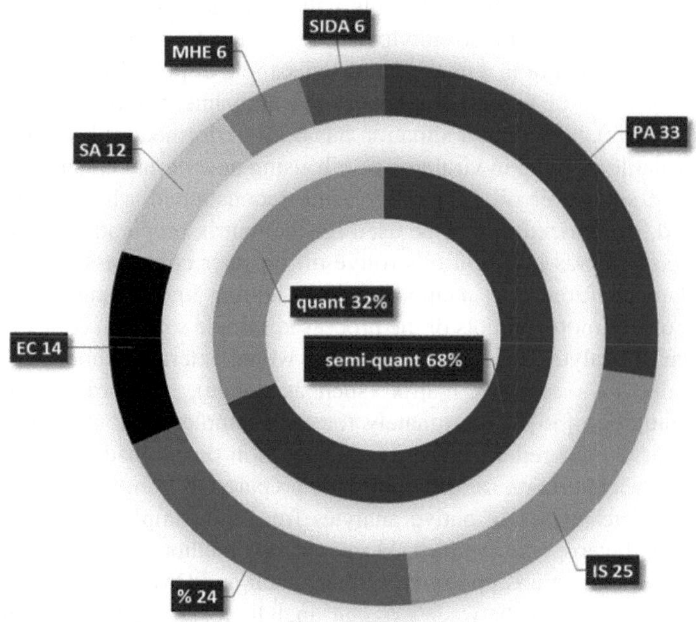

Fig. 8.5 Quantitative and semi-quantitative methods used in analysis of food flavor by SPME technique. Data from the Web of Science, years 2005–2015. *EC*—external calibration, *SA*—standard addition, *MHE*—multiple headspace extraction, *SIDA*—stable isotopes, *PA*—peak area, *%*—percent composition, *IS*—addition of an internal standard. Number of papers representing each approach shown in outer *circle*

method has been introduced in flavor chemistry in the beginning of the 1990s [43, 44], and since then it has been applied to the analysis of key odorants of many food products but mostly with solvent extraction such as performed on SAFE apparatus [45]. There are only six research papers found by authors of this chapter in the Web of Science using quantitation of flavor compounds by SPME extraction with combination of SIDA procedure (Fig. 8.5). The precision of this method has been confirmed in model experiments showing that even with the extraction yield below 10%, the results are very precise as the isotopically labeled internal standard compensated the losses almost totally [46]. The procedure using SIDA, described meticulously by Grosch in his review [47], is very simple and similar to IS method, except that it requires the calculation of a detector response factor between analyte and its labeled standard. Since its introduction, the main factor limiting its use is the need of using mass spectrometer as a detector and overall—the lack of commercially available standards of majority of flavor compounds and/or their high prices.

Internal standard (IS) quantitation is based on the addition of a chemical substance in a known amount to analyzed sample. This method is particularly recommended if multistep sample preparation procedures are required where in such case an internal standard would compensate for any losses of analytes. Clearly, this is true only if the internal standard is properly chosen, that is, it has similar chemical and physical properties as analytes. In case of SPME extraction, a complicated sample preparation procedure is eliminated nonetheless the ad/absorption process on the fiber requires to select an internal standard which would provide a comparable partition coefficient to extracted volatiles. It is particularly difficult to achieve as the composition of food flavor compounds is very varied as in chemical classes: alcohols, aldehydes, acids, ketones, carbonyls, sulfur and nitrogen, amines, aromatics, or in volatility, which usually would involve a choice of several internal standards. Therefore, an IS method is and should be generally treated as a semi-quantitative technique. With an addition of one compound to analyzed samples, it is one of the easiest methods to perform and very often used to show the differences in flavor composition between samples running in the same batch. For example, Lee and coworkers have added three different internal standards and calculated relative concentration to show differences in volatile profiles among raw and roasted almonds [48]. Similarly, in 2013, Moran et al. have shown differences in the formation of volatiles during autoxidation of lipids in lamb meat using semi-quantitative results with IS [49]. The full quantitation with IS is difficult to achieve and still may give erroneous results as there should be response factor calculated for each analyte, as described in this chapter for SIDA method. Calculation of response factor is even more important to carry out if the peak area is calculated from selected ions instead of total ion chromatogram.

For semi-quantitative analysis when performing SPME extraction, a simplified peak area abundance (PA) is mostly often used. During last 10 years, more than 30 research papers has described it and used it for food flavor analysis (Fig. 8.5). This method is even more easy than IS as no additional standards need to be added into the sample; however, to be able to compare the results, a considerable caution has to be taken in preparing the sample, for example, weighing samples and controlling extraction parameters during homogenization, so that the SPME precision can be maintained high. PA method is effectively applied for the method development and optimization of SPME extraction parameters for instance selection of fiber coating, determination of extraction time, or optimization of extraction condition such as pH, salt, or temperature. Comparison of peak areas is usually performed for analyses that are run in a single batch, using single fiber for comparative purposes.

The last one of the semi-quantitative methods used with SPME extraction is percent composition of the compound (%) calculated by measuring the corresponding peak area of the total ion chromatogram and expressed as a ratio to the total integrated peak areas. Those results often called relative percent area are more qualitative than any other mentioned methods because even comparison of the samples from the same batch can be problematic. When performing percent composition following the assumption, all flavor compounds eluted from the column have to be considered and all of them are integrated. With SPME method, sometimes additional components are showing on the chromatogram from fiber bleeding which should not be included in the flavor compounds. The most important issue in area % based semi-quantitation is the fact that SPME as a non-exhaustive extraction method based on compounds partition will provide different profiles than that obtained after exhaustive extraction. Also the % profile of obtained compounds depends on extraction time, matrix preparation, and SPME fiber used for sampling. Interpretation of percent composition results ought to be very careful because a change in the concentration of one compound affects the composition of all others.

8.3 Comparison of SPME with Other Extraction Methods

Simultaneous distillation extraction (SDE) is one of classical techniques commonly used for aroma compounds isolation. It is an exhausted extraction with the use of Likens-Nickerson and Clevenger apparatus [50]. Many publications have demonstrated that SDE has the highest amount of identified compounds of all extraction techniques [50–52]. Thus, it has great potential, and on the other hand, it requires the presence of a solvent and high temperature. Depending on the solvent used for the analysis, different compounds may be isolated. Bosch-Fuste et al. [50] showed that the use of dichloromethane caused detection of a higher amount of polar compounds (alcohols, aldehydes, and acids) in sparkling wine than in the case of SPME coated with a layer of DVB-CAR-PDMS; one needs to remember, though, that the high peaks of acids may also obscure other significant compounds. It has been shown that SDE has a greater extraction capacity for terpenes and some of furan derivatives in sparkling wine than SPME [50].

On the other hand, high temperature may influence artifacts formation [51, 53–55]. In buckwheat, aliphatic C_6–C_{10} compounds with carbonylic and hydroxylic group were identified only in the SDE extracts, which may have been caused by the use of high temperature; in potato snacks, in turn, 5 compounds were identified as artifacts [53]. Roth et al. [55] reported that the occurrence of lipid oxidation products is related to high temperature of extraction. It also has been shown that compounds soluble in water are poorly extracted by SDE [55]. In comparison with SDE, SPME showed differences in qualitative aroma profile mostly due to the absence of higher weight volatile compounds [53]. Because of the use of the high temperatures, SDE could be used to analyze volatile fractions in cooked food or to analyze compounds with high boiling temperature [51, 54]. SDE has high extraction capacity, but it is expensive because of its long duration and the use of a solvent.

Solvent-assisted flavor evaporation (SAFE) is a technique which provides gentle conditions for the isolation of volatile compounds, in combination with high vacuum and as a result, low temperature (40 °C) prevents the formation of artifacts. The use of this technique is more expensive than that of SPME (costs of solvent, liquid nitrogen as coolant, vacuum pump, and special glassware) and it is time-consuming. On the other hand, the extract is enriched with many volatiles with different polarities and broad range of molecular masses belonging to various chemical classes. Moreover, the sample may be frozen and reused; therefore, this technique is preferred in GCO analysis of key aroma compounds [45, 53, 56].

Extracts from SAFE and SPME may exhibit significant differences in aroma profiles. It has been shown that SPME extracts contain lower amounts of volatile compounds such as aldehydes, alcohols, esters, and ketones, while SAFE extracts have higher amounts of heavier molecules [52, 56–59]. Also relative amounts of extracted alcohols and terpenes were greater in SAFE than those in SPME extracts [56, 60]. This may have been caused by a higher affinity of the heavier molecules to the sorbent phase in fiber. Depending on the solvent used for SAFE, compounds with different polarities may be isolated. The use of dichloromethane caused identification of high amounts of carboxylic acid in SAFE extract; in contrast, SPME isolated only a few most abundant compounds. However; as a solvent-free technique, the SPME method has peaks of low molecular compounds such as acetaldehyde, 2,3-butanedione, and ethyl acetate that may be masked by the solvent in SAFE [55, 59]. Using solvents, one may isolate compounds from solid matrix that cannot be released without pretreatment [61]. Contrary to exhaustive methods extraction using SPME, fiber is highly dependent on fiber type, time, and temperature of extraction. Roth et al. [55] showed that at low temperatures (30 °C) and longer extraction durations, the amount of highly volatile compounds decreases and, respectively, the amount of medium- and low-volatile molecules increases. SAFE can be a more sensitive technique than SPME, but—because of the duration of isolation—SPME seems to be the more proper technique for rapid analyses of ongoing changes in the food matrix, e.g., lipid oxidation [62]. The combination of both techniques can provide sufficient data for a characterization of food flavor, useful in GCO analysis [63].

Liquid–liquid extraction (LLE) is a direct extraction based on the solubility of volatile compounds in extracting phase [64]. Solis-Solis et al. [65] have shown that LLE extracts of apricots gave a larger amount and a greater diversity of higher volatile compounds than did the SPME extracts, but both techniques were able to discriminate between different varieties of apricots. In white strawberries, LLE flavor analysis extracted the largest amount of volatile compounds, but in comparison with SPME and other techniques, it did not isolate 22 compounds that were isolated by the remaining techniques such as HS-SPME and SAFE [58]. Because of the contact of food matrix with the solvent, direct solvent extraction–high vacuum transfer (DSE-HVT) shows a larger amount of extracted volatile compounds than SPME does, but after selection of odor-active molecules, its qualitative results are similar to SPME. In addition, DSE-HVT is able to extract high molecular volatile compounds, but it can be less sensitive than SPME [66].

Hydrodistillation (HD) is a classical technique used to isolate essential oils from plants. Supercritical fluid is a solvent-free extraction based on temperature and pressure. Changes in these parameters have impact on the extraction yield of compounds with significantly different molecular weights. Use of the latter technique may lead to degradation of thermo-labile compounds; therefore, in the analysis of *Perilla frutescens*, it has been shown that only SPME isolated most of the highly volatile compounds [67].

Solid-phase extraction is widely used especially for flavor and off-flavor extraction in wines. It is based on partitioning between liquid and solid phases which allows the use of sorbents and solvents with different polarities. It has many advantages such as short time of extraction (15–30 min.), the possibility to isolate a wide range of compounds depending on their volatility and polarity, and a high repeatability [68]. Currently, a miniaturized technique based on SPE is also available: MEPS, microextraction by packed sorbent, enables the analysis of very small volumes of samples (few μl) [68]. SPE has been shown to be a more sensitive technique than MEPS and SPME, and it can isolate larger amounts of analytes. SPE and MEPS (with LiChrolut EN and C_8 as sorbent materials, respectively) concentrated higher amounts of ethyl esters, furan derivatives, and volatile fatty acids than did MEPS and SPME, while SPME mostly concentrated higher alcohols and carbonyl compounds. In comparison with SPE and SPME, MEPS extracts have the lowest amount of isolated volatile compounds [68]. An interesting result may be obtained by combining preconcentration and selectivity of SPE with subsequent extraction of volatile compounds using SPME, as was demonstrated for wine flavor compounds released after acid hydrolysis [69]. Many papers have shown that headspace techniques allow isolation of the most volatile compounds, especially ketones and sulfur compounds; there is, however, a possibility of the formation of artifacts masking low boiling compounds [70].

Purge and trap (PT) is a dynamic headspace technique based on trapping volatile compounds by an inert gas to solid or liquid matrices in a tube. Tenax®, which is widely used as sorbent in PT technique, despite its high capacity, may also absorb water, which may lead to artifact formation [54, 56, 59]. Contarini and Povolo [71] reported that the purge and trap technique isolated more low molecular compounds than did the SPME technique. In pea flour flavor analysis, purge and trap extracted a larger amount of volatile compounds, mainly alcohols (including low molecular compounds), aldehydes, and ketones than did SPME, while the latter extract contained mostly ketones and benzene derivatives [56, 59]. However, the purge and trap extract was perceived as less representative by sensory panelists than was SPME extract, which may have been caused by poor recoveries for less volatile compounds [54, 56, 72, 73]. A method in which air circulates through sorbent in a closed system (closed-looped stripping analysis—CLSA) uses charcoal as adsorbent. The use of CLSA to analyze volatile compounds in sparkling wine resulted in a similar profile to that obtained in SPME. CLSA extracted more acids, while SPME isolated larger amounts of esters. Although CLSA has greater capacity than does SPME, CLSA extraction of acids may result in the masking of smaller molecules. Moreover, CLSA takes a longer time than SPME [50].

Stir bar sorptive extraction (SBSE) is a solvent-less technique based on the sorption of analytes on a thick sorbent of PDMS. It has a larger volume of PDMS phase than does SPME, and therefore its extraction yield and sensitivity are greater [74]. On the other hand, the use of SBSE requires a thermal desorption unit or the use of a solvent. Because of its high sensitivity, SBSE is widely used for off-flavor analysis. The capacity of both techniques is dependent on pH, the presence of sugars, ethanol, etc. [75]. Compared with DI-SPME (direct immersion SPME), SBSE in wine analysis showed similar results, but higher amounts of terpenes were obtained [76]. On the other hand, the use of PDMS phase in SBSE and especially, headspace sorptive extraction (HSSE), results in a lower extraction of some compounds [51]. The use of monolithic material sorptive extraction (MMSE) with PDMS as an adsorbent resulted in a similar volatile profile to that obtained with DI-SPME, with the exception of β-damascenone [51].

8.4 SPME in Complex Profiling of Food Volatile Compounds

Complex profiling of food volatiles involves the extraction of the highest possible number of compounds, separation (in some approaches this step can be omitted) and subsequent multivariate data analysis (MVDA). The high number of data obtained in "-omics" approaches needs to be decreased without loss of useful information. With reference to the requirements of data processing, a combination of the SPME technique, as an efficient and robust extraction method, and MVDA may enable visualization, e.g., determining the volatile compounds unique to food varieties or profiling the changes caused by food processing. Because of the type and amount of data obtained by such metabolomics approach, chemometric tools are used to reduce data dimensionality [77]. Using chemometrics makes it possible to show an overview of the obtained data, which enables the demonstration of clusters and trends, the discrimination between the samples, as well as the prediction models. In analyses using SPME, methods such as PCA, PLS, and various combinations of the two tend to be employed.

Variance analysis is typically unable to show differences in the volatile profiles of many different samples [65]. In addition, a high number of identified volatile compounds may also make data treatment difficult. In this case, principal component analysis (PCA), a chemometric tool, may be used. The basis of PCA is decreasing the amount of data by normalization and data reduction (via maximizing sample variation) without the loss of information. Moreover, PCA does not require knowledge about the dataset [78]. The obtained principal components (orthogonal variables) are useful as one may compare at least two of them in order to discriminate samples. PCA in flavor analysis coupled with the SPME method is typically used to show differences between such factors as cultivars and varieties [79–81], technological processing types [75], quality [82–85], or verification of authenticity of a product [86–88]. It is also widely used in metabolomics research, which requires high amounts of data [89]. The obtained results are mainly based on the relationships between different odorants. PCA consists of grouping volatile compounds which results in the formation of various clusters unique to the selected sample groups [80, 90]. Most analyses with the use of SPME and chemometric methods are based on single-dimensional gas chromatography with mass spectrometer, but for sample profiling, comprehensive two-dimensional gas chromatography is used [87, 91]. In the latter types of analyses, large amounts of data are obtained, which calls for the selection of those significant for the composition of food; therefore, chemometric methods such as PCA, FDA (factorial discriminate analysis), and various multivariate analysis techniques are preferred. While PCA allows to establish differences between samples, FDA, in addition to this, makes it possible to predict quantitative values [65].

Hierarchical cluster analysis with PCA and SPME are tools useful for the evaluation of food authenticity. Both of them do not require knowledge about the data and enable graphic visualizations of the obtained results. Hierarchical cluster analysis (CA) is based on hierarchical grouping of samples which is dependent on their similarity. The data obtained by HCA give less information than those obtained by PCA, but they may be used in PCA for further analysis. Cluster analysis does not include information about the classes of the analyzed compounds, showing the obtained results as a dendrogram [92]. HCA in flavor analysis is used mainly for differentiation between samples, as well as detection of adulterations or differences between samples treated with various processing methods [88, 90, 93, 94]. Cagliero et al. [88] demonstrated that the results obtained by HCA with the use of enantiomers varied as a function of addition of synthetic flavors.

The data obtained by gas chromatography may be easily used for multivariate analysis such as PCA, LDA, and PLS [86]. Partial least squares regression (PLS) is the combination of principal component analysis and multiple regression and is mainly used in large datasets. It is a regression analogy of PCA that uses X and Y matrices [95]. PLS is used if one needs more information about the data than which is provided by the data obtained with PCA [96]. Depending on the Y matrix of the obtained data, which was earlier mean-centered, and whose variables were standardized to give equal variance, PLS may give qualitative (PLS-DA) or quantitative information [95, 97]. PLS is also used for fast optimization of method [86, 93, 97, 98]. The multivariate optimization design enables one to create prediction models [54, 99]. PLS allows the investigation of such issues, e.g., the impact of differences in technological processing or the impact of the cultivar on flavor profile [97], and quantitative correlation of predicted compounds [54]. Kang et al. [54] showed that the use of the PLS model allows the calculation of the concentration of fusel compounds on the basis of amino acid profile in makgeolli mash.

8.5 Conclusions

SPME in the analysis of food flavors and off-flavors has been proven to be a method of great potential in both qualitative and quantitative analyses. Though used mainly for qualitative work, its selectivity, sensitivity, and reproducibility make it an attractive tool in quantitative analysis, particularly for the determination of compounds present in trace levels. It found many uses in targeted analysis of odorants, especially compounds causing off-flavors in food. Its ease of automation enables the use of SPME for large sample sets, and its use in conjunction with such powerful separation techniques as GCxGC-ToF-MS provide a powerful device for complex "-omics" type of research.

Acknowledgment National Science Centre (Poland) for the support under a project 2012/07/B/NZ9/01634.

References

1. Nongonierma A, Voilley A, Cayot P et al (2006) Mechanisms of extraction of aroma compounds from foods, using adsorbents. Effect of various parameters. Food Rev Int 22:51–94
2. Panighel A, Flamini R (2014) Applications of solid-phase microextraction and gas chromatography/mass spectrometry (SPME-GC/MS) in the study of grape and wine volatile compounds. Molecules 19:21291–21309
3. Balasubramanian S, Panigrahi S (2011) Solid-phase microextraction (SPME) techniques for quality characterization of food products: a review. Food Bioprocess Technol 4:1–26
4. Jeleń HH, Majcher M, Dziadas M (2012) Microextraction techniques in the analysis of food flavor compounds: a review. Anal Chim Acta 738:13–26
5. Jeleń HH (2006) Solid-phase microextraction in the analysis of food taints and off-flavors. J Chromatogr Sci 44:399–415
6. Ridgway K, Lalljie SPD, Smith RM (2010) Analysis of food taints and off-flavours: a review. Food Addit Contam Part A 27:146–168
7. Stashenko EE, Martínez JR (2007) Sampling volatile compounds from natural products with headspace/solid-phase micro-extraction. J Biochem Biophys Methods 70:235–242
8. Bojko B, Cudjoe E, Gómez-Ríos GA et al (2012) SPME—quo vadis? Anal Chim Acta 750:132–151
9. Seethapathy S, Górecki T (2012) Applications of polydimethylsiloxane in analytical chemistry: a review. Anal Chim Acta 750:48–62
10. Sgorbini B, Budziak D, Cordero C et al (2010) Solvent-enhanced headspace sorptive extraction in the analysis of the volatile fraction of matrices of vegetable origin. J Sep Sci 33:2191–2199

11. Heaven MW, Nash D (2012) Recent analyses using solid phase microextraction in industries related to food made into or from liquids. Food Control 27:214–227
12. Aziz-Zanjani MO, Mehdinia A (2014) A review on procedures for the preparation of coatings for solid phase microextraction. Microchim Acta 181:1169–1190
13. Budziak D, Martendal E, Carasek E (2009) Application of an NiTi alloy coated with ZrO_2 solid-phase microextraction fiber for determination of haloanisoles in red wine samples. Microchim Acta 164:197–202
14. Nardini GS, Merib JO, Dias AN et al (2013) Determination of volatile profile of citrus fruit by HS-SPME/GC-MS with oxidized NiTi fibers using two temperatures in the same extraction procedure. Microchem J 109:128–133
15. Liang X, Liu S, Wang S et al (2014) Carbon-based sorbents: Carbon nanotubes. J Chromatogr A 1357:53–67
16. Zhang M, Zeng J, Wang Y, Chen X (2013) Developments and trends of molecularly imprinted solid-phase microextraction. J Chromatogr Sci 51:577–586
17. Rajabi Khorrami A, Narouenezhad E (2011) Synthesis of molecularly imprinted mono-lithic fibers for solid-phase microextraction of acetaldehyde from head-space of beverages stored in PET bottles. Talanta 86:58–63
18. Liu M, Peng Q, Chen Y et al (2015) A rapid space-resolved solid-phase microextraction method as a powerful tool to determine contaminants in wine based on their volatility. Food Chem 176:12–16
19. Mehdinia A, Aziz-Zanjani MO (2013) Advances for sensitive, rapid and selective extraction in different configurations of solid-phase microextraction. TrAC Trends Anal Chem 51:13–22
20. Carasek E, Pawliszyn J (2006) Screening of tropical fruit volatile compounds using solid-phase microextraction (SPME) fibers and internally cooled SPME fiber. J Agric Food Chem 54:8688–8696
21. Carasek E, Cudjoe E, Pawliszyn J (2007) Fast and sensitive method to determine chloroanisoles in cork using an internally cooled solid-phase microextraction fiber. J Chromatogr A 1138:10–17
22. Ho H-P, Lee R-J, Lee M-R (2008) Purge-assisted headspace solid-phase microextraction combined with gas chromatography–mass spectrometry for determination of chlorophenols in aqueous samples. J Chromatogr A 1213:245–248
23. Psillakis E, Yiantzi E, Sanchez-Prado L, Kalogerakis N (2012) Vacuum-assisted head-space solid phase microextraction: Improved extraction of semivolatiles by non-equilibrium headspace sampling under reduced pressure conditions. Anal Chim Acta 742:30–36
24. Psillakis E, Yiantzi E, Kalogerakis N (2013) Downsizing vacuum-assisted headspace solid phase microextraction. J Chromatogr A 1300:119–126
25. Brunton NP, Cronin DA, Monahan FJ (2001) The effects of temperature and pressure on the performance of Carboxen/PDMS fibres during solid phase microextraction (SPME) of headspace volatiles from cooked and raw turkey breast. Flavour Fragr J 16:294–302
26. O'Reilly J, Wang Q, Setkova L et al (2005) Automation of solid-phase microextraction. J Sep Sci 28:2010–2022
27. Bizkarguenaga E, Iparragirre A, Navarro P et al (2013) In-port derivatization after sorp-tive extractions. J Chromatogr A 1296:36–46
28. Pawliszyn J (2010) Theory of extraction. In: Pawliszyn J, Lord HL (eds) Handbook of sample preparation. Wiley, New York, pp 1–24
29. Lord H, Pawliszyn J (2000) Evolution of solid-phase microextraction technology. J Chromatogr A 885:153–193
30. Jiang R, Carasek E, Risticevic S et al (2012) Evaluation of a completely automated cold fiber device using compounds with varying volatility and polarity. Anal Chim Acta 742:22–29
31. Dunkel A, Steinhaus M, Kotthoff M et al (2014) Nature's chemical signatures in human olfaction: a foodborne perspective for future biotechnology. Angew Chem Int Ed 53:7124–7143

32. Lojzova L, Riddellova K, Hajslova J et al (2009) Alternative GC–MS approaches in the analysis of substituted pyrazines and other volatile aromatic compounds formed during Maillard reaction in potato chips. Anal Chim Acta 641:101–109
33. Vichi S, Romero A, Tous J et al (2008) Determination of volatile phenols in virgin olive oils and their sensory significance. J Chromatogr A 1211:1–7
34. Leksrisompong PP, Miracle RE, Drake M (2010) Characterization of flavor of whey protein hydrolysates. J Agric Food Chem 58:6318–6327
35. Peres F, Jeleń HH, Majcher MM et al (2013) Characterization of aroma compounds in Portuguese extra virgin olive oils from Galega Vulgar and Cobrançosa cultivars using GC–O and GC×GC–ToFMS. Food Res Int 54:1979–1986
36. Majcher MA, Klensporf-Pawlik D, Dziadas M, Jeleń HH (2013) Identification of aroma active compounds of cereal coffee brew and its roasted ingredients. J Agric Food Chem 61:2648–2654
37. Jeleń HH, Dziadas M, Majcher M (2013) Different headspace solid phase microextraction—gas chromatography/mass spectrometry approaches to haloanisoles analysis in wine. J Chromatogr A 1313:185–193
38. Iglesias J, Lois S, Medina I (2007) Development of a solid-phase microextraction method for determination of volatile oxidation compounds in fish oil emulsions. J Chromatogr A 1163:277–287
39. Kolb B, Ettre LS (1991) Theory and practice of multiple headspace extraction. Chromatographia 32:505–513
40. Gómez-Ariza JL, García-Barrera T, Lorenzo F, Beltrán R (2006) Use of multiple head-space solid-phase microextraction and pervaporation for the determination of off-flavours in wine. J Chromatogr A 1112:133–140
41. Bicchi C, Ruosi MR, Cagliero C et al (2011) Quantitative analysis of volatiles from solid matrices of vegetable origin by high concentration capacity headspace techniques: Determination of furan in roasted coffee. J Chromatogr A 1218:753–762
42. Costa R, Tedone L, De Grazia S et al (2013) Multiple headspace-solid-phase microextraction: an application to quantification of mushroom volatiles. Anal Chim Acta 770:1–6
43. Sen A, Laskawy G, Schieberle P, Grosch W (1991) Quantitative determination of β-damascenone in foods using a stable isotope dilution assay. J Agric Food Chem 39:757–759
44. Schieberle P (1996) Odour-active compounds in moderately roasted sesame. Food Chem 55:145–152
45. Engel W, Bahr W, Schieberle P (1999) Solvent assisted flavour evaporation–a new and versatile technique for the careful and direct isolation of aroma compounds from complex food matrices. Eur Food Res Technol 209:237–241
46. Schieberle P, Grosch W (1987) Evaluation of the flavour of wheat and rye bread crusts by aroma extract dilution analysis. Z Fuer Lebensm Unters Forsch 185:111–113
47. Grosch W (2001) Evaluation of the key odorants of foods by dilution experiments, aroma models and omission. Chem Senses 26:533–545
48. Lee J, Xiao L, Zhang G et al (2014) Influence of storage on volatile profiles in roasted almonds (*Prunus dulcis*). J Agric Food Chem 62:11236–11245
49. Bosse AK, Fraatz MA, Zorn H (2013) Formation of complex natural flavours by bio-transformation of apple pomace with basidiomycetes. Food Chem 141:2952–2959
50. Bosch-Fuste J, Riuaumatell M, Guadayol J et al (2007) Volatile profiles of sparkling wines obtained by three extraction methods and gas chromatography–mass spectrometry (GC–MS) analysis. Food Chem 105:428–435
51. Prosen H, Kokalj M, Janeš D, Kreft S (2010) Comparison of isolation methods for the determination of buckwheat volatile compounds. Food Chem 121:298–306
52. Ning L, Fu-Ping Z, Hai-Tao C et al (2011) Identification of volatile components in Chinese Sinkiang fermented camel milk using SAFE, SDE, and HS-SPME-GC/MS. Food Chem 129:1242–1252
53. Majcher M, Jeleń HH (2009) Comparison of suitability of SPME, SAFE and SDE methods for isolation of flavor compounds from extruded potato snacks. J Food Compos Anal 22:606–612

54. Kang K-M, Baek H-H (2014) Aroma quality assessment of Korean fermented red pepper paste (gochujang) by aroma extract dilution analysis and headspace solid-phase microextraction–gas chromatography–olfactometry. Food Chem 145:488–495
55. Roth M, Meiringer M, Kollmannsberger H et al (2014) Characterization of key aroma compounds in distiller's grains from wheat as a basis for utilization in the food industry. J Agric Food Chem 62:10873–10880
56. Murat C, Gourrat K, Jerosch H, Cayot N (2012) Analytical comparison and sensory representativity of SAFE, SPME, and purge and trap extracts of volatile compounds from pea flour. Food Chem 135:913–920
57. Jo Y-J, Cho IH, Song CK et al (2011) Comparison of fermented soybean paste (doen-jang) prepared by different methods based on profiling of volatile compounds. J Food Sci 76: C368–C379
58. Prat L, Espinoza MI, Agosin E, Silva H (2014) Identification of volatile compounds associated with the aroma of white strawberries (*Fragaria chiloensis*): aroma of white strawberries. J Sci Food Agric 94:752–759
59. Thomsen M, Gourrat K, Thomas-Danguin T, Guichard E (2014) Multivariate approach to reveal relationships between sensory perception of cheeses and aroma profile obtained with different extraction methods. Food Res Int 62:561–571
60. Mayuoni-kirshinbaum L, Tietel Z, Porat R, Ulrich D (2012) Identification of aroma-active compounds in "wonderful" pomegranate fruit using solvent-assisted flavour evaporation and headspace solid-phase micro-extraction methods. Eur Food Res Technol 235:277–283
61. Klensporf D, Jeleń HH (2008) Influence of the addition of raspberry seed extract on changes in the volatile pattern of stored model breakfast cereal. J Agric Food Chem 56:3268–3272
62. Havemose MS, Justesen P, Bredie WLP, Nielsen JH (2007) Measurement of volatile oxidation products from milk using solvent-assisted flavour evaporation and solid phase microextraction. Int Dairy J 17:746–752
63. Klensporf D, Jeleń HH (2008) Effect of heat treatment on the flavor of oat flakes. J Cereal Sci 48:656–661
64. Kok MF, Yong FM, Lim G (1987) Rapid extraction method for reproducible analysis of aroma volatiles. J Agric Food Chem 35:779–781
65. Solis-Solis HM, Calderon-Santoyo M, Gutierrez-Martinez P et al (2007) Discrimination of eight varieties of apricot (*Prunus armeniaca*) by electronic nose, LLE and SPME using GC–MS and multivariate analysis. Sens Actuators B Chem 125:415–421
66. Aceña L, Vera L, Guasch J et al (2010) Comparative study of two extraction techniques to obtain representative aroma extracts for being analysed by gas chromatography–olfactometry: Application to roasted pistachio aroma. J Chromatogr A 1217:7781–7787
67. Huang B, Lei Y, Tang Y et al (2011) Comparison of HS-SPME with hydrodistillation and SFE for the analysis of the volatile compounds of Zisu and Baisu, two varietal species of *Perilla frutescens* of Chinese origin. Food Chem 125:268–275
68. Mendes B, Gonçalves J, Câmara JS (2012) Effectiveness of high-throughput miniaturized sorbent- and solid phase microextraction techniques combined with gas chromatography–mass spectrometry analysis for a rapid screening of volatile and semi-volatile composition of wines—A comparative study. Talanta 88:79–94
69. Dziadas M, Jeleń HH (2010) Analysis of terpenes in white wines using SPE–SPME–GC/MS approach. Anal Chim Acta 677:43–49
70. Le Quéré J-L (2004) Cheese flavour: instrumental techniques. In: Fox PLHM PF, Cogan TM, Guinee TP (eds) Cheese: chemistry, physics and microbiology. Academic Press, New York, pp 489–510
71. Contarini G, Povolo M (2002) Volatile fraction of milk: comparison between purge and trap and solid phase microextraction techniques. J Agric Food Chem 50:7350–7355
72. Xu Y, Fan W, Qian MC (2007) Characterization of aroma compounds in apple cider using solvent-assisted flavor evaporation and headspace solid-phase microextraction. J Agric Food Chem 55:3051–3057

73. Villière A, Arvisenet G, Lethuaut L et al (2012) Selection of a representative extraction method for the analysis of odourant volatile composition of French cider by GC–MS–O and GC × GC–TOF-MS. Food Chem 131:1561–1568
74. Horák T, Čulík J, Kellner V et al (2010) Analysis of selected esters in beer: comparison of solid-phase microextraction and stir bar sorptive extraction. J Inst Brew 116:81–85
75. Forde CG, Cox A, Williams ER, Boss PK (2011) Associations between the sensory attributes and volatile composition of cabernet sauvignon wines and the volatile composition of the grapes used for their production. J Agric Food Chem 59:2573–2583
76. Gamero A, Wesselink W, de Jong C (2013) Comparison of the sensitivity of different aroma extraction techniques in combination with gas chromatography–mass spectrometry to detect minor aroma compounds in wine. J Chromatogr A 1272:1–7
77. Cevallos-Cevallos JM, Reyes-De-Corcuera JI, Etxeberria E et al (2009) Metabolomic analysis in food science: a review. Trends Food Sci Technol 20:557–566
78. Farag MA, Wessjohann LA (2012) Volatiles profiling in medicinal licorice roots using steam distillation and solid-phase microextraction (SPME) coupled to chemometrics. J Food Sci 77: C1179–C1184
79. Sousa ET, Rodrigues FdeM, Martins CC et al (2006) Multivariate optimization and HS-SPME/GC-MS analysis of VOCs in red, yellow and purple varieties of *Capsicum chinense* sp. peppers. Microchem J 82:142–149
80. Ong BT, Nazimah SAH, Tan CP et al (2008) Analysis of volatile compounds in five jack-fruit (*Artocarpus heterophyllus* L.) cultivars using solid-phase microextraction (SPME) and gas chromatography-time-of-flight mass spectrometry (GC-TOFMS). J Food Compos Anal 21:416–422
81. Lignou S, Parker JK, Oruna-Concha MJ, Mottram DS (2013) Flavour profiles of three novel acidic varieties of muskmelon (*Cucumis melo* L.). Food Chem 139:1152–1160
82. Cognat C, Shepherd T, Verrall SR, Stewart D (2012) Comparison of two headspace sampling techniques for the analysis of off-flavour volatiles from oat based products. Food Chem 134:1592–1600
83. Vaclavik L, Belkova B, Reblova Z et al (2013) Rapid monitoring of heat-accelerated reactions in vegetable oils using direct analysis in real time ionization coupled with high resolution mass spectrometry. Food Chem 138:2312–2320
84. Cheng P, Fan W, Xu Y (2013) Quality grade discrimination of Chinese strong aroma type liquors using mass spectrometry and multivariate analysis. Food Res Int 54:1753–1760
85. Poyato C, Ansorena D, Navarro-Blasco I, Astiasarán I (2014) A novel approach to monitor the oxidation process of different types of heated oils by using chemometric tools. Food Res Int 57:152–161
86. Risticevic S, Carasek E, Pawliszyn J (2008) Headspace solid-phase microextraction—gas chromatographic—time-of-flight mass spectrometric methodology for geographical origin verification of coffee. Anal Chim Acta 617:72–84
87. Cajka T, Riddellova K, Tomaniova M, Hajslova J (2010) Recognition of beer brand based on multivariate analysis of volatile fingerprint. J Chromatogr A 1217:4195–4203
88. Cagliero C, Bicchi C, Cordero C et al (2012) Fast headspace-enantioselective GC—mass spectrometric-multivariate statistical method for routine authentication of flavoured fruit foods. Food Chem 132:1071–1079
89. Settachaimongkon S, Nout MJR, Antunes Fernandes EC et al (2014) Influence of different proteolytic strains of *Streptococcus thermophilus* in co-culture with *Lactobacillus delbrueckii* subsp. bulgaricus on the metabolite profile of set-yoghurt. Int J Food Microbiol 177:29–36
90. Tu D, Li H, Wu Z et al (2014) Application of headspace solid-phase microextraction and multivariate analysis for the differentiation between edible oils and waste cooking oil. Food Anal Methods 7:1263–1270
91. Cordero C, Liberto E, Bicchi C et al (2010) Profiling food volatiles by comprehensive two-dimensional gas chromatography coupled with mass spectrometry: advanced finger-printing approaches for comparative analysis of the volatile fraction of roasted hazelnuts (*Corylus avellana* L.) from different origins. J Chromatogr A 1217:5848–5858

92. Berrueta LA, Alonso-Salces RM, Héberger K (2007) Supervised pattern recognition in food analysis. J Chromatogr A 1158:196–214
93. Welke JE, Zanus M, Lazarotto M et al (2012) Volatile characterization by multivariate optimization of headspace-solid phase microextraction and sensorial evaluation of chardonnay base wines. J Braz Chem Soc 23:678–687
94. Ligor M, Stankevičius M, Wenda-Piesik A et al (2014) Comparative gas chromatographic-mass spectrometric evaluation of hop (*Humulus lupulus* L.) essential oils and extracts obtained using different sample preparation methods. Food Anal Methods 7:1433–1442
95. Trygg J, Gullberg J, Johansson AI et al (2006) Chemometrics in metabolomics—an introduction. In: Saito PDK, Dixon PDRA, Willmitzer PDL (eds) Plant metabolomics. Springer, Berlin, pp 117–128
96. Höskuldsson A (2001) Variable and subset selection in PLS regression. Chemom Intell Lab Syst 55:23–38
97. Kebede BT, Grauwet T, Palmers S et al (2014) Effect of high pressure high temperature processing on the volatile fraction of differently coloured carrots. Food Chem 153:340–352
98. Gu S, Wang X, Tao N, Wu N (2013) Characterization of volatile compounds in different edible parts of steamed Chinese mitten crab (*Eriocheir sinensis*). Food Res Int 54:81–92
99. Samavati V, D-jomeh ZE (2013) Multivariate-parameter optimization of aroma com-pound release from carbohydrate–oil–protein model emulsions. Carbohydr Polym 98:1667–1676

Chapter 9
Applications of In Vivo and In Vitro Solid-Phase Microextraction Techniques in Plant Analysis

Fang Zhu and Guosheng Chen

Abstract Plant tissues usually consist of various complex matrices. Therefore, there is a clear and unmet need for the development of suitable systems for the collection and analysis of targeted analytes in plant, particularly for the sample pretreatment methods. Solid-phase microextraction (SPME) combines sampling, isolation, concentration, and enrichment in one step and is a simple and effective sample preparation technique. Due to the high-efficiency, cost-saving, solvent-free, and easy-to-realize automation properties, SPME is among the more promising and new, green sample preparation techniques for the determination of traces of different compounds of plants in the last decade, such as volatile organic compounds (VOCs) emitted from different organs of plants and a variety of environmental contaminants in plants. In this chapter, we summarize the recent applications of in vitro and in vivo SPME in different organs of plant and also discuss the advantages of the state-of-the-art SPME technology applied in the fields of plant analysis.

Keywords Solid-phase microextraction · Volatile organic compounds · Environmental contaminants · Plant analysis · In vivo · In vitro

9.1 Introduction

Investigation of organic compounds emitted from plants obtained great concern in the past years [1, 2]. Because many VOCs are widely used commercially, such as constituents of essential oils, their analysis has a long tradition [3]. Nearly all plants have the ability to emit volatile organic compounds (VOCs) from different organs including fruits, flowers, leaves, stems, and roots. The emission of VOC by plants

F. Zhu (✉) · G. Chen
MOE Key Laboratory of Aquatic Product Safety/KLGHEI of Environment and Energy Chemistry, School of Chemistry, Sun Yat-Sen University, Guangzhou 510275, China
e-mail: ceszhuf@mail.sysu.edu.cn

G. Ouyang and R. Jiang (eds.), *Solid Phase Microextraction*,
DOI 10.1007/978-3-662-53598-1_9

allows them to interact with other organisms at a distance. For example, many plant species emit floral volatiles which have diverse functions in pollinator's attraction and repulsion [4]. Volatiles are also released from vegetable plant organs, especially after herbivore damage. VOCs emitted from vegetables can functions as attractants for enemies of herbivores and have been suggested to serve as direct defenses against herbivores [5], as defenses against pathogens [6], as protectants against abiotic stress, and as signals in intra- and interplant communication [7]. Furthermore, a large amount of plants also emit VOCs in response to environmental or climate changes, such as light, temperature, flooding, or drought [8–10].

The translocation, accumulation, and tissue distribution of environmental pollutants in plant were also aroused extensive interesting due to the increasing concern of food safety and environment-ecology protection. The plant, particularly the esculent part of plant, can be contaminated by environmental pollutants that are present in air (by deposition), soil (by transfer), or water (by absorption and transfer), or during growth and processing [11]. In the past decade, the persistent and toxic contaminants, such as polycyclic aromatic hydrocarbons (PAHs) [12, 13], pesticides [14], pharmaceuticals, and personal care products (PCPs) [15–17], have been detected in various vegetable organs, which were grown in contaminated environment. Those contaminants may be biomagnified through food chains and threaten human health ultimately.

Since plant matrix is very complex, the extraction, separation, and quantitation of these chemicals are challenging, and there is a clear and unmet need for the development of suitable systems for the collection and analysis of the above analytes [18, 19], particularly for the sample pretreatment methods. However, the traditional sample pretreatment methods, such as solvent extraction, Soxhlet, supercritical fluid extraction (SFE), and distillative methods, including hydrodistillation (HD), stream distillation (SD), and vacuum distillation (VD) [20, 21], are laborious and time-consuming and could result in the losses of volatile analytes during sample pretreatment due to their high-volatility [22].

SPME is a solvent-free sample preparation technique and combines sampling, analyte isolation, and enrichment into one step. In this approach, microquantities of the solid sorbent or liquid polymer inappropriate format are exposed to the sample. Quantification is based on the amount of analyte extracted at appropriate conditions [23]. Because of its convenience and speed, SPME was implemented by researchers worldwide in environmental, food, aromatic, metallic, forensic, biologic, and pharmaceutical samples [24–27] and especially played a significant role in the volatile and nonvolatile organic compound analysis of plant [19]. As time goes on, more and more researchers use solid-phase microextraction (SPME) techniques for rapid extraction of VOCs and other environmental contaminants from plants [19].

9.2 Applications

9.2.1 Fruits

Fruit aroma is an important signal for assessing the quality of fruits. However, a large amount of aroma volatiles emitted from fruits are unstable and may be thermally decomposed and degraded during thermal extraction or distillation [28]. Owing to the convenience and speed features of SPME, SPME technique has became an efficient tool to detect the aromatic compounds from fruits.

SPME technique was used to identify and monitor the VOCs emitted from berry fruit [29]. In total, 74 volatile compounds and 24 aroma-active compounds were detected and characterized by HS-SPME within only 20-min headspace extraction time, using DVB/CAR/PDMS as an extraction coating. Similar method was adopted for the measurement of VOCs emitted from apricot [30], plums [31], and tomato [32]. In addition, the effect of salinity on yield of 2-acetyl-1-pyrroline content in the grains was studied through HS-SPME [33] and the effect of vineyard training system and location on the free flavor compounds emitted from Traminette grape [34]. In these studies, a large number of VOCs were identified in a short analysis time using SPME sampling technique. Such efficient detection is not realizable using conventional methodologies such as solvent extraction or Soxhlet.

Beyond the fast extraction property, high extraction capacity is another merit of SPME technique. A series of researches demonstrated that SPME possesses superior extraction capacity toward volatile aroma compounds than other traditional techniques. For example, Xie et al. found that HS-SPME coupled with GC/MS could be useful and appropriate for the rapid extraction and qualitative analysis of volatile components from plants at analytical level, compared with steam distillation and solvent extraction [35]. Furthermore, SPME-GC-MS can provide mass data for principal component analysis (PCA), quantitative descriptive analysis (QDA), and sensory analysis [36–38]. These unique merits facilitate the well understanding of the contribution of VOCs to the overall aroma and flavor, which play guiding roles in agricultural cultivation and relative industry.

Most plant species contain different varieties after long-term natural evolution or artificial cultivation. In addition, fruits may vary in color, flavor, size, and shape and belong to diverse cultivation. Flavor is one of the most appreciated characteristics of fruit, which can be used to distinguish the types or cultivars of the fruits. HS-SPME has been successfully applied in the discrimination of volatiles from flowers at three blooming stages of nine citrus cultivars [39] and the volatile profiling of Ficus carica varieties [40]. In addition, HS-SPME-GC-MS was used for the determination of free and bond volatile compounds in six varieties of citrus. The detection result showed that free volatile contents in sweet orange were the most abundant followed by those in grapefruits and mandarins and bound volatiles also significantly differed among cultivars [41]. Furthermore, qualitative ripening-dependent changes of VOCs were studied via headspace solid-phase microextraction, in which various VOCs emitted from white grapes in different maturity stages were detected using SPME method [42].

Besides the VOC analysis of fruits, analysis of contaminant residues in fruits also takes an important part, as environmental pollution has become a global problem. A self-made SPME fiber based on molecularly imprinted polymer technology was applied to the determination of triazine herbicides in rice and onion [43]. Chai et al. evaluate the pesticide residue contents in strawberry and cucumber through HS-SPME [44]. Acetic acid solution was suggested to be the most effective solution in removing the pesticide residue. In addition, herbicides have also been detected by SPME-GC-MS successfully. Except for contaminant analysis, endogenous substances of plant were successfully detected by SPME recently. For instance, a multilayer interparticle linking hybrid MOF-199 SPME fiber was prepared for plant hormone ethylene analysis in grapes, wampees, blueberries, and durian husks [45]. Besides the analysis of organic, SPME was also used for the determination of metal contents in rice samples. For instance, Zarrin et al. determined lead, cadmium, and copper in rice samples using SPME coupled with differential pulse anodic stripping voltammetry (DPASV). Rice sample was digested to an aqueous solution. A self-made solgel multiwalled carbon nanotube SPME fiber was used for the extraction of the solution under the protection of a hollow fiber [46].

Employing HS-SPME, one effective way to release the analytes from their matrices is to heat the sample to a high temperature. However, due to the exothermic nature of the adsorption, it can adversely decrease their partition coefficients and subsequent extraction from headspace to the fiber. In order to overcome the difficulties mentioned above, another novel SPME configuration, termed cold-fiber SPME (CF-SPME), has been developed to achieve higher extraction efficiencies [47, 48] and used in the determination of flavor profile of rice samples successfully. Ghiasvand et al. determine the flavor profile in Iranian fragrant rice samples successfully with cold-fiber SPME-GC-TOFMS [49]. In another research, Chai et al. analyze organophosphorous pesticides in tomatoes with circulating cooling SPME [50]. It is demonstrated that the CF-SPME enhances the extraction efficiency and sensitivity compared with the traditional SPME.

In the studies mentioned above, SPME was only combined with one-dimensional GC. As the development of GC, high-resolution comprehensive two-dimensional gas chromatography (GC×GC) is clearly preferable, in which compounds coeluting from the first column undergo additional separation on the second one. As a result, it was feasible to enhance the separation potential compared to the one-dimensional GC. In terms of its selective separation and increased resolution power, Pedroso et al. used a dynamic headspace solid-phase microextraction (DHS-SPME) for sample extraction that was coupled to GC×GC with flame ionization detection (FID) [51]. Qualitative ripening-dependent changes of pineapple complex volatiles were studied by Schmarr et al. via HS-SPME-GC×GC-qMS [42].

All of the above applications are in vitro SPME methods. In contrast, in vivo SPME sampling approach can indicate or predict the processes occurring in a complex living system more accurately [52, 53]. Even though Flamini et al. have studied the volatiles emitted in vivo by the fruit of grapefruit (*Citrus paradisi* L.) [54], the fruits were still required to be picked from the plants and pericarp needed to be removed from fruits before sampling, which cannot be referred to as in vivo sampling. Hence, the nondestructive in vivo sampling will be a promising development direction for the application of SPME in fruits' analysis.

Table 9.1 presents some other applications of SPME for the analysis of VOCs released from different fruits.

Table 9.1 Applications of SPME in fruit studies in recent years

Analytes	Samples	Device used	Extraction mode/time/temperature	References
Plant aroma	Common tomato, cherry tomatoes, longan, durian, Chinese mango varieties, and Allium varieties	75 μm CAR/PDMS, 65 μm PDMS/DVB	HS/60 or 30 min	[54]
Black pepper aroma and flavor	Shiraz grape berries	100 μm PDMS	HS/10 min/25 °C	[125]
Aroma volatiles	Tomato	65 μm CW/DVB	HS/55 min/40 °C	[126]
VOCs	Tomato	100 μm PDMS	HS/30 min	[127]
VOCs	Dry bean	50/30 μm DVB/CAR/PDMS	HS/1 h/50 °C	[128]
Aroma volatile	Mountain papaya fruit	65 μm PDMS/DVB	HS/60 min/20 °C	[129]
VOCs	*Juniperus communis* L. berries	50/30 μm DVB/CAR/PDMS	HS/30 min/50 °C	[130]
VOCs	Fruits of *Peucedanum alsaticum* L.	30 μm PDMS	HS/30 min/room temperature	[131]
VOCs	Longan	75 μm CAR/PDMS	HS/1 h/room temperature	[132]
VOCs	Melon fruit	50/30 μm DVB/CAR/PDMS	HS/30 min/35 °C	[133]
VOCs	Peanuts	50/30 μm DVB/CAR/PDMS	HS/30 min/40 °C	[134]
VOCs	Durian	30 μm DVB/CAR	HS/10 min/28 °C	[135]
VOCs	Hot peppers *Capsicum annuum* L.	50/30 μm DVB/CAR/PDMS	HS/15 min/40 °C	[136]
VOCs	Cacao beans	65 μm PDMS/DVB	HS/10 min/60 °C	[137]
VOCs	Apples	50/30 μm DVB/CAR/PDMS	HS/30 min/50 °C	[138]

(continued)

Table 9.1 (continued)

Analytes	Samples	Device used	Extraction mode/time/temperature	References
VOCs	Cucumber fruit	75 μm CAR/PDMS	HS/30 min/50 °C	[139]
VOCs	Litchi	50/30 μm DVB/CAR/PDMS	HS/30 min/40 °C	[140]
VOCs	*Fragaria chiloensis* fruits	65 μm PDMS/DVB	HS/30 min/20 °C	[141]
VOCs	Peaches and nectarines	65 μm PDMS/DVB	HS/30 min/45 °C	[142]
VOCs	Apples	100 μm PDMS 65 μm CW/DVB	HS/30 min/25 °C	[143]
VOCs	Juice and peel of oranges	50/30 μm DVB/CAR/PDMS	HS/40 min/40 °C	[144]
VOCs	*Peucedanum cervaria* fruits	30 μm PDMS	HS/30 min	[145]
Active compounds	Tomato	50/30 μm DVB/CAR/PDMS	HS/40 min/55 °C	[146]
VOCs	Strawberry fruits		HS/40 min/room temperature	[147]
VOCs	Asturian cider apples	65 μm PDMS/DVB	HS/10 min/60 °C	[148]
VOCs	Different plant parts of grapefruit	100 μm PDMS	HS/15 min/room temperature	[149]
VOCs	Leaves and fruits	65 μm PDMS/DVB	HS/20 min/40 °C	[150]
VOCs	Leaves and fruits	65 μm PDMS/DVB	HS/60 min/40 °C	[151]
Aroma compounds	Buckwheat	50/30 μm DVB/CAR/PDMS	HS/15 min/50 °C	[152]
VOCs	Soursop	75 μm CAR/PDMS	HS/30 min/25 °C	[153]
VOCs	Tomato cultivars	100 μm PDMS	HS/30 min/30 °C	[154]
VOCs	Capsicum fruits	50/30 μm DVB/CAR/PDMS	HS/40 min/40 °C	[155]
VOCs and polyphenols	"Wanmi" peaches	65 μm PDMS/DVB	HS/30 min/45 °C	[156]
VOCs	Apricot fruit	50/30 μm DVB/CAR/PDMS	HS/15 min/35 ± 0.5 °C	[157]
VOCs	Green-ripe banana fruit	60 μm PDMS/DVB	HS/4 min/20 °C	[158]
VOCs	Three table grapes puree	65 μm PDMS/DVB	HS/20 min/40 °C	[159]
VOCs	Fresh red and white guava fruit meat	100 μm PDMS	HS/15 min/45 °C	[160]

(continued)

Table 9.1 (continued)

Analytes	Samples	Device used	Extraction mode/time/temperature	References
VOCs	Pulp and skin of *Vitis vinifera* L. grape varieties	50/30 μm DVB/CAR/PDMS	HS/45 min/60 °C	[161]
Aroma volatiles	Muskmelon fruits	100 μm PDMS	HS/15 min	[162]
Aroma potential	Andean blackberry purees	65 μm PDMS/DVB	HS/20 min/60 °C	[163]
Aroma volatiles	Rices of Indian scented rice cultivars	50/30 μm DVB/CAR/PDMS	HS/10–30 min/70–100 °C	[164]
Aroma profile	Purees of plant matrices	50/30 μm DVB/CAR/PDMS	HS/45 min/50 °C	[165]
VOCs	Stoned table olives	65 μm PDMS/DVB	HS/20 min/40 °C	[166]
VOCs	Garlic	85 μm PA	HS/15 min/30 °C	[167]
VOCs	Garlic	75 μm CAR/PDMS	HS/30 min/30 °C	[168]
VOCs	Leaves and fruits of *Mangifera indica* var. *coquinho*	35 μm NiTi-ZrO2/PDMS	HS/45 min/40 °C	[169]
VOCs	Strawberry fruits	85 μm CAR/PDMS	HS/10–20 min	[170]
VOCs	Chili pepper fruits	100 μm PDMS 50/30 μm DVB/CAR/PDMS	HS/30 min/40 °C	[171]
VOCs	Apple	75 μm DVB/CAR/PDMS	HS/10 min/40 °C	[172]
VOCs	Cantaloupe fruits	100 μm PDMS	HS/15 min/30 °C	[173]
VOCs	Apples	65 μm PDMS/DVB	HS/25 min/room temperature	[174]
VOCs	Two fresh pineapple varieties	65 μm PDMS/DVB	HS/40 min/25 °C	[175]
VOCs	Peach fruit	65 μm PDMS/DVB	HS/10 min/50 °C	[176]
VOCs	Fruit of *Pyrus ussuriensis* cultivars	65 μm PDMS/DVB	HS/30 min/40 °C	[177]
VOCs	Banana	65 μm PDMS/DVB	HS/60 min/50 °C	[178]
VOCs	*Vitis vinifera* L. grapes	50/30 μm DVB/CAR/PDMS	HS/45 min/60 °C	[179]
VOCs	Fruit	50/30 μm DVB/CAR/PDMS	HS/40 min/40 °C	[180]
Sesquiterpene	Grape	85 μm PA	HS/10 min/45 °C	[181]

(continued)

Table 9.1 (continued)

Analytes	Samples	Device used	Extraction mode/time/temperature	References
VOCs	Fruits of roasted plantains	50/30 μm DVB/CAR/PDMS	HS/10 min/40 °C	[182]
VOCs	New passion fruit	100 μm PDMS	HS/10 min/30 °C	[183]
VOCs	Apricot	75 μm CAR/PDMS	HS/30 min/40 °C	[30]
VOCs	Tomato	75 μm CAR/PDMS	HS/20 min/60 °C	[184]
VOCs	Thapsia	50/30 μm DVB/CAR/PDMS	HS/40 min/60 °C	[185]
VOCs	Plums	65 μm PDMS/DVB	HS/20 min/40 °C	[31]
Hormones	Tomato	β-cyclodextrin (β-CD)-modified carbon nanotubes (CNTs) and a hollow fiber	1-octanol/60 min/1150 rpm	[32]
Essential oil and antioxidants	Fruit	PDMS-DVB	HS/30 min/50 °C	[186]
VOCs	Malaysian durians	50/30 μm DVB/CAR/PDMS	HS/30 min/40 °C	[187]

9.2.2 Flowers

The flowers' fragrance plays an important role in almost branches of present-day life and industry, such as foods, aroma therapy, pharmaceutical, cosmetics, and household products. Due to the volatile characteristic of aroma compounds and the complexity of matrices, SPME is currently the predominant sample preparation method for the analysis of different volatile compounds as well as environmental contaminants in aromatic or medicinal flowers owing to its superior performance [13, 55, 56].

HS-SPME coupled with GC-MS was widely applied to the analysis of aroma-active compounds. Cullere et al. found that safranal is the most important aroma compounds of the saffron sample by HS-SPME-GC-MS [57]. The VOCs emitted from flowers of different plants, such as *Patrinia scabiosifolia* [56], *Tecucrium massiliense* L. [58], *Catharanthis roseus* (L.) G. Don [59], *Jasminum sambac* (L.) Ait. [60], *Murrarya paniculata* (Linn.) Jack [61], and *Chrysanthemum* [62], were also identified by SPME approach. In these cases, alcohols, esters, aromatics, and especially terpenoids, such as monoterpenes, sesquiterpenes, and their oxygenated compounds, were the major components found in the flowers.

The variations of VOCs from flowers could be influenced by many factors, such as abiotic factors (season, temperature, water, etc.), cultivars, plant species, or different stages of flowering. A simple and solvent-free method using HS-SPME (85 μm CAR/PDMS) and GC-MS was developed for the determination of more than 90 volatile compounds from fresh flowers and nectar of male sterile and fertile lines of *Allium cepa* L. The result showed that a great difference was found for the number of bee visits and seed yield among the onion lines analyzed [55]. The distinctions of the volatile profile of rose petals under different treatment temperatures were measured by Rizzolo et al. The PDMS/DVB implemented at 60 °C was the most suitable extraction phase to sample the rose VOCs, not only from raw petal, but also from processed rose petals and the food product, compared with PDMS, CAR/PDMS, and DVB/CAR/PDMS fibers [63]. Also, the components' difference of VOCs between the different part of flowers [64], liquid essential oils and flowers [65], and the flowers from different cultivars [66] was compared by HS-SPME and between the flowers from different cultivars [66]. In addition, the reliable SPME technique was also employed to further confirm the accuracy of newly developed approach for VOCs determination from lilac blossom, named surface acoustic wave sensor (SAW) [67].

Besides the applications on flower VOCs analysis, a research on environmental contaminants in flower was reported [13]. In this study, three types of solgel-based SPME fibers were fabricated to extract PAHs from saffron samples under direct immersion (DI) mode. The results showed that the DI-SPME method exhibited enumerated advantages to the determination of the PAHs at trace level, by using the self-made fibers, such as relative short extraction time and low detection limits. Furthermore, a new poly(acrylamide-co-ethylene glycol dimethacrylate) (poly (AM-co-EGDMA) coating was prepared for the preconcentration of 24-epibrassinolide (24-epiBL), which represents a new sixth class of plant hormones with wide occurrence in plant samples, from pollen samples [68].

Not only in vitro SPME technique, but also the recently developed in vivo SPME technique is used for the determination of VOCs in living plants, which causes a more accurate result and enables the capture of more representative metabolites, presenting opportunities to detect the low-abundance, short-lived, and unstable compounds [69]. A commonly used device for in vivo sampling of flower scent with SPME is shown in Fig. 9.2 [70]. A glass funnel was placed around a developing flower, and aluminum foil was placed as an enclosure to limit the air movement. The SPME device was fixed besides the glass funnel, with the hollow SPME needle stuck through the aluminum foil, exposing the fiber close to the flower. Through the study of scent production by Petunia Mitchell flowers, Verdonk et al. demonstrated that SPME coupled to GC-MS is an outstanding tool to measure the volatile production by flowers in vivo, and the similar result was confirmed by Manzo et al. [71]. Song et al. analyzed the VOCs emitted from living daffodil flowers with in vivo SPME by the similar device [72]. Based on the multiple fibers, Caceres et al. maximized the discovery of VOCs in tomato flower

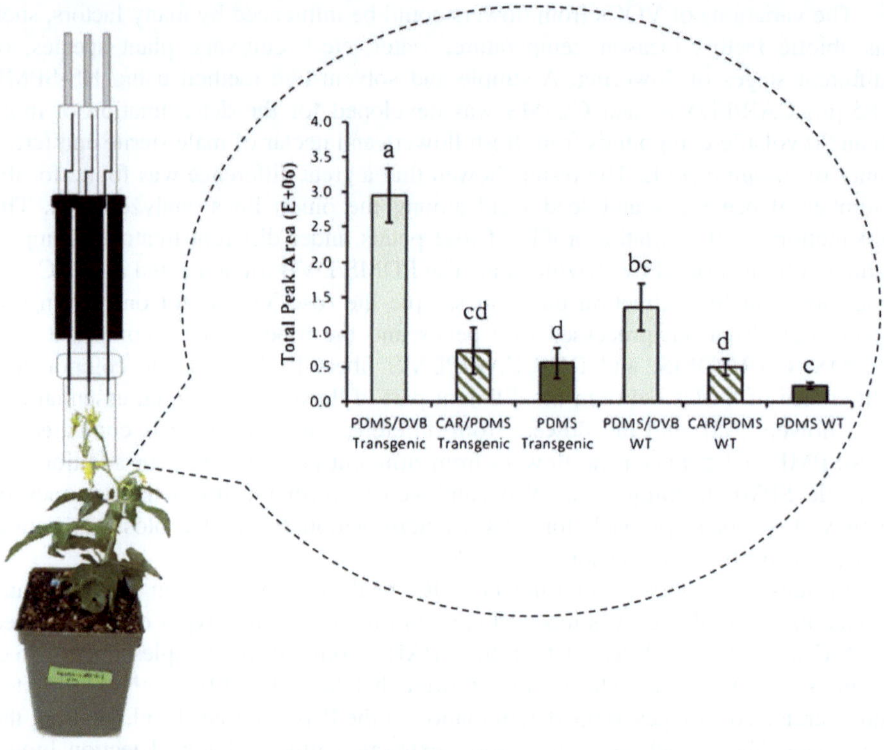

Fig. 9.1 Experimental setup for the in vivo sampling of Micro-Tom flowers. VOCs were accumulated in the vial for 24 h prior to the introduction of three SPME fibers (extraction time was 20 min). Reprinted with the permission from Ref. [73].Copyright 2015 NRC Research Press

[73]. The three fibers used in this work were polydimethylsiloxane (PDMS), PDMS/divinylbenzene (DVB), and carboxen (CAR)/PDMS (Fig. 9.1). The largest amounts of volatile organic compounds (VOCs) were observed with the PDMS/DVB fiber for both wild-type and transgenic plants, but the CAR/PDMS and PDMS fibers contributed to the detection of selective compounds. Data revealed the presence of 45 VOCs from transgenic plants and 35 from the wild-type plants when all three fibers were used together.

As mentioned above, SPME becomes a perfect method for the analysis of volatile and nonvolatile organic compounds in flowers owing to its convenient and fast sampling procedure. The combination of in vivo SPME and metabolic engineering or genetic engineering in the analysis of targeted compounds would improve the research efficiency significantly and bring about more detail at the molecular level simultaneously.

Table 9.2 presents the relatively recent applications of SPME for flower studies.

Table 9.2 Applications of SPME for flower studies in recent years

Analytes	Samples	Device used	Extraction mode/time/temperature	References
VOCs	*Murrarya paniculata* (Linn.) Jack flowers	100 μm PDMS	HS/90 min/room temperature/in vivo	[61]
VOCs	Chamomile flower heads	50/30 μm CAR/DVB/PDMS 100 μm PDMS	HS/30 min/80 °C	[188]
VOCs	*Chrysanthemum* flowers	100 μm PDMS	HS/15 min/70 °C	[62]
VOCs	Rose petals	65 μm PDMS/DVB	HS/30 min/60 °C	[63]
VOCs	*Chimonanthus praecox* flowers	75 μm PDMS/CARk	HS/30 min/15 °C	[69]
VOCs	Chinese daffodil flowers	85 μm PDMS/CAR	HS/15 min/22 ± 2 °C/in vivo	[70]
VOCs	*Jasminum sambac* (L.) Ait. (Oleaceae) flowers	100 μm PDMS	HS/60 min/30 ± 2 °C	[60]
VOCs	Lilac blossom	50/30 μm DVB/CAR/PDMS	HS/60 min/23 °C	[67]
VOCs	*Teucrium flavum* L. subsp flavum	100 μm PDMS 65 μm PDMS/DVB 50/30 μm DVB/CAR/PDMS	HS/30 min/60 °C	[64]
VOCs	*Catharanthus roseus* (L.) G. Don	50/30 μm DVB/PDMS	HS/60 min/50 °C	[59]
VOCs	*Delphinium elatum* L. "Blue Bird"	100 μm PDMS	HS/8 h/room temperature/in vivo	[189]
VOCs	*Allium cepa* L.	85 μm PDMS/CAR	HS/30 min/30 °C	[55]
VOCs	*Teucrium massiliense* L.	30 μm DVB/CAR/PDMS	HS/30 min/70 °C	[58]
VOCs	Flowering rice panicles	100 μm PDMS	HS/30 min/22 ± 1 °C	[190]
VOCs	*Brunfelsia calycina* petals	65 μm PDMS/DVB	HS/30 min/room temperature	[73]
VOCs	Flowers of two kinds of roses	75 μm CAR/PDMS	HS/6 h	[8]
VOCs	The inflorescences of Brazilian chamomile	65 μm CW/DVB	HS/60 min/40 ± 0.1 °C	[9]
VOCs	Flowers of *Patrinia scabiosifolia*	PDMS/DVB	HS/30 min/80 °C	[56]
VOCs	Spanish saffron	100 μm PDMS	HS/20 min/36 °C	[57]
VOCs	Saffron	8 μm PDMS 25 μm PEG 40 μm PEG/CNTs	DI/PDMS and PEG fibers 60 min and PEG/CNTs 40 min/20 ± 1 °C	[13]

(continued)

Table 9.2 (continued)

Analytes	Samples	Device used	Extraction mode/time/temperature	References
VOCs	Hallabong (*Citrus kiyomi* × *Citrus ponkan*) Blossom	100 μm PDMS 85 μm CAR/PDMS	HS/30 min/30 °C	[65]
VOCs	Exocarpium Citri Grandis	50/30 μm DVB/CAR/PDMS	HS/80 min/40 °C	[35]
VOCs	Citrus	50/30 μm DVB/CAR/PDMS	HS/30 min/40 °C	[39]

9.2.3 Leaves

The screen and profile of volatiles emitted from leaf were capable of assessing the effective components [74], distinguishing subspecies [75], explaining the insect attracting and expelling effects [76], and exploring environmental effects on volatile emission [77–80]. Just like the application of SPME in any other parts of a plant, a satisfactory efficiency should be ensured prior to the application of SPME to leaves. Microwave-assisted SPME was conducted to extract the volatile compounds more sufficiently [81]. Multiple fibers were evaluated to be a more powerful approach for extracting a wide range of polarities [82]. On most occasions, PDMS/DVB and PDMS/carboxen were satisfactory choices. Schossler et al. demonstrated that SPME is a more suitable tool to capture VOCs emitted from leafs compared with conventional hydrodistillation method, in which the hydrodistillation method was more likely to cause the volatile component degradation due to the high-temperature treatment [83]. Wakte et al. identified 31 volatiles in Pandanus amaryllifolius under optimized extraction parameters, of which 21 compounds were reported for the first time, which demonstrate the superiority of SPME [84]. To enhance the extraction capacity, the extraction temperature needs to be optimized when HS-SPME is used. When the DI-SPME is used, stirring is also useful to facilitate the extraction if the powdered leaf sample is preliminarily immersed in solvent [78]. Temperature optimization and stirring are also used for the analysis of other plant organs by SPME.

The leaves of different plant species and subspecies could release various volatile compounds. Many works studied the differences of the released volatile fingerprints, to explore the different potential applications or to distinguish each species especially on some occasions that the difference in morphology is inconspicuous. A HS-SPME technique combined with gas chromatography mass spectrometry is reported to distinguish the volatile metabolomic profile of fresh leaves from *Mentha* × *piperita* L. and *Mentha spicata* L. aromatic plants. The results suggested that the HS-SPME/GC-qMS methodology could be used to establish the similarities and differences between both investigated Lamiaceae plants and could probably be used for plant differentiation based on the volatile metabolomic profile [85]. The other efforts to distinguish the indigenous and cultivated citrus species in

Taiwan [75], the cultivar Anamed (A3), and *Artemisia annua* L. wild-type species [86] were reported using HS-SPME techniques. Furthermore, three different Chamaecyparis species (Cupressaceae) grown in Taiwan could be distinguished based on the volatile differences from their mature leaves through the GC-MS analysis [87].

The volatile compounds released from plant leaves are influenced by many factors, such as infection, health state, and seasons (temperature). Smith et al. measured the VOC emissions of five species of plants in the subtribe Centaureinae (Asteraceae). Leaves of each plant were punctured with a needle, and the VOCs were collected by solid-phase microextraction (SPME) periodically over 48 h and analyzed by GC–MS. The result demonstrated that damage caused an immediate increase of 200–600% in the composition of VOCs emitted from each plant species, and the amounts generally remained high for at least 48 h [77]. In this study reported by Köllner et al., HS-SPME with PDMS coating was adopted to study the patterns of volatile sesquiterpene emission throughout the damaged leaf and in neighboring leaves. Sesquiterpene volatiles were released at high rates from damaged leaves, but at much lower rates from neighboring leaves [88]. The composition of volatile organic compounds emitted by in vitro shoots of *Agastache rugosa* (Fischer & C.A. Meyer) O. Kuntze (Lamiaceae) was studied using HS-SPME-GC-MS and compared to those emitted by adult plants and in vitro-germinated seedlings [78]. The volatile organic compounds occurring in plant leaves to detect plant stress before visible sign were monitored by HS-SPME, in which 47 metabolites were confirmed with authentic standards and of which 19 were described for the first time [79]. Commercial CAR/DVB/PDMS fiber equipped on an autosampler was used to conduct SPME. In another research, the influence of heating temperature and duration on volatile compounds of *P. pubescens* shoots was studied through SPME. The results showed that the major volatile compound in winter bamboo shoots when heated at 100 °C is n-heneicosane, whose relative content rises with increasing heating duration. On the other hand, spring bamboo shoots when heated at 100 °C are composed mainly of hydrocarbon, aldehyde, and alcohol compounds, totaling 80.59%. After 60-min heating, benzyl salicylate becomes the major volatile compound [80].

In most of the analysis mentioned above, the leaves were chopped or ground, or even preliminarily extracted by hydrodistillation prior to extraction with SPME fibers. However, such pretreatments may cause the loss or degradation of the VOCs [83]. For example, Huang et al. found that the volatile components of powdered leaves and fresh leaves of *Perilla frutescens* were different. And a different number of volatiles and a different content of main components were also detected for powdered leaves (treated with liquid nitrogen) and fresh leaves [89]. Yassaa and Williams successfully analyzed the enantiomeric monoterpene emissions from natural Scots pine through HS-SPME and a gas exchange system [90]. A significant difference of enantiomeric ratios in natural emissions was observed compared with detached needles from Scots pine. The work provided evidence that in vitro analysis may cause unreal profiles of live vegetation, and in vivo analysis should be more desirable. Ten *W. glabra* plants from Nepal were sampled in vivo in order to evaluate the volatile organic compound (VOC) fingerprint emitted by living plants, which were enclosed in a customized 11 cm × 21 cm Teflon cage [91]. Similarly, SPME device with a customized collection bag was applied to in situ collect volatiles from intact leaves of *Centaurea solstitialis* and mechanically damaged leaves, and the different VOCs profiles were captured [92].

The recently established in vivo SPME was proposed to sample VOCs emitted from leaves of plants, aimed to obtain the more accurate profiles. Chen et al. used 65 μm PDMS/DVB fiber for the detection of volatile compounds from leaves of living Calocedrus macrolepis [93]. Branch was enclosed in Teflon film and a gas flow flowed through the chamber. The branch chamber was shaded with aluminum foil to investigate light and temperature dependencies of primary emissions. Compared with traditional air sampling and analysis methods, the developed protocol was thought superior since it eliminated the loss of these compounds to the walls of sampling lines and instrument parts and also eliminated the need for solvents (Fig. 9.2). To measure the carbamate pesticides in the leaf, PDMS/DVB fiber was directly inserted in the leaf of live jade plant [20]. Compared with microdialysis, in vivo SPME was demonstrated to be more sensitive, precise, and accurate. For VOC analysis, SPME was successfully applied for in vivo contaminant or endogenous nonvolatile component tracing in plant. Chen et al. used 45 μm PDMS fiber for long-time continuous monitoring of synthetic musks in different organs of aloe leaf, and the feasibility was demonstrated by comparing with traditional extraction method [94]. Furthermore, an effort focused on studying the effect of nanomaterials on contaminants' environmental fate in plant leaf was made using in vivo SPME [95]. Importantly, the novel SPME probe based on phenylboronic acid-functionalized carbon nanotubes was proposed or ultrasensitive carbohydrate determination in living plant leaf without any expensive enzymes or tedious pretreatment procedure (Fig. 9.3) [96].

Table 9.3 presents some recent applications of SPME for leaf studies.

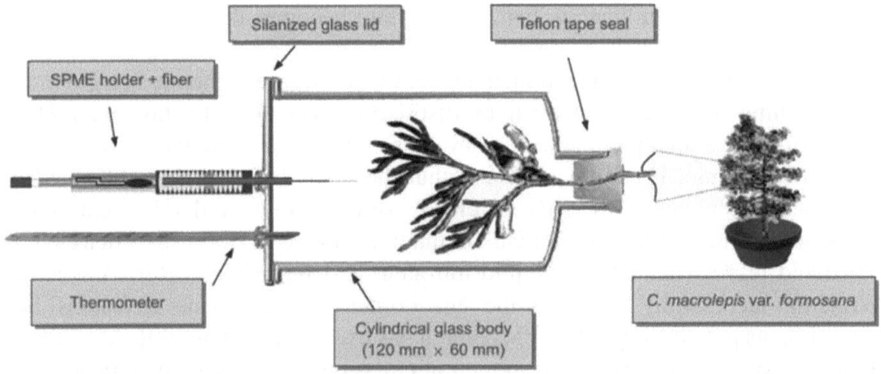

Fig. 9.2 Schematic diagram of SPME sampling of volatiles emitted from *C. macrolepis* var. *formosana* live leaves. Reprinted from Ref. [93]. With a kind permission from Springer Science +Business Media

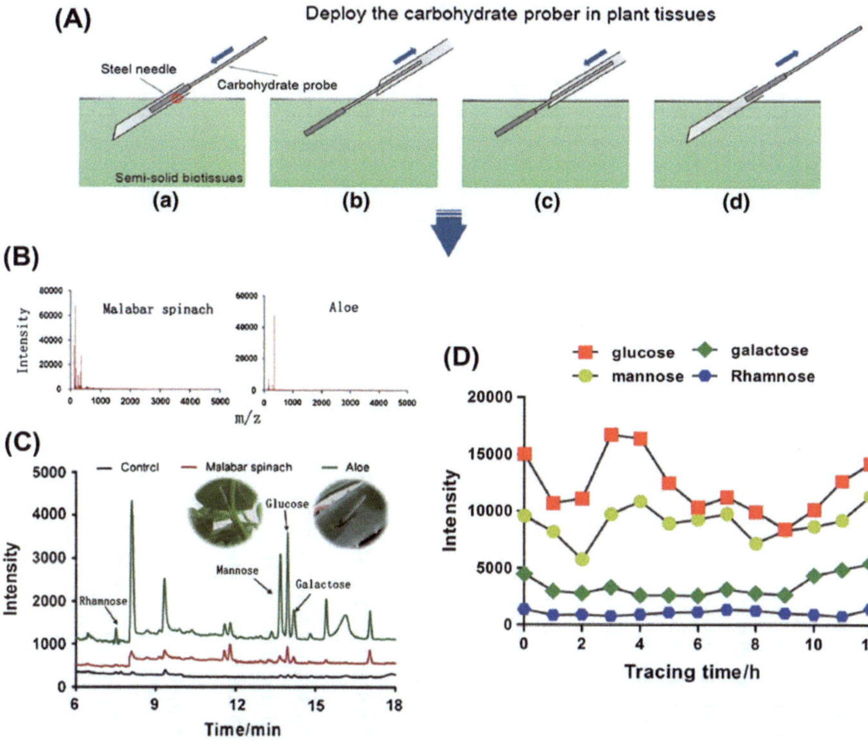

Fig. 9.3 **A** The in vivo sampling procedure in plant tissues. The carbohydrate probe is deployed under the guidance of a steel needle (*a*), removal of the steel needle and exposure of the probe to the carbohydrate in plant tissues (*b*), the steel needle is carefully put back in the plant tissue at the end of sampling (*c*), the carbohydrate probe is removed (*d*). The probe was readily inserted into or removed from plant tissues; **B** biologic macromolecule analysis in the eluent with MALDI-TOF MS; **C** carbohydrate assay in the stem of Malabar spinach and leaf of aloe. The signals of the corresponding carbohydrate were apparent; **D** in vivo continuous carbohydrate monitoring in aloe leaf using the proposed probe. Reproduced from Ref. [96] with the permission from the Royal Society of Chemistry

9.2.4 Stems, Roots, and Seeds

The applications of SPME technique, focusing on the stems, roots, and seeds, were relatively absent compared with the analysis of volatile compounds from fruits, flowers, and leaves.

Similar to the cases described above, a satisfactory efficiency should be ensured prior to the application of SPME to stems, roots, and seeds. Wang et al. compared the performance of HS-SPME method and hydrodistillation for volatile compound extraction from cinnamon bark. The result showed that the extracted components of the two methods were significantly different [97]. More complete compositions were extracted by hydrodistillation. However, SPME method was thought to only

Table 9.3 Applications of SPME for leaf studies in recent years

Analytes	Samples	Device used	Extraction mode/time/temperature	References
Volatiles	Leaves	PDMS 100 μm	HS/30 min	[74]
(±)-Methyl jasmonate	Fresh leaf material of Jasminum	100 μm PDMS	HS/15 min/60 °C	[191]
Volatiles	Fresh leaves	100 μm PDMS	HS/15 min/70 °C	[75]
VOCs	Seeds of *Artemisia annua* cultivars	100 μm PDMS	HS/5 min/35 °C	[86]
VOCs	Leaves of three Chamaecyparis species	65 μm PDMS/DVB	HS/15 min/50 °C	[87]
VOCs	Allium plant leaves	100 μm PDMS	HS/10 min/room temperature	[192]
Terpenoid volatiles	Eastern hemlock branch	100 μm PDMS	HS/15 min/50 °C	[193]
VOCs	Pine needles	100 μm PDMS	HS/45 min/room temperature	[76]
BVOCs	Leaves	7 μm PDMS	HS/30 min/30 °C	[83]
VOCs	Leaves of five species	100 μm PDMS/in vivo	HS/55 min/room temperature	[77]
VOCs	Shoots of *Agastache rugosa* (Fischer & C. A. Meyer) O. Kuntze	50/30 μm DVB/CAR/PDMS	HS/15 min/60 °C	[78]
VOCs	Leaves of *Vitis vinifera*	50/30 μm DVB/CAR/PDMS	HS/60 min/90 °C	[79]
VOCs	*Phyllostachys pubescens* shoots	65 μm PDMS/DVB	HS/30 min	[80]
VOCs	*Toona sinensis* (A. Juss.) Roem complete plantlet	65 μm PDMS/DVB	MAE-HS	[81]
VOCs	Maté leaves and twigs	Various types of fibers	HS/60 min/80 °C	[82]
2-Acetyl-1-pyrroline	Leaves	50/30 μm CAR/DVB/PDMS	HS/35 min/80 °C	[84]
VOCs	Zisu and Baisu powder	75 μm CAR/PDMS	HS/20 min/45 °C	[89]
Enantiomeric monoterpene	Scots pine branch	65 μm PDMS/DVB	HS/1 min/0–5 °C	[90]

(continued)

Table 9.3 (continued)

Analytes	Samples	Device used	Extraction mode/time/temperature	References
VOCs	Centaurea	100 μm PDMS	HS/60 min	[92]
BVOCs	Leaves of live *C. macrolepis* var. *formosana*	PDMS/DVB	HS/15 min/21.8 °C	[93]
Carbofuran, carbaryl	Leaf of the jade plant	PDMS	DI/30 min/room temperature	[20]
Aloe	Synthetic musks	45 μm PDMS	DI/20 min/room temperature	[94]
Leaf mustard	Eight contaminants	45 μm PDMS	DI/20 min/room temperature	[95]
Aloe	Carbohydrate	Phenylboronic acid-functionalized carbon nanotubes	DI/40 min/room temperature	[96]
Isothiocyanates	Leaves of Lepidium	65 μm PDMS−DVB	HS/35 °C	[194]
Acidic phytohormones	Aloe leaf	Polyaniline-sheathed electrospun nanofiber	DI/20 min	[195]
VOCs	Eucalyptus	100 μm PDMS	HS/30 min/41.5 °C	[196]
28-epihomobrassinolide	*Arabidopsis thaliana*	In tube SPME with a poly (MAA-co-EDMA) monolith	5 min/PH 6.0	[197]
VOCs	Plant clone	50/30 μm DVB/CAR/PDMS	HS/30 min/45 °C	[198]
VOCs	Leaves of Lamiaceae	50/30 μm DVB/CAR/PDMS	HS/60 min/40 °C	[85]
VOCs	Maize leaves	100 μm PDMS	HS/60 min/40 °C	[199]
Essential oil	Plinia cerrocampanensis leaf	50/30 μm DVB/CAR/PDMS	HS/10 min/49 °C	[200]

extract the readily volatile compounds. In another research, the composition of essential oil extracted from *Rhaponticum acaule* L. roots growing wild in Algeria was studied by hydrodistillation and by HS-SPME, and the results showed that HS-SPME extraction could be considered as an alternative technique for the isolation of volatiles from plant [98]. Zhang et al. optimized SPME conditions from the aspects of extraction temperature, extraction duration, and selection of SPME fibers and then compared the components extracted by the developed HS-SPME and steam distillation (SD) [99]. The study demonstrated that HS-SPME method used less sample amount, shorter extraction time, and simpler procedure than SD, and 73 compounds were detected by HS-SPME, while 61 compounds were detected by SD.

As the extraction performance was satisfied, more and more researchers adopt SPME technique to VOC analysis in stems, roots, or seeds. Distillation-extraction with mixed solvent and HS-SPME were compared by Riu-Aumatell et al. through the extraction of volatile compounds in ginseng [100]. HS-SPME was used to extract volatile compounds from *Panax ginseng* C.A. Meyer cultured with 3 different cultivation methods [101]. It was concluded that the organic and good agriculture practice cultivation can increase the ginseng characteristic volatile compounds. In other works, Yu et al. used microwave distillation-assisted SPME to extract volatile constituents in ginger [102], and volatile constituents from *G. glabra*, *G. inflata*, and *G. echinata* roots were profiled using steam distillation and solid-phase microextraction by Farag et al. [103].

Mebazaa et al. studied the volatile compounds in fenugreek seeds with two methods, solvent extraction and static HS-SPME. For static HS-SPME protocol, the seeds were ground and two kinds of fibers were compared. DVB/carboxen/PDMS fiber was confirmed to possess high extraction efficiency to the odorous compounds of fenugreek seeds [104]. Hashemi et al. functionalized SBA-15 with 3-[bis (2-hydroxyethyl)amino]propyl-triethoxysilane and then glued the functionalized particles to copper wire to develop a new SPME fiber to study the essential oil composition of cumin seeds collected from different climates. The homemade fiber was demonstrated to be superior to commercial PDMS fiber in the work [105].

Assisted by the SPME-GC-MS method, Azarnia and coworkers studied the impacts of crop year and processing [106] and storage conditions [107] on volatile flavor profiles of pea. Tamogami et al. investigated the biosynthetic pathway of methyl (E)-2-hexenoate in Achyranthes bidentata plant and suggested that Achyranthes bidentata possess a specific pathway for the production of methyl (E)-2-hexenoate (1) from (Z)-2-hexenol in response to methyl jasmonate [108]. In another work, Min used SPME-GC-MS method and resazurin reagent to study ethanol released from aged radish seeds [109].

With sampling rate calibrated in vivo SPME, Chen et al. successfully traced the uptake and depuration of synthetic musks in the root and stem of aloe plant [94]. In the work, a PDMS fiber was directly inserted in the root and stem of aloe plant. After the duration of 20 min, the fiber was removed and rinsed with deionized water, dried using a Kimwipe tissue, and then assembled into a recycled SPME fiber to be directly introduced to GC-MS for analysis. Meanwhile, the accuracy of

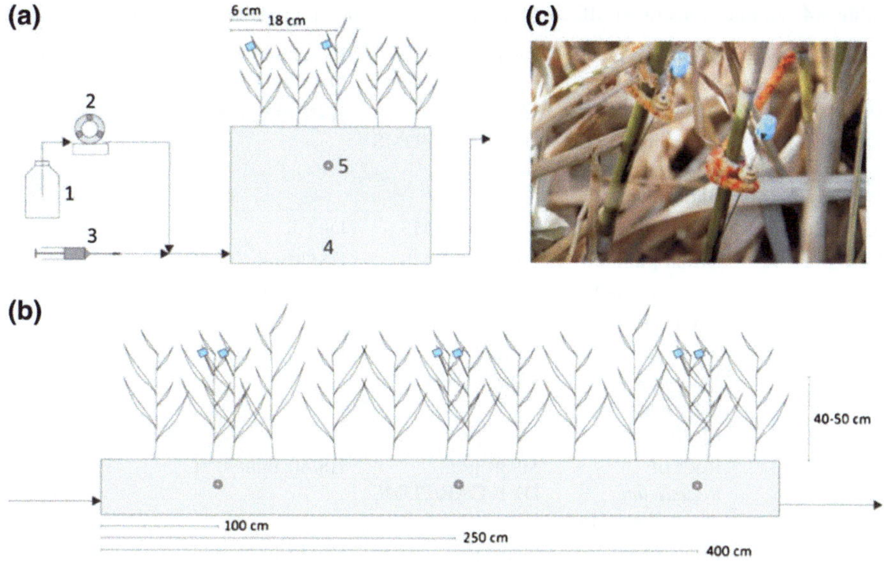

Fig. 9.4 Schemes of the experimental setup. **a** Laboratory setup(*1* nutrient solution reservoir, *2* peristaltic pump, *3* MTBE dosage by syringe pump, *4* laboratory microcosm, and *5* pore water sampling port) and in vivo SPME sampling locations. **b** Field setup and in vivo SPME and pore water sampling locations. **c** Inserted SPME assemblies during exposition in the field plot constructed wetland. Reprinted from Ref. [110] with the kind permission from Springer Science +Business Media

the proposed method was demonstrated by comparing with traditional extraction method. By this means, the concentration of synthetic musks was traced during a long time. Furthermore, in vivo SPME was applied to in suit monitor methyl tert-butyl ether (MTBE) in the common reed (*Phragmites australis*) from a phytoremediation site highly polluted with MTBE, with the SPME fiber directly introduced into the aerenchyma of the plant stem (Fig. 9.4) [110]. SPME was demonstrated to be a satisfactory method to observe the toxic effects on plants. Recently, SPME has been developed to be a powerful tool for phytoscreening, which is termed as a method to map subsurface contaminants by the analysis of contaminants in plants. Limmer et al. [111] and Holm et al. [112] detected the collected cores of trunk with HS-SPME approach, and abundant target compounds [tetrachloroethene and trichloroethene, and chlorinated volatile organic compounds (CVOCs)] were successfully captured with PDMS fibers. Their results suggested that SPME decreases the detection level considerably and increases the sensitivity of phytoscreening.

Table 9.4 presents some recent applications of SPME for stem, root, and seed studies.

Table 9.4 Applications of SPME for stem, root, and seed studies in recent years

Analytes	Samples	Device used	Extraction mode/time/temperature	References
VOCs	Cinnamon (*Cinnamomum Cassia* Presl) Bark powder	50/30 μm DVB/CAR/PDMS	HS/120 min/50 °C	[96]
Allelochemical	Tomato	100 μm PDMS	DI/1 h	[110]
VOCs	Urban plume	100 μm PDMS	HS/30 min/35 °C	[111]
CVOC	Tree core and birch sap	85 μm CAR/PDMS	HS/30 min/35 °C	[201]
VOCs	Roots of Echinops kebericho Mesfin	65 μm PDMS/DVB	HS/30 min/60 °C	[202]
VOCs	Root of *Ligusticum chuanxiong* Hort	50/30 μm DVB/CAR/PDMS	HS/40 min/90 °C	[98]
VOCs	Roots of white salsify	50/30 μm DVB/CAR/PDMS	HS/75 min/80 °C	[99]
VOCs	Panax	65 μm PDMS	HS/40 min/40 °C	[100]
VOCs	Ginger rhizome	65 μm PDMS/DVB	HS/2 min	[101]
VOCs	Seeds	50/30 μm DVB/CAR/PDMS	HS/60 min/22 ± 2 °C	[103]
VOCs	Seeds	100 μm PDMS	HS/28 min/70 °C	[104]
VOCs	Seeds of selected field pea cultivars	75 μm CAR/PDMS 75 μm DVB/CAR/PDMS	HS/60 min/40 °C	[105]
VOCs	Seeds of pea cultivars	75 μm CAR/PDMS	HS/30 min/50 °C	[106]
VOCs	Aged radish (*Raphanus sativus* L.) seeds	65 μm CAR/DVB	HS/24 h/25 °C	[108]
VOCs	Purees of plant matrices	50/30 μm DVB/CAR/PDMS	HS/10–60 min/5–60 ° C	[203]
Essential oil	*Rhaponticum acaule* L. roots	30 μm DVB/CAR/PDMS	HS/30 min/70 °C	[97]
VOCs	Jatropha ribifolia	CW-DVB, PDMS-DVB, and PDMS	HS/15 s/25 °C	[204]
Diterpenes	Leaves and stems of Stevia rebaudiana Bertoni	85 μm polyacrylate	HS/10 min/45 °C	[205]

9.2.5 Multiple Plant Organs and Whole Plants

SPME has also been used for volatile compound analysis in multiple plant organs, with the objective to study the difference of VOC profiles of different organs. Shanmugam et al. used HS-SPME to detect the composition of the main constituents from different plant parts (rhizome, stems, leaves, and flowers) of cultivated *Hedychium coronarium* Koenig, and the enantiomeric ratio of some of these was successfully measured [113]. Yamani et al. used optimized HS-SPME/GC-MS to identify the volatile bioactive compounds in the leaves, flower spikes, and, for the first time, the flower nectar of Australian-grown *A. rugosa*. Methyl chavicol was the predominant headspace volatile compound in the flowers with nectar, flower spikes, and leaves, with a total of 97.16, 96.74, and 94.35%, respectively. The results indicate that HS-SPME/GC-MS could be a useful tool for screening estragole concentration in herbal products [114]. Vázquez et al. demonstrated that HS-SPME method can achieve comparable results to those obtained by essential oil analysis, by using very fewer samples, a shorter extraction time, and a much simpler procedure [115]. In addition, HS-SPME was demonstrated to be a more desirable approach compared with hydrodistillation since HS-SPME was able to capture the volatiles not only emitted by specific plant parts, but also emitted by different parts within a single flower [64]. Table 9.5 lists the other applications of SPME method for the analysis of VOCs in multiple plant organs.

Dried or fresh (homogenized) plant samples were generally adopted for whole plant analysis, but living whole plants were also analyzed by SPME directly in some cases. For instance, a novel SPME configuration, namely electromembrane-surrounded solid-phase microextraction (EM-SPME), was developed for acidic herbicide monitoring in wheat plant tissues [116]. Only 20-min time frame was needed to reach the highest extraction efficiency of the analytes from a 24 mL alkaline sample solution across the organic liquid membrane and into the aqueous acceptor phase through a 50 V electrical field and to their final adsorption on a carbonaceous anode. In addition to high sample cleanup, which made the proposed method appropriate for the analysis of acidic compounds in a complicated media (plant tissue) (Fig. 9.5). In situ HS-SPME coupled to GC-MS approach was developed to measure sesquiterpene emission rates from gray pine and ponderosa pine. In this approach, a plastic bag was used to package the pine during sampling. The effects of sample linear velocity and sample relative humidity on sesquiterpene absorption were investigated with a capillary diffusion system [117]. Based on the in situ SPME sampling technique, Limmer et al. developed an *in planta* sampling method to obtain high-frequency measurements of chlorinated ethenes in oak (*Quercus rubra*) and bald cypress (*Taxodium distichum*) trees growing above a contaminated plume during a 4-year trial, which addressed the unexplained variability of contaminant concentrations in trees [118].

Table 9.5 Applications of SPME for multiple plant organs and whole plant studies in recent years

Analytes	Samples	Device used	Extraction mode/time/temperature	References
VOCs	Different plant parts of Citrus limon (Rutaceae)	100 μm PDMS	HS/20 min/room temperature	[206]
VOCs	Aerial medicinal plant	Anodized aluminum fiber	HS/30 min/70 °C	[115]
Sesquiterpenes	Plants	100 μm PDMS	HS/10 min/room temperature	[117]
VOCs	Flowers, leaves, and stems	50/30 μm DVB/PDMS	HS/60 min/50 °C	[207]
VOCs	Flowers, leaves, and stems	30 μm DVB/CAR/PDMS	HS/15, 30, and 45 min	[208]
VOCs	Rice	100 μm PDMS	HS/30 min/22 ± 1 °C	[209]
VOCs	Different organs and pollen of Cistus albidus	100 μm PDMS	HS/15 min/room temperature	[210]
VOCs	Flowering aerial parts, leaves, stems, and flowers	100 μm PDMS 65 μm PDMS/DVB 50/30 μm DVB/CAR/PDMS	HS/30 min/60 °C	[14]
VOCs	Leaves and fruits of *Mangifera indica* var. *coquinho*	35 μm NiTi-ZrO2/PDMS	HS/45 min/40 °C	[211]
VOCs	Chamomile flower heads, sage leaves	50/30 μm CAR/DVB/PDMS 100 μm PDMS	HS/30 min/30 °C	[212]
2-acetyl-1-pyrroline	Leaves, stems, and panicles	50/30 μm DVB/CAR/PDMS	HS/20 min/80 °C	[213]
VOCs	Stems and leaves	65 μm PDMS/DVB	HS/3 min/250 °C	[214]
VOCs	*Adenostyles briquetii* Gamisa	30 μm DVB/CAR/PDMS	HS/varies in time and temperature	[215]

(continued)

Table 9.5 (continued)

Analytes	Samples	Device used	Extraction mode/time/temperature	References
VOCs	Peppermint	50/30 μm DVB/CAR/PDMS	HS/30 min/27 °C	[216]
VOCs	Seeds	75 μm CAR/PDMS 65 μm CW/DVB 65 μm PDMS/DVB	HS/20 min/45 °C	[217]
VOCs	Leaves	65 μm PDMS/DVB	HS/5–60 min/25 °C	[218]
VOCs	Aerial plant parts of bastard balm	100 μm PDMS	HS/30 min/40 °C	[219]
Organophosphorus pesticides	Pakchoi	65 μm homemade OH-TSO/DVB	HS/50 min/75 °C	[220]
VOCs	Cultivated dumebuchu	100 μm PDMS 85 μm CAR/PDMS	HS/40 min/30 °C	[221]
Pyrethroids	Vegetables (whole plants)	IL-linked fiber	Immerse/50 min/50 °C	[222]
VOCs	Seeds of cultivated *Tagetes minuta* L	100 μm PDMS	HS/15,40 min/70 °C	[223]
VOCs	*Elsholtzia splendens*	65 μm PDMS	HS/40 min/40 °C	[224]
VOCs	*Thymus vulgaris* L.	30 μm CAR/PDMS	HS/30 min/25 °C	[225]
Organochlorine pesticides	Seaweed samples	65 μm PDMS/DVB	DI/60 min/room temperature	[226]
VOCs	Rosemary	7 μm PDMS	HS/45 s/25 °C	[227]
VOCs	Ground hay samples	70 μm CW/DVB	HS/8 min/50 °C	[228]
VOCs	BRC, BT, BRM, and BC	100 μm PDMS 65 μm DVB 75 μm CAR	HS/60 min/60 °C	[229]
VOCs	Subsp of *Melittis melissophyllum* L. subsp *melissophyllum* and subsp *albida* (Guss) P. W. Ball (Lamiaceae)	100 μm PDMS	HS/30 min/40 °C	[230]

(continued)

Table 9.5 (continued)

Analytes	Samples	Device used	Extraction mode/time/temperature	References
VOCs	Packaged alfalfa sprouts	75 μm CAR/PDMS.	HS/15 min/20 °C	[231]
Polycyclic aromatic hydrocarbons	Vegetables (whole plants)	70 μm benzoxy-C/OH-TSO	HS/40 min/60 °C	[232]
Pesticides	Lettuce	100 μm PDMS 85 μm PA 65 μm PDMS/DVB 50 μm CW/TPR	Immerse/30 min/room temperature	[13]
VOCs	Liverwort	65 μm PDMS/DVB	HS/10 min/40 °C	[233]
VOCs	Micropropagated plants of *Salvia stenophylla*	50/30 μm DVB/CAR/PDMS	HS/15 min/80 °C	[234]
VOCs	Full aerial parts and separated organs of fresh *T. marum*	30 μm CAR/PDMS	HS/30 min/70 °C	[235]
VOCs	Aerial parts of *Teucrium polium* L.	Nanoporous silica polyaniline SPME fiber	MA-HS/450 W/4 min	[236]
VOCs	Fresh fruits, dried seeds, wood, bark, and leaves of *Bixa orellana*	50/30 μm CAR/PDMS/DVB	HS/3 h/30 °C	[237]
Flavonoids	Aerial parts of *D. graveolens*	PDMS	HS/10 min/25 °C	[238]
VOCs	Aerial parts of *Tagetes filifolia* Lag	Not mentioned	HS/30 min/40 °C	[239]
VOCs	Whole plants	100 μm PDMS	HS/15 min/25 °C	[240]
VOCs	Leaves, flower spikes, and nectar	PA	HS/50 min/40 °C	[113]
VOCs	Flowering individuals of *C. odorum*	DVB-CAR-PDMS	HS/10 min/40 °C	[241]
VOCs	Whole Arabidopsis plants	65 μm PDMS/DVB	HS	[242]

(continued)

Table 9.5 (continued)

Analytes	Samples	Device used	Extraction mode/time/temperature	References
VOCs	Herbal plants	75 μm PDMS/CAR	HS/30 min/40 °C	[243]
Acidic herbicides	Whole wheat tissue	/	Electromembrane-surrounded SPME	[116]
Geosmin	Cores, roots, and leaves of *Beta vulgaris* ssp. *vulgaris*	PDMS/DVB	HS/10 min/60 °C	[244]
VOCs	Faba bean	50/30 μm DVB-CAR-PDMS	HS/60 min/50 °C	[245]
VOCs	Aerial parts of *E. platyloba*	Anodized aluminum SPME fiber	HS/30 min/70 °C	[119]

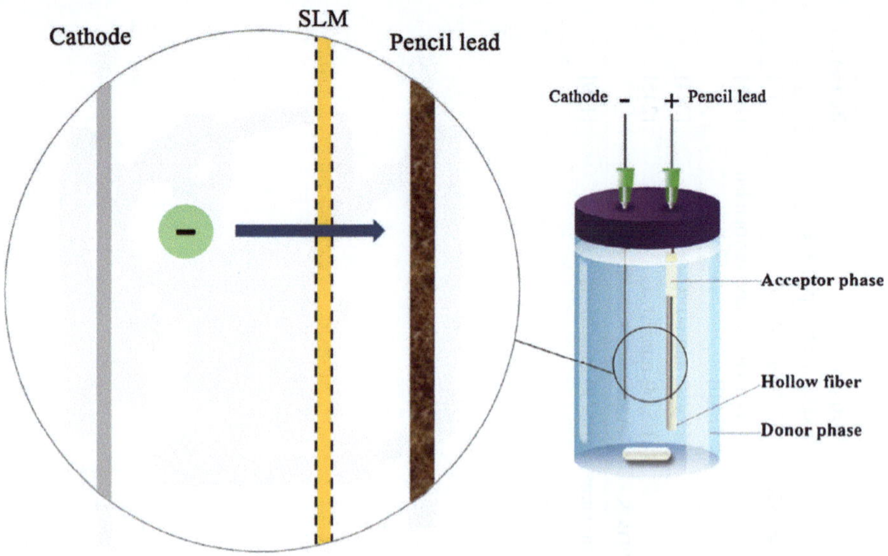

Fig. 9.5 Equipment used for the EM-SPME method and mechanism of transportation across liquid–liquid–liquid–solid boundaries. Reprinted with the permission from Ref. [116]. Copyright 2014 American Chemical Society

9.3 Conclusion

Solid-phase microextraction, combining sampling, isolation, concentration, and enrichment in one step, is a simple and effective sample preparation technique. It possessed advantages over conventional procedures, such as simple operation, low cost, low solvent consumption, speed, high enrichment, and ease-to-realize automation. Owing to the high flexibility, SPME is widely used in the analysis of volatile/nonvolatile compounds in complex matrix, such as plant tissue samples. As microextraction is one of the most important research and development frontiers in modern sample pretreatment, it could be expected that the applications of SPME technique on plant analysis will be extended to more and more endogenous and exogenous substances in the future.

In vivo SPME sampling was an emerging area of research in the past decades. Although in vitro method can address the requirement of most plant analysis, some researches demonstrated that in vivo sampling method could provide more accurate and reliable information of the distribution of target organic compounds, offering a more reliable profile in living biologic system [120–124]. Also, this method could long time trace the change of VOCs under different conditions, such as different seasons, temperatures, different stages of flowering, or caused by insect stimulating or bacteria infecting, without sacrifice of specimen. Some VOCs emitted from plants are important pheromones, which not only act as direct defense compounds but may also signal alarm to neighboring plants. By coupling with portable GC or

GC-MS, in vivo SPME sampling method could provide more abundant real-time information of plant pheromones, which was unable to be achieved by other methods.

Acknowledgement This research was supported by projects of National Natural Science Foundation of China (21377172, 21225731, 21477166, 21527813), and the NSF of Guangdong Province (S2013030013474)

We acknowledge financial support from the projects of NNSFC (Grants 21377172, 21677182)

References

1. Schiestl FP (2015) Ecology and evolution of floral volatile-mediated information transfer in plants. New Phytol 206:571–577
2. Oikawa PY, Lerdau MT (2013) Catabolism of volatile organic compounds influences plant survival. Trends Plant Sci 18:695–703
3. Bicchi C (2004) Special issue: analysis of flavors and fragrances. J Chromatogr Sci 42:401–401
4. Unsicker SB, Kunert G, Gershenzon J (2009) Protective perfumes: the role of vegetative volatiles in plant defense against herbivores. Curr Opin Plant Biol 12:479–485
5. Huang M, Sanchez-Moreiras AM, Abel C et al (2012) The major volatile organic compound emitted from Arabidopsis thaliana flowers, the sesquiterpene (E)-beta-caryophyllene, is a defense against a bacterial pathogen. New Phytol 193:997–1008
6. D'Alessandro M, Erb M, Ton J et al (2014) Volatiles produced by soil-borne endophytic bacteria increase plant pathogen resistance and affect tritrophic interactions. Plant, Cell Environ 37:813–826
7. Maffei ME, Gertsch J, Appendino G (2011) Plant volatiles: production, function and pharmacology. Nat Prod Rep 28:1359–1380
8. Yan HJ, Zhang H, Wang QG et al (2011) Isolation and identification of a putative scent-related gene RhMYB1 from rose. Mol Biol Rep 38:4475–4482
9. Petronilho S, Maraschin M, Delgadillo I et al (2011) Sesquiterpenic composition of the inflorescences of Brazilian chamomile (*Matricaria recutita* L.): impact of the agricultural practices. Ind Crops Prod 34:1482–1490
10. Kaewtathip T, Charoenrein S (2012) Changes in volatile aroma compounds of pineapple (*Ananas comosus*) during freezing and thawing. Int J Food Sci Technol 47:985–990
11. Wenzl T, Simon R, Kleiner J et al (2006) Analytical methods for polycyclic aromatic hydrocarbons (PAHs) in food and the environment needed for new food legislation in the European Union. Trends Anal Chem 25:716–725
12. Lei FF, Huang JY, Zhang XN et al (2011) Determination of polycyclic aromatic hydrocarbons in vegetables by headspace SPME-GC. Chromatographia 74:99–107
13. Sarafraz-Yazdi A, Ghaemi F, Amiri A (2012) Comparative study of the sol-gel based solid phase microextraction fibers in extraction of naphthalene, fluorene, anthracene and phenanthrene from saffron samples extractants. Microchim Acta 176:317–325
14. Melo A, Aguiar A, Mansilha C et al (2012) Optimisation of a solid-phase microextraction/HPLC/diode array method for multiple pesticide screening in lettuce. Food Chem 130:1090–1097
15. Macherius A, Eggen T, Lorenz WG et al (2012) Uptake of galaxolide, tonalide, and triclosan by carrot, barley, and meadow fescue plants. J Agric Food Chem 60:7785–7791
16. Herklotz PA, Gurung P, Vandenheuvel B et al (2010) Uptake of human pharmaceuticals by plants grown under hydroponic conditions. Chemosphere 78:1416–1421

17. Wu X, Conkle JL, Ernst F et al (2014) Treated wastewater irrigation: uptake of pharmaceutical and personal care products by common vegetables under field conditions. Environ Sci Technol 48:11286–11293
18. Materic D, Bruhn D, Turner C et al (2015) Methods in plant foliar volatile organic compounds research. Appl Plant Sci 3:1500044
19. Zhu F, Xu J, Ke Y et al (2013) Applications of in vivo and in vitro solid-phase microextraction techniques in plant analysis: a review. Anal Chim Acta 794:1–14
20. Zhou SN, Ouyang GF, Pawliszyn J et al (2008) Comparison of microdialysis with solid-phase microextraction for in vitro and in vivo studies. J Chromatogr A 1196:46–56
21. Bicchi C, Cordero C, Liberto E et al (2008) Headspace sampling of the volatile fraction of vegetable matrices. J Chromatogr A 1184:220–233
22. Stashenko EE, Martinez JR (2008) Sampling flower scent for chromatographic analysis. J Sep Sci 31:2022–2031
23. Zhang Z, Yang MJ, Pawliszyn J (1994) Solid-phase microextraction, a for preparation. Anal Chem 66:844A–853A
24. Ouyang G, Vuckovic D, Pawliszyn J (2011) Nondestructive sampling of living systems using in vivo solid-phase microextraction. Chem Rev 111:2784–2814
25. Souza-Silva EA, Jiang R, Rodriguez-Lafuente A et al (2015) A critical review of the state of the art of solid-phase microextraction of complex matrices I. Environmental analysis. Trends Anal Chem 71:224–235
26. Xu C, Chen G, Xiong Z et al (2016) Applications of solid-phase microextraction in food analysis. Trends Anal Chem 80:12–29
27. Souza-Silva EA, Reyes-Garces N, Gomez-Rios GA et al (2015) A critical review of the state of the art of solid-phase microextraction of complex matrices III. Bioanalytical and clinical applications. Trends Anal Chem 71:249–264
28. Kimbaris AC, Siatis NG, Daferera DJ et al (2006) Comparison of distillation and ultrasound-assisted extraction methods for the isolation of sensitive aroma compounds from garlic (*Allium sativum*). Ultrason Sonochem 13:54–60
29. Kraujalyte V, Leitner E, Venskutonis PR et al (2013) Characterization of *Aronia melanocarpa* volatiles by headspace-solid-phase microextraction (HS-SPME), simultaneous distillation/extraction (SDE), and gas chromatography-olfactometry (GC-O) methods. J Agric Food Chem 61:4728–4736
30. Gokbulut I, Karabulut I (2012) SPME-GC-MS detection of volatile compounds in apricot varieties. Food Chem 132:1098–1102
31. Chai Q, Wu B, Liu W et al (2012) Volatiles of plums evaluated by HS-SPME with GC-MS at the germplasm level. Food Chem 130:432–440
32. Song X, Ha W, Chen J et al (2014) Application of beta-cyclodextrin-modified, carbon nanotube-reinforced hollow fiber to solid-phase microextraction of plant hormones. J Chromatogr A 1374:23–30
33. Gay F, Maraval I, Roques S et al (2010) Effect of salinity on yield and 2-acetyl-1-pyrroline content in the grains of three fragrant rice cultivars (*Oryza sativa* L.) in Camargue (France). Field Crops Res 117:154–160
34. Ji T, Dami IE (2008) Characterization of free flavor compounds in traminette grape and their relationship to vineyard training system and location. J Food Sci 73:C262–C267
35. Xie Z, Liu Q, Liang Z et al (2013) The GC/MS analysis of volatile components extracted by different methods from Exocarpium Citri Grandis. J Anal Methods Chem 2013:918406
36. Guzmán-Gerónimo RI, López MG, Dorantes-Alvarez L (2008) Microwave processing of avocado: volatile flavor profiling and olfactometry. Innov Food Sci Emerg 9:501–506
37. Vázquez-Araújo L, Chambers E, Funk DB (2011) References for "musty" odor notes in sensory analysis of grain sorghum. J Cereal Sci 54:460–466
38. Kraujalytė V, Leitner E, Venskutonis PR (2012) Chemical and sensory characterisation of aroma of *Viburnum opulus* fruits by solid phase microextraction-gas chromatography-olfactometry. Food Chem 132:717–723

39. Azam M, Song M, Fan F et al (2013) Comparative analysis of flower volatiles from nine citrus at three blooming stages. Int J Mol Sci 14:22346–22367
40. Silva BM, Santos RP, Mendes LS et al (2011) *Dracaena draco* L. fruit: phytochemical and antioxidant activity assessment. Food Res Int 44:2182–2189
41. Ren J, Tai Y, Dong M et al (2015) Characterisation of free and bound volatile compounds from six different varieties of citrus fruits. Food Chem 185:25–32
42. Steingass CB, Carle R, Schmarr HG (2015) Ripening-dependent metabolic changes in the volatiles of pineapple (*Ananas comosus* (L.) Merr.) fruit: I. Characterization of pineapple aroma compounds by comprehensive two-dimensional gas chromatography-mass spectrometry. Anal Bioanal Chem 407:2591–2608
43. Djozan D, Ebrahimi B (2008) Preparation of new solid phase micro extraction fiber on the basis of atrazine-molecular imprinted polymer: application for GC and GC/MS screening of triazine herbicides in water, rice and onion. Anal Chim Acta 616:152–159
44. Chai M, Tan G (2010) Headspace solid-phase microextraction for the evaluation of pesticide residue contents in cucumber and strawberry after washing treatment. Food Chem 123:760–764
45. Zhang Z, Huang Y, Ding W et al (2014) Multilayer interparticle linking hybrid MOF-199 for noninvasive enrichment and analysis of plant hormone ethylene. Anal Chem 86:3533–3540
46. Es'haghi Z, Khalili M, Khazaeifar A et al (2011) Simultaneous extraction and determination of lead, cadmium and copper in rice samples by a new pre-concentration technique: hollow fiber solid phase microextraction combined with differential pulse anodic stripping voltammetry. Electrochim Acta 56:3139–3146
47. Zhang Z, Pawliszyn J (1995) Quantitative extraction using an internally cooled solid phase microextraction device. Anal Chem 67:34–43
48. Chen Y, Pawliszyn J (2006) Miniaturization and automation of an internally cooled coated fiber device. Anal Chem 78:5222–5226
49. Ghiasvand AR, Setkova L, Pawliszyn J (2007) Determination of flavour profile in Iranian fragrant rice samples using cold-fibre SPME-GC-TOF-MS. Flavour Fragr J 22:377–391
50. Chai X, Jia J, Sun T et al (2008) Suitability of a novel circulating cooling SPME for analysis of organophosphorous pesticides in tomatoes. Chromatographia 67:309–313
51. Pedroso M, Ferreira E, Hantao L et al (2011) Identification of volatiles from pineapple (*Ananas comosus* L.) pulp by comprehensive two-dimensional gas chromatography and gas chromatography/mass spectrometry. J Sep Sci 34:1547–1554
52. Vuckovic D, de Lannoy I, Gien B et al (2011) In vivo solid-phase microextraction: capturing the elusive portion of metabolome. Angew Chem Int Ed 50:5344–5348
53. Vuckovic D, Risticevic S, Pawliszyn J (2011) In vivo solid-phase microextraction in metabolomics: opportunities for the direct investigation of biological systems. Angew Chem Int Ed 50:5618–5628
54. Zhang Z, Li G (2007) A preliminary study of plant aroma profile characteristics by a combination sampling method coupled with GC-MS. Microchem J 86:29–36
55. Soto VC, Maldonado IB, Jofré VP et al (2015) Direct analysis of nectar and floral volatile organic compounds in hybrid onions by HS-SPME/GC-MS: relationship with pollination and seed production. Microchem J 122:110–118
56. Kang WY, Wang JM, Tian PY (2011) Analysis of volatiles in the flowers of *Patrinia scabiosifolia* by HS-SPME-GC-MS. Chem Nat Compd 47:101–102
57. Cullere L, San-Juan F, Cacho J (2011) Characterisation of aroma active compounds of Spanish saffron by gas chromatography-olfactometry: quantitative evaluation of the most relevant aromatic compounds. Food Chem 127:1866–1871
58. Djabou N, Paolini J, Desjobert JM et al (2010) Qualitative and quantitative analysis of volatile components of *Teucrium massiliense* L.—identification of 6-methyl-3-heptyl acetate as a new natural product. Flavour Fragr J 25:475–487
59. De Pinho PG, Goncalves RF, Valentao P et al (2009) Volatile composition of *Catharanthus roseus* (L.) G. Don using solid-phase microextraction and gas chromatography/mass spectrometry. J Pharm Biomed 49:674–685

60. Edris AE, Chizzola R, Franz C (2008) Isolation and characterization of the volatile aroma compounds from the concrete headspace and the absolute of *Jasminum sambac* (L.) Ait. (Oleaceae) flowers grown in Egypt. Eur Food Res Technol 226:621–626
61. Rout PK, Rao YR, Sree A et al (2007) Composition of essential oil, concrete, absolute, wax and headspace volatiles of *Murraya paniculata* (Linn.) Jack flowers. Flavour Fragr J 22:352–357
62. Dong L, Wang J, Deng C et al (2007) Gas chromatography-mass spectrometry following pressurized hot water extraction and solid-phase microextraction for quantification of eucalyptol, camphor, and borneol in *Chrysanthemum* flowers. J Sep Sci 30:86–89
63. Bianchi G, Nuzzi M, Leva AA et al (2007) Development of a headspace-solid phase micro extraction method to monitor changes in volatile profile of rose (*Rosa hybrida*, cv David Austin) petals during processing. J Chromatogr A 1150:190–197
64. Maggi F, Bilek T, Cristalli G et al (2009) Comparison of the characterisation of the fruit-like aroma of *Teucrium flavum* L. subsp flavum by hydrodistillation and solid-phase micro-extraction. J Sci Food Agric 89:2505–2518
65. Chung MS (2012) Volatile compounds of the Hallabong (*Citrus kiyomi* x *Citrus ponkan*) blossom. Food Sci Biotechnol 21:285–290
66. Vázquez-Araújo L, Rodríguez-Solnan R, Cortés-Diéguez SM et al (2013) Use of hydrodistillation and headspace solid-phase microextraction to characterize the volatile composition of different hop cultivars. J Sci Food Agric 93:2568–2574
67. Oh SY, Du Shin H, Kim SJ et al (2008) Rapid determination of floral aroma compounds of lilac blossom by fast gas chromatography combined with surface acoustic wave sensor. J Chromatogr A 1183:170–178
68. Pan J, Hu Y, Liang T et al (2012) Preparation of solid-phase microextraction fibers by in-mold coating strategy for derivatization analysis of 24-epibrassinolide in pollen samples. J Chromatogr A 1262:49–55
69. Bojko B, Cudjoe E, Gómez-Ríos GA et al (2012) SPME—quo vadis? Anal Chim Acta 750:132–151
70. Verdonk JC, Ric de Vos CH, Verhoeven HA et al (2003) Regulation of floral scent production in petunia revealed by targeted metabolomics. Photochemistry 62:997–1008
71. Manzo A, Panseri S, Vagge I et al (2014) Volatile fingerprint of Italian populations of orchids using solid phase microextraction and gas chromatography coupled with mass spectrometry. Molecules 19:7913–7936
72. Song G, Xiao J, Deng C et al (2007) Use of solid-phase microextraction as a sampling technique for the characterization of volatile compounds emitted from Chinese daffodil flowers. J Anal Chem 62:674–679
73. Cáceres LA, McDowell TW, Scott IM et al (2015) In vivo extraction of volatile organic compounds (VOCs) from Micro-Tom tomato flowers with multiple solid phase microextraction (SPME) fibers. Can J Chem 93:143–150
74. Cheng WW, Lin CT, Chu FH et al (2009) Neuropharmacological activities of phytoncide released from *Cryptomeria japonica*. J Wood Sci 55:27–31
75. Lin SY, Roan SF, Lee CL et al (2010) Volatile organic components of fresh leaves as indicators of indigenous and cultivated citrus species in Taiwan. Biosci Biotechnol Biochem 74:806–811
76. Paiva MR, Mateus E, Santos MH et al (2011) Pine volatiles mediate host selection for oviposition by *Thaumetopoea pityocampa* (Lep., Notodontidae). J Appl Entomol 135:195–203
77. Smith L, Beck JJ (2015) Duration of emission of volatile organic compounds from mechanically damaged plant leaves. J Plant Physiol 188:19–28
78. Zielinska S, Piatczak E, Kalemba D et al (2011) Influence of plant growth regulators on volatiles produced by in vitro grown shoots of *Agastache rugosa* (Fischer & C.A.Meyer) O. Kuntze. Plant Cell Tissue Organ Culture 107:161–167

79. Weingart G, Kluger B, Forneck A et al (2012) Establishment and application of a metabolomics workflow for identification and profiling of volatiles from leaves of *Vitis vinifera* by HS-SPME-GC-MS. Phytochem Anal 23:345–358
80. Chung MJ, Cheng SS, Lin CY et al (2012) Profiling of volatile compounds of *Phyllostachys pubescens* shoots in Taiwan. Food Chem 134:1732–1737
81. Mu RM, Wang XR, Liu SX et al (2007) Rapid determination of volatile compounds in *Toona sinensis* (A. Juss.) Roem. by MAE-HS-SPME followed by GC-MS. Chromatographia 65:463–467
82. Araujo HC, Lacerda MEG, Lopes D et al (2007) Studies on the aroma of mate (*Ilex paraguariensis* St. Hil.) using headspace solid-phase microextraction. Phytochem Anal 18:469–474
83. Schossler P, Schneider GL, Wunsch D et al (2009) Volatile compounds of *Baccharis punctulata*, *Baccharis dracunculifolia* and *Eupatorium laevigatum* obtained using solid phase microextraction and hydrodistillation. J Braz Chem Soc 20:277–287
84. Wakte KV, Thengane RJ, Jawali N et al (2010) Optimization of HS-SPME conditions for quantification of 2-acetyl-1-pyrroline and study of other volatiles in *Pandanus amaryllifolius* Roxb. Food Chem 121:595–600
85. Silva CL, Camara JS (2013) Profiling of volatiles in the leaves of Lamiaceae species based on headspace solid phase microextraction and mass spectrometry. Food Res Int 51:378–387
86. Reale S, Fasciani P, Pace L et al (2011) Volatile fingerprints of artemisinin-rich *Artemisia annua* cultivars by headspace solid-phase microextraction gas chromatography/mass spectrometry. Rapid Commun Mass Spectrom 25:2511–2516
87. Lin C, Chen Y, Cheng S et al (2011) Rapid differentiation of three Chamaecyparis species (Cupressaceae) Grown in Taiwan using solid-phase microextraction-gas chromatography/mass spectrometry, cluster analysis, and principal component analysis. J Agric Food Chem 59:10854–10859
88. Kollner TG, Lenk C, Schnee C et al (2013) Localization of sesquiterpene formation and emission in maize leaves after herbivore damage. BMC Plant Biol 13:15
89. Huang B, Lei Y, Tang Y et al (2011) Comparison of HS-SPME with hydrodistillation and SFE for the analysis of the volatile compounds of Zisu and Baisu, two varietal species of *Perilla frutescens* of Chinese origin. Food Chem 125:268–275
90. Yassaa N, Williams J (2007) Enantiomeric monoterpene emissions from natural and damaged Scots pine in a boreal coniferous forest measured using solid-phase microextraction and gas chromatography/mass spectrometry. J Chromatogr A 1141:138–144
91. Giorgi A, Panseri S, Mattara MS et al (2013) Secondary metabolites and antioxidant capacities of *Waldheimia glabra* (Decne.) Regel from Nepal. J Sci Food Agric 93:1026–1034
92. Beck JJ, Smith L, Merrill GB (2008) In situ volatile collection, analysis, and comparison of three Centaurea species and their relationship to biocontrol with herbivorous insects. J Agric Food Chem 56:2759–2764
93. Chen Y, Cheng S, Chang S (2010) Monitoring the emission of volatile organic compounds from the leaves of *Calocedrus macrolepis* var. *formosana* using solid-phase micro-extraction. J Wood Sci 56:140–147
94. Chen G, Jiang R, Qiu J et al (2015) Environmental fates of synthetic musks in animal and plant: an in vivo study. Chemosphere 138:584–591
95. Chen G, Qiu J, Liu Y et al (2015) Carbon nanotubes act as contaminant carriers and translocate within plants. Sci Rep 5:15682
96. Chen G, Qiu J, Xu J et al (2016) A novel probe based on phenylboronic acid functionalized carbon nanotubes for ultrasensitive carbohydrate determination in biofluids and semi-solid biotissues. Chem Sci 7:1487–1495
97. Wang R, Wang R, Yang B (2011) Comparison of volatile compound composition of cinnamon (*Cinnamomum cassia* Presl) bark prepared by hydrodistillation and headspace solid phase microextraction. J Food Process Eng 34:175–185

98. Benyelles B, Allali H, El Amine Dib M et al (2014) Essential oil from *Rhaponticum acaule* L. roots: Comparative study using HS-SPME/GC/GC-MS and hydrodistillation techniques. J Saudi Chem Soc 18:972–976
99. Zhang C, Qi M, Shao Q et al (2007) Analysis of the volatile compounds in *Ligusticum chuanxiong* Hort. using HS-SPME-GC-MS. J Pharm Biomed 44:464–470
100. Riu-Aumatell M, Vargas L, Vichi S et al (2011) Characterisation of volatile composition of white salsify (*Tragopogon porrifolius* L.) by headspace solid-phase microextraction (HS-SPME) and simultaneous distillation-extraction (SDE) coupled to GC-MS. Food Chem 129:557–564
101. Lee KS, Kim GH, Kim HH et al (2012) Volatile compounds of *Panax ginseng* CA Meyer cultured with different cultivation methods. J Food Sci 77:C805–C810
102. Yu Y, Huang T, Yang B et al (2007) Development of gas chromatography-mass spectrometry with microwave distillation and simultaneous solid-phase microextraction for rapid determination of volatile constituents in ginger. J Pharm Biomed 43:24–31
103. Farag MA, Wessjohann LA (2012) Volatiles profiling in medicinal licorice roots using steam distillation and solid-phase microextraction (SPME) coupled to chemometrics. J Food Sci 77:C1179–C1184
104. Mebazaa R, Mahmoudi A, Fouchet M et al (2009) Characterisation of volatile compounds in Tunisian fenugreek seeds. Food Chem 115(2009):1326–1336
105. Hashemi P, Shamizadeh M, Badiei A et al (2009) Study of the essential oil composition of cumin seeds by an amino ethyl-functionalized nanoporous SPME fiber. Chromatographia 70:1147–1151
106. Azarnia S, Boye JI, Warkentin T et al (2011) Volatile flavour profile changes in selected field pea cultivars as affected by crop year and processing. Food Chem 124:326–335
107. Azarnia S, Boye JI, Warkentin T et al (2011) Changes in volatile flavour compounds in field pea cultivars as affected by storage conditions. Int J Food Sci Technol 46:2408–2419
108. Tamogami S, Noge K, Agrawal GK et al (2015) Methyl jasmonate elicits the production of methyl (E)-2-hexenoate from (Z)-2-hexenol via (Z)-2-hexenal in *Achyranthes bidentata* plant. FEBS Lett 589:390–395
109. Min TG (2012) Detection of ethanol released from aged radish (*Raphanus sativus* L.) seeds using resazurin. Hortic Environ Biotechnol 53:66–71
110. Reiche N, Mothes F, Fiedler P et al (2013) A solid-phase microextraction method for the in vivo sampling of MTBE in common reed (*Phragmites australis*). Environ Monit Assess 185:7133–7144
111. Limmer MA, Balouet JC, Karg F et al (2011) Phytoscreening for chlorinated solvents using rapid in vitro SPME sampling: application to urban plume in Verl, Germany. Environ Sci Technol 45(2011):8276–8282
112. Holm O, Rotard W (2011) Effect of radial directional dependences and rainwater influence on CVOC concentrations in tree core and birch sap samples taken for phytoscreening using HS-SPME-GC/MS. Environ Sci Technol 45:9604–9610
113. Shanmugam PV, Yadav A, Chanotiya CS et al (2015) Enantiomer differentiation of key volatile constituents from leaves, stems, rhizome and flowers of cultivated *Hedychium coronarium* Koenig from India. J Essent Oil Res 27:101–106
114. Yamani H, Mantri N, Morrison PD et al (2014) Analysis of the volatile organic compounds from leaves, flower spikes, and nectar of Australian grown *Agastache rugosa*. BMC Complement Altern Med 14:495
115. Vázquez AM, Aimar ML, Demmel GI et al (2014) Identification of volatile compounds of Clinopodium odorum (Lamiaceae): a comparison between HS-SPME and classic hydrodistillation. Bol Latinoam Caribe Plantas Med Aromát 13:285–296
116. Rezazadeh M, Yamini Y, Seidi S et al (2014) Electromembrane surrounded solid phase microextraction followed by injection port derivatization and gas chromatography-flame ionization detector analysis for determination of acidic herbicides in plant tissue. J Agric Food Chem 62(2014):3134–3142

117. Baker B, Sinnott M (2009) Analysis of sesquiterpene emissions by plants using solid phase microextraction. J Chromatogr A 1216:8442–8451
118. Limmer MA, Holmes AJ, Burken JG et al (2014) Phytomonitoring of chlorinated ethenes in trees: a four-year study of seasonal chemodynamics in planta. Environ Sci Technol 48:10634–10640
119. Gholivand MB, Piryaei M, Abolghasemi M (2011) Anodized aluminum wire as a solid-phase microextraction fiber for rapid determination of volatile constituents in medicinal plant. Anal Chim Acta 701:1–5
120. Togunde OP, Oakes KD, Servos MR et al (2012) Optimization of solid phase microextraction for non-lethal in vivo determination of selected pharmaceuticals in fish muscle using liquid chromatography-mass spectrometry. J Chromatogr A 1261:99–106
121. Yeung JCY, de Lannoy I, Gien B et al (2012) Semi-automated in vivo solid-phase microextraction sampling and the diffusion-based interface calibration model to determine the pharmacokinetics of methoxyfenoterol and fenoterol in rats. Anal Chim Acta 742:37–44
122. Zhang X, Oakes KD, Wang S et al (2012) In vivo sampling of environmental organic contaminants in fish by solid-phase microextraction. Trends Anal Chem 32:31–39
123. Wang S, Oakes KD, Bragg LM et al (2011) Validation and use of in vivo solid phase micro-extraction (SPME) for the detection of emerging contaminants in fish. Chemosphere 85:1472–1480
124. Ouyang G, Oakes KD, Bragg L et al (2011) Sampling-rate calibration for rapid and nonlethal monitoring of organic contaminants in fish muscle by solid-phase microextraction. Environ Sci Technol 45:7792–7798
125. Parker M, Pollnitz AP, Cozzolino D et al (2007) Identification and quantification of a marker compound for 'Pepper' aroma and flavor in shiraz grape berries by combination of chemometrics and gas chromatography-mass spectrometry. J Agric Food Chem 55:5948–5955
126. Marković K, Vahčić N, Ganić KK et al (2007) Aroma volatiles of tomatoes and tomato products evaluated by solid-phase microextraction. Flavour Fragr J 22:395–400
127. Ortiz-Serrano P, Gil JV (2007) Quantitation of free and glycosidically bound volatiles in and effect of glycosidase addition on three tomato varieties (*Solanum lycopersicum* L.). J Agric Food Chem 55:9170–9176
128. Oomah BD, Liang LS, Balasubramanian P (2007) Volatile compounds of dry beans (*Phaseolus vulgaris* L.). Plant Food Hum Nutr 62:177–183
129. Balbontín C, Gaete-Eastman C, Vergara M et al (2007) Treatment with 1-MCP and the role of ethylene in aroma development of mountain papaya fruit. Postharvest Biol Technol 43:67–77
130. Vichi S, Riu-Aumatell M, Mora-Pons M et al (2007) HS-SPME coupled to GC/MS for quality control of *Juniperus communis* L. berries used for gin aromatization. Food Chem 105:1748–1754
131. Malm A, Głowniak K, Łoś R et al (2008) Variation of the volatile content of the fruits of *Peucedanum alsaticum* L. Acta Chromatogr 20:119–133
132. Zhang Z, Zeng D, Li G (2008) Study of the volatile profile characteristics of longan during storage by a combination sampling method coupled with GC/MS. J Sci Food Agric 88:1035–1042
133. Obando-Ulloa JM, Moreno E, García-Mas J et al (2008) Climacteric or non-climacteric behavior in melon fruit—1. Aroma volatiles. Postharvest Biol Technol 49:27–37
134. Greene JL, Sanders TH, Drake MA (2008) Characterization of volatile compounds contributing to naturally occurring fruity fermented flavor in peanuts. J Agric Food Chem 56:8096–8102
135. Chin ST, Nazimah SAH, Quek SY et al (2008) Changes of volatiles' attribute in durian pulp during freeze- and spray-drying process. LWT Food Sci Technol 41:1899–1905

136. Ziino M, Condurso C, Romeo V et al (2009) Volatile compounds and capsaicinoid content of fresh hot peppers (*Capsicum annuum* L.) of different Calabrian varieties. J Sci Food Agric 89:774–780
137. Humston EM, Zhang Y, Brabeck GF et al (2009) Development of a GC x GC-TOFMS method using SPME to determine volatile compounds in cacao beans. J Sep Sci 32:2289–2295
138. Ferreira L, Perestrelo R, Caldeira M et al (2009) Characterization of volatile substances in apples from Rosaceae family by headspace solid-phase microextraction followed by GC-qMS. J Sep Sci 32:1875–1888
139. Zawirska-Wojtasiak R, Goslinski M, Szwacka M et al (2009) Aroma evaluation of transgenic, thaumatin II-producing cucumber fruits. J Food Sci 74:C204–C210
140. Wu Y, Pan Q, Qu W et al (2009) Comparison of volatile profiles of nine litchi (*Litchi chinensis* Sonn.) cultivars from Southern China. J Agric Food Chem 57:9676–9681
141. Gonzalez M, Gaete-Eastman C, Valdenegro M et al (2009) Aroma development during ripening of *Fragaria chiloensis* fruit and participation of an alcohol acyltransferase (FcAAT1) gene. J Agric Food Chem 57:9123–9132
142. Wang Y, Yang C, Li S et al (2009) Volatile characteristics of 50 peaches and nectarines evaluated by HP-SPME with GC-MS. Food Chem 116:356–364
143. Reis SFAR, Rocha SM, Barros AS et al (2009) Establishment of the volatile profile of 'Bravo de Esmolfe' apple variety and identification of varietal markers. Food Chem 113:513–521
144. Fan G, Qiao Y, Yao X et al (2009) Free and bound volatile compounds in juice and peel of Jincheng oranges. Eur Food Res Technol 229:571–578
145. Skalicka-Wozniak K, Los R, Glowniak K et al (2009) Volatile compounds in fruits of *Peucedanum cervaria* (LAP.) L. Chem Biodivers 6:1087–1092
146. Rodriguez-Lafuente A, Nerin de la Puerta C, Batlle R (2009) Determination of fifteen active compounds released from paraffin-based active packaging in tomato samples via microextraction techniques. Anal Bioanal Chem 395:203–211
147. Zhang Y, Wang G, Dong J et al (2009) Analysis of volatile components in strawberry cultivars Xingdu 1 and Xingdu 2 and their Parents. Agric Sci China 8:441–446
148. Abrodo PA, Llorente DD, Corujedo SJ et al (2010) Characterisation of Asturian cider apples on the basis of their aromatic profile by high-speed gas chromatography and solid-phase microextraction. Food Chem 121:1312–1318
149. Flamini G, Cioni PL (2010) Odour gradients and patterns in volatile emission of different plant parts and developing fruits of grapefruit (*Citrus paradisi* L.). Food Chem 120:984–992
150. Oliveira AP, Silva LR, Andrade PB et al (2010) Determination of low molecular weight volatiles in *Ficus carica* using HS-SPME and GC/FID. Food Chem 121:1289–1295
151. Oliveira AP, Silva LR, Guedes de Pinho P et al (2010) Volatile profiling of *Ficus carica* varieties by HS-SPME and GC-IT-MS. Food Chem 123:548–557
152. Prosen H, Kokalj M, Janeš D et al (2010) Comparison of isolation methods for the determination of buckwheat volatile compounds. Food Chem 121:298–306
153. Cheong KW, Tan CP, Mirhosseini H et al (2010) Equilibrium headspace analysis of volatile flavor compounds extracted from soursop (*Annona muricata*) using solid-phase microextraction. Food Res Int 43:1267–1276
154. Ortiz-Serrano P, Gil JV (2010) Quantitative comparison of free and bound volatiles of two commercial tomato cultivars (*Solanum lycopersicum* L.) during ripening. J Agric Food Chem 58:1106–1114
155. Rodriguez-Burruezo A, Kollmannsberger H, Gonzalez-Mas M et al (2010) HS-SPME comparative analysis of genotypic diversity in the volatile fraction and aroma-contributing compounds of capsicum fruits from the annuum-chinense-frutescens complex. J Agric Food Chem 58:4388–4400
156. Wang Y, Yang C, Liu C et al (2010) Effects of bagging on volatiles and polyphenols in "Wanmi" peaches during endocarp hardening and final fruit rapid growth stages. J Food Sci 75:S455–S460

157. Lo Bianco R, Farina V, Indelicato SG et al (2010) Fruit physical, chemical and aromatic attributes of early, intermediate and late apricot cultivars. J Sci Food Agric 90:1008–1019
158. Yang X, Song J, Fillmore S et al (2011) Effect of high temperature on color, chlorophyll fluorescence and volatile biosynthesis in green-ripe banana fruit. Postharvest Biol Technol 62:246–257
159. Yang C, Wang Y, Wu B et al (2011) Volatile compounds evolution of three table grapes with different flavour during and after maturation. Food Chem 128:823–830
160. Thuaytong W, Anprung P (2011) Bioactive compounds and prebiotic activity in Thailand-grown red and white guava fruit (*Psidium guajava* L.). Food Sci Technol Int 17:205–212
161. Perestrelo R, Barros AS, Rocha SM et al (2011) Optimisation of solid-phase microextraction combined with gas chromatography-mass spectrometry based methodology to establish the global volatile signature in pulp and skin of *Vitis vinifera* L. grape varieties. Talanta 85:1483–1493
162. Oh SH, Lim BS, Hong SJ et al (2011) Aroma volatile changes of netted muskmelon (*Cucumis melo* L.) fruit during developmental stages. Hortic Environ Biotechnol 52:590–595
163. Meret M, Brat P, Mertz C et al (2011) Contribution to aroma potential of Andean blackberry (*Rubus glaucus* Benth.). Food Res Int 44:54–60
164. Mathure SV, Wakte KV, Jawali N et al (2011) Quantification of 2-acetyl-1-pyrroline and other rice aroma volatiles among Indian scented rice cultivars by HS-SPME/GC-FID. Food Anal Method 4:326–333
165. Martendal E, De Souza Silveira CD, Nardini GS et al (2011) Use of different sample temperatures in a single extraction procedure for the screening of the aroma profile of plant matrices by headspace solid-phase microextraction. J Chromatogr A 1218:3731–3736
166. Malheiro R, De Pinho PG, Casal S et al (2011) Determination of the volatile profile of stoned table olives from different varieties by using HS-SPME and GC/IT-MS. J Sci Food Agr 91:1693–1701
167. Ma Y, Song D, Wang Z et al (2011) Effect of ultrahigh pressure treatment on volatile compounds in garlic. J Food Process Eng 34:1915–1930
168. Kim NY, Park MH, Jang EY et al (2011) Volatile distribution in garlic (*Allium sativum* L.) by solid phase microextraction (SPME) with different processing conditions. Food Sci Biotechnol 20:775–782
169. Gebara SS, De Oliveira Ferreira W, Ré-Poppi N et al (2011) Volatile compounds of leaves and fruits of *Mangifera indica* var. *coquinho* (Anacardiaceae) obtained using solid phase microextraction and hydrodistillation. Food Chem 127:689–693
170. Cha DH, Loeb GM, Linn CE et al (2011) Electrophysiological and behavioral identification of a volatile blend involved in host location of female strawberry sap beetle, *Stelidota geminata*. Entomol Exp Appl 140:153–162
171. Junior SB, De Marchi Tavares de Melo A, Zini CA et al (2011) Optimization of the extraction conditions of the volatile compounds from chili peppers by headspace solid phase micro-extraction. J Chromatogr A 1218:3345–3350
172. Aprea E, Gika H, Carlin S et al (2011) Metabolite profiling on apple volatile content based on solid phase microextraction and gas-chromatography time of flight mass spectrometry. J Chromatogr A 1218:4517–4524
173. Beaulieu JC, Ingber BF, Lea JM (2011) Physiological, volatile, and SEM surface effects resulting from cutting and dipping treatments in cantaloupe. J Food Sci 76:S415–S422
174. Fong SS, Sagi-Kiss V, Brereton RG (2011) Self-organizing maps and support vector regression as aids to coupled chromatography: illustrated by predicting spoilage in apples using volatile organic compounds. Talanta 83:1269–1278
175. Zheng LY, Sun GM, Liu YG et al (2012) Aroma volatile compounds from two fresh pineapple varieties in China. Int J Mol Sci 13:7383–7392
176. Sanchez G, Besada C, Badenes ML et al (2012) A non-targeted approach unravels the volatile network in peach fruit. PLoS ONE 7:e38992

177. Qin G, Tao S, Cao Y et al (2012) Evaluation of the volatile profile of 33 Pyrus ussuriensis cultivars by HS-SPME with GC-MS. Food Chem 134:2367–2382
178. Pontes M, Pereira J, Câmara JS (2012) Dynamic headspace solid-phase microextraction combined with one-dimensional gas chromatography-mass spectrometry as a powerful tool to differentiate banana cultivars based on their volatile metabolite profile. Food Chem 134:2509–2520
179. Perestrelo R, Caldeira M, Camara JS (2012) Solid phase microextraction as a reliable alternative to conventional extraction techniques to evaluate the pattern of hydrolytically released components in *Vitis vinifera* L. grapes. Talanta 95:1–11
180. Moreno E, Fita A, González-Mas MC et al (2012) HS-SPME study of the volatile fraction of Capsicum accessions and hybrids in different parts of the fruit. Sci Hortic 135:87–97
181. May B, Wüst M (2012) Temporal development of sesquiterpene hydrocarbon profiles of different grape varieties during ripening. Flavour Frag J 27:280–285
182. Lasekan O (2012) Headspace solid-phase microextraction gas chromatography-mass spectrometry (HS-SPME-GC-MS) determination of volatile compounds in roasted plantains (French sombre and Dwarf Kalapua). LWT Food Sci Technol 46:536–541
183. Goldenberg L, Feygenberg O, Samach A et al (2012) Ripening attributes of new passion fruit line featuring seasonal non-climacteric behavior. J Agric Food Chem 60:1810–1821
184. Farneti B, Cristescu SM, Costa G et al (2012) Rapid tomato volatile profiling by using proton-transfer reaction mass spectrometry (PTR-MS). J Food Sci 77:C551–C559
185. Drew DP, Rasmussen SK, Avato P et al (2012) A comparison of headspace solid-phase microextraction and classic hydrodistillation for the identification of volatile constituents from thapsia spp. provides insights into guaianolide biosynthesis in apiaceae. Phytochem Anal 23:44–51
186. Chizzola R, Saeidnejad AH, Azizi M et al (2014) *Bunium persicum*: variability in essential oil and antioxidants activity of fruits from different Iranian wild populations. Genet Resour Crop Evol 61(2014):1621–1631
187. Chin ST, Nazimah SAH, Quek SY et al (2007) Analysis of volatile compounds from Malaysian durians (*Durio zibethinus*) using headspace SPME coupled to fast GC-MS. J Food Compos Anal 20:31–44
188. Bicchi C, Cordero C, Liberto E et al (2007) Reliability of fibres in solid-phase microextraction for routine analysis of the headspace of aromatic and medicinal plants. J Chromatogr A 1152:138–149
189. Bar-Akiva A, Ovadia R, Rogachev I et al (2010) Metabolic networking in *Brunfelsia calycina* petals after flower opening. J Exp Bot 61:1393–1403
190. Fujii T, Hori M, Matsuda K (2010) Attractants for rice leaf bug, *Trigonotylus caelestialium* (Kirkaldy), are emitted from flowering rice panicles. J Chem Ecol 36:999–1005
191. Del Castillo MLR, Blanch GP (2007) Enantiomeric purity of (±)-methyl jasmonate in fresh leaf samples and commercial fragrances. J Sep Sci 30:2117–2122
192. Hori M (2007) Onion aphid (*Neotoxoptera formosana*) attractants, in the headspace of Allium fistulosum and A. tuberosum leaves. J Appl Entomol 131:8–12
193. Lagalante AF, Montgomery ME, Calvosa FC et al (2007) Characterization of terpenoid volatiles from cultivars of eastern hemlock (*Tsuga canadensis*). J Agric Food Chem 55:10850–10856
194. Sansom CE, Jones VS, Joyce NI et al (2015) Flavor, glucosinolates, and isothiocyanates of Nau (Cook's scurvy grass, *Lepidium oleraceum*) and other rare New Zealand Lepidium species. J Agric Food Chem 63:1833–1838
195. Wu Q, Wu D, Guan Y (2014) Polyaniline sheathed electrospun nanofiber bar for in vivo extraction of trace acidic phytohormones in plant tissue. J Chromatogr A 1342:16–23
196. Hantao LW, Aleme HG, Passador MM et al (2013) Determination of disease biomarkers in Eucalyptus by comprehensive two-dimensional gas chromatography and multivariate data analysis. J Chromatogr A 1279:86–91

197. Wang X, Ma Q, Li M et al (2013) Automated and sensitive analysis of 28-epihomobrassinolide in *Arabidopsis thaliana* by on-line polymer monolith microextraction coupled to liquid chromatography-mass spectrometry. J Chromatogr A 1317:121–128
198. Hantao LW, Toledo BR, De Lima Ribeiro FA et al (2013) Comprehensive two-dimensional gas chromatography combined to multivariate data analysis for detection of disease-resistant clones of Eucalyptus. Talanta 116:1079–1084
199. Kollner TG, Lenk C, Schnee C et al (2013) Localization of sesquiterpene formation and emission in maize leaves after herbivore damage. BMC Plant Biol 13:15
200. Durant AA, Rodriguez C, Herrera L et al (2014) Anti-malarial activity and HS-SPME-GC-MS chemical profiling of Plinia cerrocampanensis leaf essential oil. Malaria J 13:18
201. Engelberth J, Alborn HT, Schmelz EA et al (2004) Airborne signals prime plants against insect herbivore attack. Proc Natl Acad Sci USA 101:1781–1785
202. Hymete A, Rohloff J, Iversen TH et al (2007) Volatile constituents of the roots of Echinops kebericho Mesfin. Flavour Fragr J 22:35–38
203. Martendal E, Silveira CDD, Nardini GS et al (2011) Use of different sample temperatures in a single extraction procedure for the screening of the aroma profile of plant matrices by headspace solid-phase microextraction. J Chromatogr A 1218:3731–3736
204. Da Silva CEL, CostaWF Da, Minguzzi S et al (2013) Assessment of volatile chemical composition of the essential oil of Jatropha ribifolia (Pohl) Baill by HS-SPME-GC-MS using different fibers. J Anal Methods Chem 2013:352606
205. Wölwer-Rieck U, May B, Lankes C et al (2014) Methylerythritol and mevalonate pathway contributions to biosynthesis of mono-, sesqui-, and diterpenes in glandular trichomes and leaves of *Stevia rebaudiana* bertoni. J Agric Food Chem 62:2428–2435
206. Flamini G, Tebano M, Cioni PL (2007) Volatiles emission patterns of different plant organs and pollen of *Citrus limon*. Anal Chim Acta 589:120–124
207. De Pinho PG, Goncalves RF, Valentao P et al (2009) Volatile composition of *Catharanthus roseus* (L.) G. Don using solid-phase microextraction and gas chromatography/mass spectrometry. J Pharm Biomed 49:674–685
208. Djabou N, Paolini J, Desjobert JM et al (2010) Qualitative and quantitative analysis of volatile components of *Teucrium massiliense* L.—identification of 6-methyl-3-heptyl acetate as a new natural product. Flavour Fragr J 25:475–487
209. Fujii T, Hori M, Matsuda K (2010) Attractants for Rice Leaf Bug, *Trigonotylus caelestialium* (Kirkaldy), are emitted from flowering rice panicles. J Chem Ecol 36:999–1005
210. Maccioni S, Baldini R, Cioni PL et al (2007) In vivo volatiles emission and essential oils from different organs and pollen of *Cistus albidus* from Caprione (Eastern Liguria, Italy). Flavour Fragr J 22:61–65
211. Gebara SS, Ferreira ED, Re-Poppi N et al (2011) Volatile compounds of leaves and fruits of *Mangifera indica* var. *coquinho* (Anacardiaceae) obtained using solid phase microextraction and hydrodistillation. Food Chem 127:689–693
212. Bicchi C, Cordero C, Liberto E et al (2007) Reliability of fibres in solid-phase microextraction for routine analysis of the headspace of aromatic and medicinal plants. J Chromatogr A 1152:138–149
213. Maraval I, Sen K, Agrebi A et al (2010) Quantification of 2-acetyl-1-pyrroline in rice by stable isotope dilution assay through headspace solid-phase microextraction coupled to gas chromatography-tandem mass spectrometry. Anal Chim Acta 675:148–155
214. Yang ZB, Mao HL, Kang WY et al (2010) Rapid determination of volatile compounds in *Gymnotheca involucrata* Pei. by MAE-HS-SPME followed by GC-MS. J Am Oil Chem Soc 87:737–745
215. Paolini J, Nasica E, Desjobert JM et al (2008) Analysis of volatile constituents isolated by hydrodistillation and headspace solid-phase microextraction from *Adenostyles briquetii* Gamisans. Phytochem Anal 19:266–276

216. Mucciarelli M, Camusso W, Maffei M et al (2007) Volatile terpenoids of endophyte-free and infected peppermint (*Mentha piperita* L.): chemical partitioning of a symbiosis. Microb Ecol 54:685–696
217. Fernandes F, De Pinho PG, Valentao P et al (2009) Volatile constituents throughout *Brassica oleracea* L. Var. *acephala* germination. J Agric Food Chem 57:6795–6802
218. Courtois EA, Painec CET, Blandinieres PA et al (2009) Diversity of the volatile organic compounds emitted by 55 species of tropical trees: a survey in French Guiana. J Chem Ecol 35:1349–1362
219. Maggi F, Martonfi P, Conti F et al (2011) Volatile components of whole and different plant parts of bastard balm (*Melittis melissophyllum* L., Lamiaceae) collected in Central Italy and Slovakia. Chem Biodivers 8:2057–2079
220. Wang YL, Zeng ZR, Liu MM et al (2008) Determination of organophosphorus pesticides in pakchoi samples by headspace solid-phase microextraction coupled with gas chromatography using home-made fiber. Eur Food Res Technol 226:1091–1098
221. Chung MS (2010) Volatile compounds of the cultivated dumebuchu (*Allium senescens* L. var. *senescens*). Food Sci Biotechnol 19:1679–1682
222. Zhang Y, Wang X, Cp Lin et al (2012) A novel SPME fiber chemically linked with 1-vinyl-3-hexadecylimidazolium hexafluorophosphate ionic liquid coupled with GC for the simultaneous determination of pyrethroids in vegetables. Chromatographia 75:789–797
223. Ghiasvand AR, Nasseri M, Farsizaeh S et al (2011) Chemical characterization of cultivated *Tagetes minuta* L. by use of ultrasound-assisted head space SPME and GC-MS. Chromatographia 73:1031–1035
224. Choi HS, Min KC (2008) Aroma-active compounds of *Elsholtzia splendens* using AEDA and HS-SPME-GC-O dilution analysis. Flavour Fragr J 23:58–64
225. Dawidowicz AL, Rado E, Wianowska D et al (2008) Application of PLE for the determination of essential oil components from *Thymus vulgaris* L. Talanta 76:878–884
226. Moreno DV, Ferrera ZS, Rodriguez JJS (2007) SPME and SPE comparative study for coupling with microwave-assisted micellar extraction in the analysis of organochlorine pesticides residues in seaweed samples. Microchem J 87:139–146
227. Tigrine-Kordjani N, Chemat F, Meklati BY et al (2007) Relative characterization of rosemary samples according to their geographical origins using microwave-accelerated distillation, solid-phase microextraction and Kohonen self-organizing maps. Anal Bioanal Chem 389:631–641
228. Valtiner SM, Bonn GK, Huck CW (2008) Characterisation of different types of hay by solid-phase micro-extraction-gas chromatography mass spectrometry and multivariate data analysis. Phytochem Anal 19:359–367
229. de Oliveira ALL, da Silva DB, Turatti ICC et al (2009) Volatile constituents of Brazilian Bostrychia species (Rhodomelaceae) from mangrove and rocky shore. Biochem Syst Ecol 37:761–765
230. Maggi F, Conti F, Cristalli G et al (2011) Chemical differences in volatiles between *Melittis melissophyllum* L. subsp melissophyllum and subsp albida (Guss) P. W. Ball (Lamiaceae) determined by solid-phase microextraction (SPME) coupled with GC/FID and GC/MS. Chem Biodivers 8:325–343
231. Siripatrawan U, Harte BR (2007) Solid phase microextraction/gas chromatography/mass spectrometry integrated with chemometrics for detection of *Salmonella typhimurium* contamination in a packaged fresh vegetable. Anal Chim Acta 581:63–70
232. Lei FF, Huang JY, Zhang XN et al (2011) Determination of polycyclic aromatic hydrocarbons in vegetables by headspace SPME-GC. Chromatographia 74:99–107
233. Bukvicki D, Gottardi D, Veljic M et al (2012) Identification of volatile components of liverwort (*Porella cordaeana*) extracts using GC/MS-SPME and their antimicrobial activity. Molecules 17:6982–6995
234. Musarurwa HT, Koegelenberg L, Makunga NP (2012) Chemical variation in essential oil profiles detected using headspace solid-phase microextraction gas chromatography

spectrometry in response to potassium, nitrogen, and water available to micropropagated plants of *Salvia stenophylla* (Burch. ex Benth.). J Plant Growth Regul 31:207–220

235. Djabou N, Andreani S, Varesi L et al (2013) Analysis of the volatile fraction of *Teucrium marum* L. Flavour Fragr J 28:14–24
236. Gholivand MB, Piryaei M, Abolghasemi MM et al (2013) Rapid analysis of volatile components from *Teucrium polium* L. by nanoporous silica-polyaniline solid phase microextraction fibre. Phytochem Anal 24:69–74
237. Giorgi A, Marinis PD, Granelli G, et al (2013) Secondary metabolite profile, antioxidant capacity, and mosquito repellent activity of *Bixa orellana* from Brazilian Amazon region. J Chem 409826
238. Alvarez-Ospina H, Cruz IR, Duarte G et al (2013) HPLC determination of the major active flavonoids and GCMS Analysis of volatile components of *Dysphania graveolens* (Amaranthaceae). Phytochem Anal 24:248–254
239. Vázquez AM, Aimar ML, Demmel GI et al (2013) Determination of volatile organic compounds of *Tagetes filifolia* Lag. (Asteraceae) from Cordoba (Argentina) using HS-SPME analysis. Bol Latinoam Caribe Plantas Med Aromát 12:143–149
240. Pistelli L, Noccioli C, Dangiolillo F et al (2013) Composition of volatile in micropropagated and field grown aromatic plants from Tuscany Islands. Acta Biochim Pol 60:43–50
241. Vázquez AM, Aimar ML, Demmel GI et al (2014) Identification of volatile compounds of *Clinopodium odorum* (Lamiaceae): A comparison between HS-SPME and classic hydrodistillation. Bol Latinoam Caribe Plantas Med Aromát 13:285–296
242. Truong D-H, Delory BM, Vanderplanck M et al (2014) Temperature regimes and aphid density interactions differentially influence VOC emissions in Arabidopsis. Arthropod Plant Interact 8:317–327
243. Dagdelen S, Bilenler T, Durmaz G et al (2014) Volatile composition, antioxidant and antimicrobial activities of herbal plants used in the manufacture of van herby (OTLU) cheese. J Food Process Preserv 38:1716–1725
244. Freidig AK, Goldman IL (2014) Geosmin (2 beta,6 alpha-dimethylbicyclo [4.4.0]decan-1 beta-ol) production associated with *Beta vulgaris* ssp. *vulgaris* is cultivar specific. J Agric Food Chem 62:2031–2036
245. Oomah BD, Razafindrainibe M, Drover JCG (2014) Headspace volatile components of Canadian grown low-tannin faba bean (*Vicia faba* L.) genotypes. J Agric Food Chem 94:473–481

Chapter 10
Solid-Phase Microextraction in Binding Studies

Dajana Vuckovic

Abstract This chapter briefly discusses how solid-phase microextraction (SPME) can be successfully implemented to accurately measure ligand–receptor binding constants, % plasma protein binding, blood-to-plasma distribution ratios, and free versus total analyte concentrations. The main practical considerations and assumptions to successfully apply SPME to binding studies are discussed in detail, including the effect of significant versus negligible depletion, extraction time, pH, and fiber fouling. The binding results obtained with SPME are compared to literature values and to other techniques used for binding determinations, and the advantages and disadvantages of different methods are clearly discussed. Recent applications of SPME in binding studies are highlighted, with the primary focus on binding studies in biological fluids and tissues since this application area is where SPME has been most extensively applied. Highlighted examples show that SPME has been successfully applied for ligands with both low and high binding affinities, for ligands with different affinities to multiple sites on a given receptor, to study differences in inter-species binding as well as partitioning coefficients for various tissue types. The importance of studying inter-individual variability of free concentrations in future studies is highlighted and can contribute to the development of personalized medicine.

Keywords Ligand–receptor binding · Binding constant · Negligible depletion · Fiber fouling · Biological fluid · Tissues · Free concentration · Total concentration · Plasma protein binding

---

D. Vuckovic (✉)
Department of Chemistry and Biochemistry and PERFORM Centre, Concordia University, Montreal, Canada
e-mail: dajana.vuckovic@concordia.ca

G. Ouyang and R. Jiang (eds.), *Solid Phase Microextraction*,
DOI 10.1007/978-3-662-53598-1_10

10.1 Introduction

In blood, primary binding of drugs and metabolites occurs to highly abundant carrier proteins such as serum albumin (acidic, basic, and neutral drugs) and α-1-acidglycoprotein (basic drugs, and steroids) while some drugs can also bind globulins or lipoproteins (few lipophilic and basic drugs) [1, 2]. The binding of a ligand to protein dictates its biological availability because ligand–protein complex cannot cross cell membranes or bind to receptors. As such, bound ligand cannot reach tissue or undergo liver metabolism or renal excretion. Therefore, for highly bound drugs, the binding of drug to transport proteins creates drug reservoir, and thus impacts important parameters such as elimination half-life, biologically active concentration, metabolism, and clearance. Therefore, the evaluation of the extent of binding plays a key role in pharmacology and toxicology. The competition of drugs and other ligands for the same binding site on a given protein can also be responsible for adverse drug–drug interactions. Beyond these applications, ligand–receptor binding can play an important role in environmental analysis and few of these types of applications including binding to dissolved organic matter and insecticide uptake as highlighted in applications section of this chapter.

Solid-phase microextraction is one of the methods that can be used to determine the extent of plasma protein binding (% PPB), binding constants between individual proteins and their ligands or free concentrations of analytes in complex matrices. Other methods that can be used for this purpose include equilibrium dialysis, ultrafiltration, affinity chromatography, and ultracentrifugation to name just a few [2]. The majority of the methods used to date relies on one of two main principles: (1) separation of free and bound ligand or (2) measurable change in a property of ligand or target upon binding. Among all of these methods, equilibrium dialysis and ultrafiltration are the most popular and rely on the separation between bound and unbound ligand. Equilibrium dialysis is the gold standard technique for this application but suffers from long times required to reach equilibrium, typically 4–28 h [3] as well as issues such as volume shifts which are not accounted for in many studies. Due to long equilibration times, equilibrium dialysis is also not suitable to determine binding of unstable drugs and drugs that have very poor aqueous solubility. Ultrafiltration reduces the time required to perform the binding experiment, but can suffer from protein leakage through the device which adversely affects the accuracy of the determination. Another issue that is commonly faced by both equilibrium dialysis and ultrafiltration is non-specific binding which can cause significant errors in the measurement. SPME can address both of these major issues by increasing experimental throughput and improving accuracy of binding data, making it a valuable addition to the arsenal of methods for free concentration determinations.

The main principle that permits the use of SPME in binding studies is that the amount of analyte extracted by SPME is proportional to the free concentration of the analyte. Thus, in SPME, it is not necessary to physically separate the free and bound analyte. The amount of analyte extracted by the device will depend on the free concentration of analyte, which in turn will depend on the extent of analyte binding to receptors or carrier proteins present in the matrix under study. Therefore, for accurate binding studies, it is critical not to disturb the equilibrium between ligand and receptor by the SPME or to compensate correctly for any introduced disturbance.

Any non-covalent reversible binding equilibrium with 1:1 stoichiometry can be described according to the equation below:

$$L + R \rightleftarrows LR$$

The equilibrium constant, K_a for this equilibrium can be described according to the expression

$$K_a = \frac{[LR]}{[L][R]}$$

where $[LR]$ is the concentration of the ligand–receptor complex at equilibrium, $[L]$ is the concentration of free ligand at equilibrium, and $[R]$ is the equilibrium concentration of the receptor. Receptor is the term used for any protein that can bind the ligand. SPME can then be used to determine $[L]$ according to the equation below [4]:

$$[L] = C_{\text{free}} = \frac{n}{f_c}$$

where n is the number of moles of analyte extracted and f_c is the fiber constant that represents the product of K_{fs} (or K_A if using adsorptive coating far from saturation) and the volume of the fiber, if the amount extracted is independent of the volume, which occurs when sample volume $\gg V_f K_{fs}$ [5]. If the latter assumption is not met, then f_c incorporates all the terms from full SPME equation. The above equation holds regardless of the extent of depletion, but the coating selected must not extract receptor molecules or receptor–ligand complex. The fiber constant can easily be determined by performing external calibration using standard solutions that contain no receptor molecules. Typically, multiple points are used to build the calibration curve and obtain an accurate determination of f_c although we have recently shown that single-point calibration can also work very well when using 96-well automated SPME [6].

10.2 Design of SPME Methods to Evaluate Ligand–Receptor Binding

Figure 10.1 shows the main workflow and steps required to perform binding study to determine binding constants and/or stoichiometry using SPME. The main steps of the workflow are (1) development of SPME method, (2) selection of appropriate

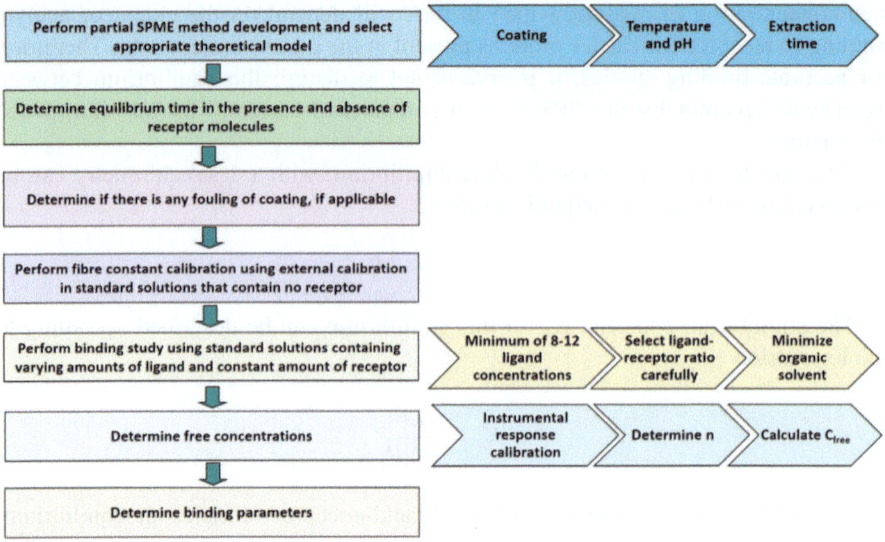

Fig. 10.1 Workflow of the main steps required to perform an accurate binding study using SPME for studies of ligand binding with a single receptor to determine binding constants. Adapted by permission from Nature Publishing Group: Ref. [7], copyright 2010

model/experiment design and the verification of the assumptions required for the model to be applied, (3) fiber constant calibration (4) execution of binding study and determination of free concentration using an appropriate analytical method, and (5) determination of binding parameters. The most common experimental design uses multiple receptor solutions for the binding study. However, one of the advantages of SPME is that even a single receptor solution can be used to obtain binding parameters. The details of SPME set-up and model calculations for such situations where receptor availability is limited are proposed and discussed elsewhere [4].

The main parameters to consider during SPME development for binding studies are coating, extraction time, pH, and temperature. To obtain the most biologically relevant results, temperature and pH are typically selected to match physiological conditions of 37 °C and pH of 7.4. For binding applications, steps that are often used to optimize analytical sensitivity of SPME but that can lead to disruption of ligand–receptor binding such as extreme pH adjustment, addition of solvents and/or salts are typically not explored because they would lead to incorrect determination of free concentration(s) of interest. Additional considerations that should be addressed during SPME method development specifically for binding studies are discussed in more detail below. In addition, step-by-step procedure for ligand–receptor binding studies of non-volatile analytes using automated 96-fiber SPME is available [7].

SPME binding studies can be performed under two different paradigms. The first and simplest approach is **negligible depletion** when the amount of ligand extracted by the fiber is less than 5% of free (unbound) amount [8]. In this case, the amount of analyte extracted is so small it does not significantly disturb the binding equilibrium between analyte and the binding matrix. The free concentration of analyte is determined using SPME and appropriate matrix-free calibration. From practical perspective, negligible depletion conditions can be achieved most easily by using very thin coatings and/or using coatings that do not have very high K_{fs} values for the analytes of interest. Pre-equilibrium SPME can also be applied to limit depletion and increase sample throughput; however, mass uptake kinetics need to be carefully studied in this approach [9]. An excellent review by Heringa and Hermens [8] covers all aspects of negligible depletion SPME in detail.

Binding studies can also be accurately carried out with significant depletion provided that (1) the amount of analyte extracted by the fiber is small in comparison to the amount of analyte bound to binding matrix, (2) ligand–protein complex has fast off rates (important both for kinetic and equilibrium sampling), and (3) same analyte uptake occurs in presence and absence of matrix or appropriate correction for uptake rates is applied (kinetic SPME) [10]. In this case, the calculations must account for depletion as described in detail elsewhere [4, 5]. Table 10.1 shows example results demonstrating that both negligible depletion and significant depletion experimental designs can yield equivalent binding constants provided that the above described theoretical constraints are met by the system under study. The results shown are for the binding of diazepam to bovine serum albumin (BSA), a well-studied system. In addition, Table 10.1 also shows that pre-equilibrium SPME can be successfully employed to yield accurate binding constants. Within experimental error, all the results shown agree well with reported literature values of log K = 3.85–4.51 with the exception of 5 mL sample tested at pre-equilibrium where deviation of up to 0.2 log units was found.

Another important parameter to consider when selecting **SPME coating** for binding studies is adsorption of proteins to SPME fiber, also known as fouling. If fouling occurs, it can affect analyte uptake and yield erroneous binding constants. Fouling of the fiber can affect the binding data in two possible ways: (i) decreasing uptake of the analyte into the coating which would cause an overestimation of

Table 10.1 Influence of depletion and pre-equilibrium sampling on calculated binding affinities of diazepam to BSA

Sample volume (mL)	30 min exposure		Equilibrium exposure	
	Log $K \pm$ SD	% Depletion	Log $K \pm$ SD	% Depletion
1.5	4.42 ± 0.02	8.5	4.42 ± 0.003	47
5	4.67 ± 0.02	1.7	4.49 ± 0.04	19
10	4.56 ± 0.10	1.1	4.44 ± 0.03	11
20	4.37 ± 0.05	0.8	4.38 ± 0.004	6.5
35	4.36 ± 0.04	0.4	4.38 ± 0.01	3.8

Diazepam was spiked in different vial sizes at a total concentration of 5 μM. BSA concentration was 150 μM in all vials. Exposure time was either 30 min or at equilibrium (24 h). Free concentrations were calculated using a time-matched fiber calibration in phosphate-buffered saline (PBS). Reprinted from Ref. [10], Copyright 2015, with permission from Elsevier binding and is problematic if pre-equilibrium SPME is used or (ii) increasing concentration of analyte in the fiber which would cause an underestimation of binding [10]. The use of biocompatible coatings such as biocompatible C18 commercial fibers or polymers such as polypyrrole and polyacrylontirile is preferred to minimize such adsorptive effects. Peltenburg et al. examined the extent of adsorption of BSA on biocompatible C18/SCX fibers commercially available from Supelco and found concentration dependent adsorption of serum albumin with 0.4–0.6 nmol BSA adsorbed for 6–72 h of exposure [10]. However, even with these values, the authors found that the influence on binding constants was negligible with errors up to 0.02 log units on binding constants depending on the ligand tested and if the analyte occupancy of protein is <10%. The latter requirement is generally necessary to ensure total protein concentration and free protein concentration can be assumed to be equivalent in binding calculations. It is also physiologically relevant for many studies as concentrations of carrier proteins such as albumin tend to be very high. Unfortunately, the authors did not test very short exposures of 1–10 min that are typically used for in vivo studies, pre-equilibrium SPME and equilibrium SPME using thin coatings. Also, it must be emphasized that the extent of fouling depends strongly on fiber type, with previous studies reporting higher fouling for PA and ion-exchange types of coatings [11] while no fouling was detected for PDMS within limit-of-detection of Bradford assay [12]. Therefore, for binding studies, it is highly recommended to evaluate extent of fouling for the exact coating and extraction time combination used in the experimental design and to preferentially select biocompatible coatings for these types of experiments. If the analytes of interest are volatile or semi-volatile, then headspace SPME can be applied, thus completely eliminating the possibility of fouling. Another option that has been successfully employed to date is to protect non-biocompatible coatings using an appropriate membrane as shown for paclitaxel in Table 10.2.

After the selection of coating, the second key parameter to consider during SPME method development is **extraction time**, and whether pre-equilibrium or equilibrium SPME will be used. Equilibrium SPME is more commonly used as long as the time required to reach equilibrium is not too excessive for the application of interest. In this case, it is important to ensure that the extraction time selected is sufficient to establish two equilibria: (1) equilibrium between sample and coating and (2) binding equilibrium between ligand and receptor. In many cases, 1 h or overnight incubation is sufficient to establish equilibrium between drugs and proteins, but this needs to be verified experimentally for the system of interest as many exceptions do exist. In particular, systems with very low K_d and/or very low k_{off} and k_{on} rates will require longer time to reach equilibrium. At low receptor occupancy, if LR complex half-life is known ($t_{1/2}$), equilibration time should be set at least 5 times $t_{1/2}$ to reach equilibrium [13]. If SPME extraction time is shorter than the time required to reach ligand–receptor binding equilibrium, then either the longer time can be selected for the extraction time or pre-incubation of ligand and receptor must be incorporated into SPME procedure. If kinetic SPME is to be used, the key idea is that equilibrium between protein and ligand must be established prior to SPME sampling. The second important consideration if using pre-equilibrium SPME is that uptake dynamics (or

Table 10.2 Example applications of SPME for ligand–receptor binding and partitioning determinations

Analyte [Refs.]	Application	Equilibration time	SPME coating	Experimental value
Pharmaceutical/biological				
Paclitaxel [17]	Free drug concentration in liposome formulation	Pre-equilibrium 1 h	Membrane-protected carbowax/template resin (CW/TPR)	0.36 μg/mL
Ibuprofen [19]	% PPB	10 min	Polypyrrole	99.8 ± 0.05%
Warfarin [19]	% PPB	10 min	Polypyrrole	99.7 ± 0.2%
Verapamil [19]	% PPB	10 min	Polypyrrole	85.3 ± 2.4%
Propranolol [19]	% PPB	10 min	Polypyrrole	75.4 ± 5.3%
Caffeine [19]	% PPB	10 min	Polypyrrole	67.0 ± 5.5%
Diazepam [5]	Binding affinity and number of sites to HSA	8 min	Restricted access C18	1.76×10^6 M^{-1}, 3.31×10^4 M^{-1}, 8.1×10^3 M^{-1} three binding sites with different affinity
Lindane, parathion, paraoxon [27]	Fat/blood, muscle/blood and liver/blood partition coefficients	24 h	Polyacrylate	Fat/blood 53.5 parathion 122 lindane 11.3 paraoxon
Isosorbidedinitrate [5]	Binding affinity and number of sites to HSA	8 min	Restricted access C18	K = 2.78×10^4 M^{-1}, three identical binding sites
Carbamazepine [16]	Binding affinity and number of sites to HSA and % PPB in vivo and ex vivo	50 min Pre-equilibrium 2-min in vivo	Biocompatible C18	3.38×10^4 M^{-1} 7.50×10^3 M^{-1} two binding sites 96.1 ± 0.6% (in vivo)
Estradiol [28]	Binding affinity and number of sites to HSA	3 h	Polyacrylate	8.9×10^4 M^{-1}

(continued)

Table 10.2 (continued)

Analyte [Refs.]	Application	Equilibration time	SPME coating	Experimental value
Chlorhexidine [5]	Binding to saliva proteins	15 min	CW/TPR	$K = 3.51 \times 10^5\ M^{-1}$, one binding site
Propranolol [24, 25]	% PPB and whole blood binding	2 min	Polystyrene/collagen core–sheath nanofiber	66–87% PPB 99.4% whole blood
Aripiprazole [24]	% PPB and whole blood binding	2 min	Polystyrene/collagen core–sheath nanofiber	76% PPB 93.9% whole blood
Desipramine [24, 25]	% PPB and whole blood binding	2 min	Polystyrene/collagen core–sheath nanofiber	81–91% PPB 99.4% whole blood
Aroma compounds 2-methyl-butanal, methional, 2-pentanone, 2-heptanone and nonanal [29]	Binding to myofibrillar proteins and dependence on protein oxidation	Headspace 30 min	CAR/PDMS	Influence of protein oxidation on binding successfully evaluated
2-nonanone [30]	Binding to milk proteins including binding constants and number of binding sites and comparison of protein affinity	Headspace 5 min	PDMS	Albumin > β lactoglobulin > α lactoglobulin > α casein > β casein
Environmental				
Alachlor, lindane, *p,p′*-DDE, PCB-15, PCB-138, atrazine, ametryn, atraton, terbuthylazine, terbutryn, terbumeton [31]	Sorption to humic substances and lime soil/model minerals—determination of free fraction and binding constants	Pre-equilibrium 20–30 min	PDMS	K_{HA} calculated for different types of humic acids (HA)
N-phenyl-2-naphtylamine, fluoranthene, pyrene, PCB 101, lindane, methoxychlor, benzo[a] anthracene, [32]	Free concentration determination and algae partitioning coefficients	Pre-equilibrium 5 min	PA	Log *K* range 1.3–4.5 and good correlation with log K_{OW}

(continued)

Table 10.2 (continued)

Analyte [Refs.]	Application	Equilibration time	SPME coating	Experimental value
Methyl parathion, atrazine, prometryn, *p′p′*-DDD, anthraquinone [32]	Free concentration determination and algae partitioning coefficients	Pre-equilibrium 5 min	DVB/CAR/PDMS	Log K range 1.3–4.5 and good correlation with log K_{OW}
Chlorfenvinphos [33]	Insecticide uptake and inter-insect variability in free concentration	Pre-equilibrium 45 min	PDMS	Internal dose found to be 84% of contact dose 78% free
Other				
Phenetole, toluene, *m*-xylene, naphthalene, biphenyl, fluorine, phenanthrene [34]	Determination of stoichiometry and binding constants with cyclodextrin	120 min	PDMS/DVB	Stoichiometry of 1:1 or 1:2 depending on analyte 111.3–3510 M^{-1} for 1:1 binding 7.30×10^5–9.03×10^6 M^{-2} for 1:2 binding

time required to reach equilibrium) must be the same in presence and absence of binding matrix. If these two conditions are not met, then matrix-free calibration will not yield the correct free concentration, which will in turn impact the accuracy of binding constants significantly. Recently, Alam et al. [14] have proposed a detailed mathematical model/computational simulation to predict whether equilibration time will be affected by presence of binding matrix if physicochemical properties of analyte, matrix, and SPME coating are known. This model applies to the situation where only single binding matrix component is present, with further on-going efforts to expand this computational approach to multi-compartment models such as binding to plasma or whole blood. Zhang et al. [9] have also recently shown that kinetic calibration methods that use deuterated internal standards preloaded on the fiber can be used to correct for mass transfer variation due to protein binding and/or matrix tortuosity in order to yield correct free concentration measurement. Figure 10.2 shows an example extraction time profile of chlorpromazine in the absence and presence of bovine serum albumin [15]. It shows visually there was no significant difference detected in the time required to reach equilibrium, and this was further confirmed statistically by calculation of $t_{95\%}$. The figure also illustrates that, as expected, the amount of analyte extracted decreases in the presence of the receptor due to the decrease of its free concentration because of ligand–receptor binding.

Solvent percentage in samples used for the determination of binding should be kept ≤1% v/v to ensure binding is not disturbed by the presence of organic solvent. For instance, the presence of 1, 2, and 3% methanol for binding study between carbamazepine and human serum albumin showed no effect on binding constants calculated for high-affinity site, but showed significant differences (2.5 fold difference in K_a) for low-affinity binding site which was more exposed and accessible to solvent [16]. For plasma protein binding determinations, even smaller percentage of organic modifier is preferred (<0.01%) to obtain most physiologically relevant results [19]. Also, it is important to emphasize that the presence of organic solvent generally lowers K_{fs} values which in turn decreases the amount of analyte extracted

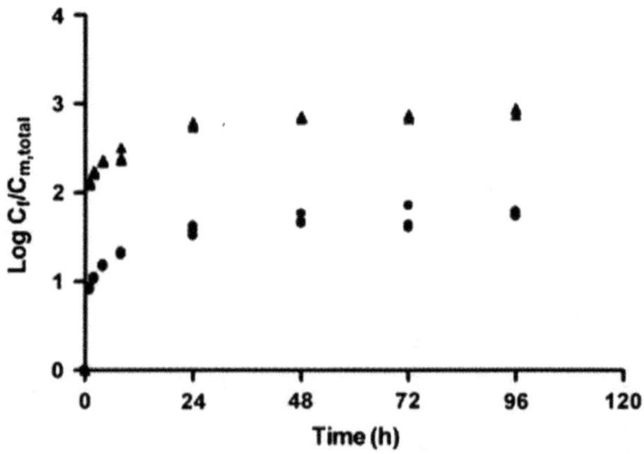

Fig. 10.2 Absorption profile of 100 μM chlorpromazine with (*filled circle*) and without (*filled triangle*) 600 μM bovine serum albumin (individual data points of triplicate experiments are shown, but in several cases, the replicates overlap). Reprinted from Ref. [15], Copyright 2011, with permission from Elsevier

by the coating. Therefore, it is important to keep percentage of solvent very low and equal in all samples used for binding studies.

pH control is crucial to obtain highly relevant physiological binding values especially for basic drugs. In a seminal work on the topic, Musteata et al. [19], showed significant influence of pH on % PPB and recommended incubation with 10% CO_2 or dilution 1:10 with isotonic PBS as optimal methods of pH control. Simple adjustment of pH using concentrated phosphate or citrate buffer resulted in irreproducible % PPB values and is not recommended. If dilution approach is used, PPB for undiluted sample can be calculated as follows if assumption that $C_{\mathrm{drug}} \ll C_{\mathrm{p}}$ holds:

$$\%\,\mathrm{PPB}_{\mathrm{diluted}} = \frac{100}{1 + \frac{d}{K \times C_{\mathrm{p}}}}$$

$$\%\,\mathrm{PPB} = \frac{100}{1 + \frac{1}{K \times C_{\mathrm{p}}}}$$

where d is dilution factor, C_{p} is concentration of protein, and K is the binding constant between ligand and the binding matrix (same as K_{a} defined in Sect. 10.1). $K \times C_{\mathrm{p}}$ is therefore determined from the diluted binding data, and those values are then used to find % PPB for undiluted sample. The assumption that drug concentration is low enough in comparison to protein concentration should be scrutinized during study design. Thus, the selection of ligand to receptor ratio plays a critical role for both (1) physiological relevance and (2) selecting and applying appropriate model.

Temperature is another parameter that requires strict control when performing binding studies with SPME. The temperature has been shown to affect the extent of drug–protein binding significantly with even 50% differences found when comparing room temperature and physiologically relevant 37 °C [18].

10.3 Determination of Binding Constants and Number of Binding Sites

Peltenburg et al. [10] determined K_a for interaction of BSA with four different drugs across different concentrations. Experimental log K values ranged between 4.36 and 4.56 for amitriptyline (literature values 3.91–4.80), between 2.74 and 3.40 for amphetamine (literature values 3.03–3.15), and between 3.80–4.40 for diazepam (literature values 3.85–4.51). The authors also determined binding of the same drugs to human serum albumin with good agreement to literature data. For instance, human serum binding to diazepam was determined to have log K value of 5.22 ± 0.06 in excellent agreement with literature values of 4.18–6.04. Overall, the study showed good ability to distinguish species-related differences in affinity when

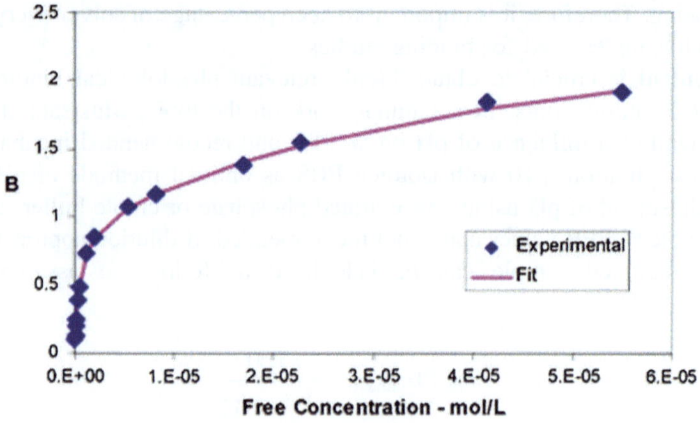

Fig. 10.3 Binding curve obtained by SPME for diazepam binding to human serum albumin and used to determine three binding constants shown in Table 10.2. Reprinted with the permission from Ref. [4]. Copyright 2005 American Chemical Society

using SPME for binding studies. In an earlier study on the topic, Musteata and Pawliszyn showed three different methods can be used to accurately determine binding constants: method of multiple standard solutions, method of successive additions, or method of successive extractions. The latter two options are invaluable for studies with limited sample availability. Figure 10.3 shows the binding curve obtained using typical method of multiple receptor solutions to study binding of diazepam to human serum albumin. Three binding sites with different binding affinities could be successfully distinguished as shown in Table 10.2 and match well with expected literature values. As shown in Figs. 10.1 and 10.3, when constructing binding curves to determine binding constants and stoichiometry, it is crucial to collect many points across wide concentration range. This will permit to get a reliable fit of the model and will allow to verify the applicability of the model. For instance, 1:1 binding stoichiometry will yield hyperbolic binding curves as shown in Fig. 10.3 example. Any deviations from the model fit should be scrutinized carefully as they likely indicate that the model applied is not appropriate.

10.4 Determination of Plasma Protein Binding (% PPB)

The total extent of % PPB for an analyte can be determined using the equation below:

$$\%\text{Plasma protein binding}\,(\%\,\text{PPB}) = \frac{C_{\text{total}} - C_{\text{free}}}{C_{\text{total}}} \times 100\%$$

where C_{total} is the total concentration of analyte in plasma and C_{free} is the free concentration of the analyte in plasma measured using SPME [19, 20]. If not using negligible depletion design, where C_{total} and C_0 can be considered approximately equal, C_{total} can be calculated using the equation below:

$$C_{total} = C_{0\ plasma} - \frac{n_{plasma}}{V_s}$$

where $C_{0\ plasma}$ is the initial concentration of analyte spiked in plasma, n_{plasma} is the amount of analyte extracted by SPME, and V_s is the sample volume of plasma used for the experiment. C_{free} is calculated in the same way as described previously, using fiber constant.

Experimentally, it is recommended to perform plasma protein binding experiments at minimum of three concentrations with triplicate determinations at each concentration [1, 21] although many studies in literature to date only report one concentration. The use of multiple concentrations helps to determine any changes in binding that depend on drug concentration, perhaps due to saturation of binding sites.

Musteata et al. [19] systematically investigated the parameters that can affect the determination of plasma protein binding using five common drugs ibuprofen, warfarin, verapamil, propranolol, and caffeine, which span the range from low to high binding. The authors found that the dilution of plasma with isotonic PBS or incubation with 10% CO_2 in the atmosphere was necessary to maintain physiological pH and obtain correct % PPB binding values for ionisable drugs. pH control is particularly important for basic drugs, where small change of pH, from physiological pH of 7.4 can significantly affect the degree of ionization of analyte. With the use of pH control, the data obtained correlated well with average values of plasma protein binding found in the literature. Table 10.2 shows that the SPME approach applies well to both low-binding and very highly bound drugs. However, SPME would not be a suitable choice, if the ligand binding is extremely low so that the amount of analyte extracted by SPME fiber in the presence and absence of receptor changes only slightly to an extent that this small change cannot be accurately detected during sample analysis within experimental error.

SPME was also recently used to determine percent plasma protein binding for several eicosanoids in human plasma [22]. The values of 99.3% for 12-hydroxyeicosatetraenoic acid (12-HETE), 99.6% for arachidonic acid, and 99.9% for docosahexaenoic acid were obtained. The authors showed that diffusion-based calibration in and without matrix was able to yield accurate % protein binding levels in agreement with literature values. This is in agreement with other studies where both kinetic and equilibrium SPME were used for accurate determination of binding constants such as presented in Table 10.1. In another study, SPME was applied to measure human serum albumin binding constants and % PPB of polyunsaturated fatty acids [23]. The affinity constants (K_a) of individual polyunsaturated fatty acids to human serum albumin ranged from 9.2×10^4 to 4.3×10^5 M^{-1}, while % PPB ranged from 98.0 to 99.7%. These results are in agreement with literature and show the same trends as previously established for fatty acids: (1) K_a increases with increase in fatty acid chain length and (2) K_a increases with decrease in the number of double bonds present. This shows the accuracy of SPME is excellent in binding studies and permits accurate detection of small differences in the extent of binding. Wu et al. proposed polystyrene/collagen core–sheath nanofiber membrane within microfluidic device as a new biocompatible coating type that can be used to achieve equilibrium rapidly (within 2 min) [24]. They used the newly designed device to evaluate plasma protein binding for several drugs. The results in this study showed generally good agreement between experimental and literature values; however, some discrepancies depending on the exact device and study design were observed as shown in Table 10.2 for propranolol and desipramine. For propranolol, % PPB values reported ranged from 66.3 to 87% between the two studies [24, 25], which merits additional investigation. The observed difference could possibly originate from different device designs used in the two studies. If there were significant adsorptive losses to microfluidic device, the accuracy of free concentration determination would be compromised, similar to what is often observed in membrane dialysis and ultrafiltration methods. Other possible contributing factors could include different lots of plasma or differences in plasma pH if this parameter was not strictly controlled between the two studies.

Peltenburg et al. also clearly show the importance of performing binding studies in complex matrix because human serum albumin binding values even at physiologically relevant concentrations were not predictive of in vivo situation due to competition for binding sites and the contribution of proteins other than albumin to the binding, especially alpha-1-acid glycoprotein which is known to bind cationic drugs. As such, European Bioanalysis Forum recently recommended that in vitro binding studies should be performed in (1) human plasma, (2) the animal species used in toxicology and pharmacology studies, (3) human serum albumin, and (4) human a1-acid glycoprotein in order to better understand binding characteristics of compound under study [21]. The inclusion of positive controls whenever performing binding studies is also recommended. Suggested positive control compounds are warfarin for high to very high binding and atenolol or ketamine for low to medium binding during drug development phase [21].

10.5 Determination of Blood-to-Plasma Concentration Ratios

In addition to % PPB, SPME can also be used to determine partitioning of drugs to red blood cells and blood-to-plasma concentration ratios. The blood-to-plasma concentration ratio significantly differs from unity as a result of either exclusion of drug from red blood cells (ratio <1) or the partitioning into or binding of drug to red blood cells (ratio >1). Experimentally, as shown in Fig. 10.4, this type of study is performed by spiking drug at various concentrations into whole blood, incubating for sufficient time to establish partitioning equilibrium and then dividing each blood sample into two aliquots. One aliquot is used to perform equilibrium SPME

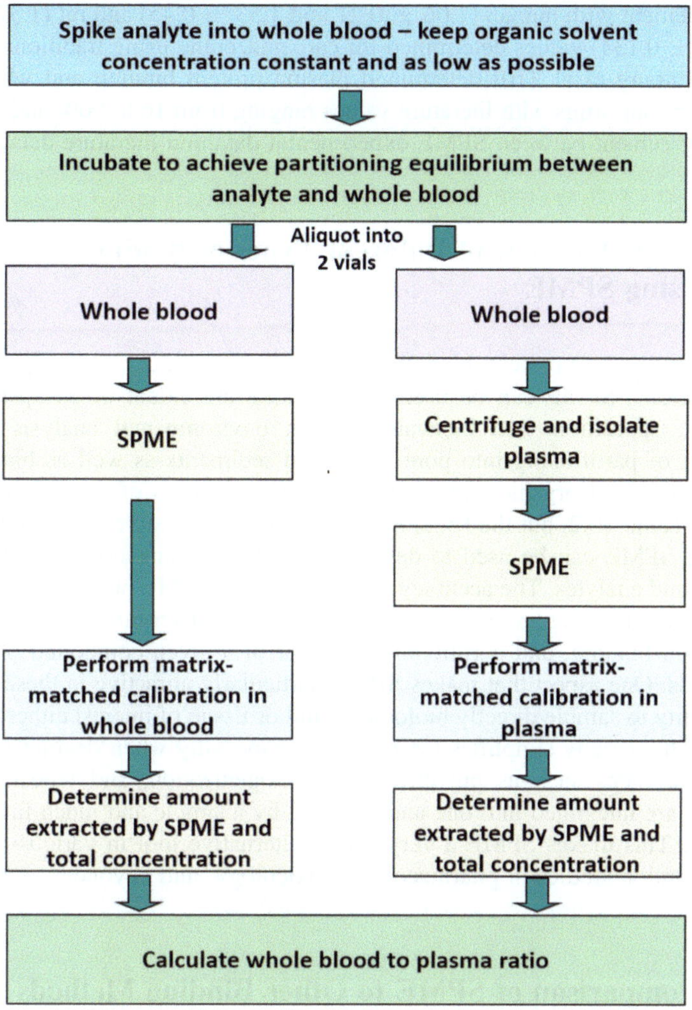

Fig. 10.4 Schematic of the procedure for the determination of whole blood-to-plasma ratio

sampling directly in the spiked whole blood. Second aliquot is used to isolate plasma after centrifugation, and then equilibrium SPME is performed in the resulting plasma sample. Matrix-matched calibration curves are constructed in both matrices (plasma and whole blood) and used to determine the total analyte concentrations in the two aliquots. Blood-to-plasma concentration ratio is then calculated as the ratio of the concentration found in blood versus plasma. This experiment should be performed at minimum of three concentrations, and preferably using three replicates at each concentration level. For instance, this approach was used to determine blood-to-plasma ratio for carbamazepine (1.1 ± 0.2) and its metabolite carbamazepine-10,11-epoxide (1.3 ± 0.4) for mice, data which is in good agreement with human (1.06 ± 0.21 and 1.53 ± 0.45) and rat (1.37 ± 0.125 and 1.32 ± 0.134) values determined for carbamazepine using traditional methods [26]. Peltenburg et al. [10] determined plasma protein binding and whole blood binding for four drugs with literature values ranging from 16 to 96% and with good general agreement between SPME experimental data and literature data.

10.6 Brief Overview of Different Types of Binding Studies Using SPME

In addition to drug-protein binding studies, % PPB and blood-to-plasma ratio determinations, SPME can be used to determine the extent of encapsulation by liposomes, speciation, and bioavailability in environmental analysis including evaluation of partitioning into pore water and sediments as well as binding constants to dissolved organic carbon. Few of different types of possible studies are shown in Table 10.2, but the list is not meant to be exhaustive. Overall, the results show that SPME can be used to determine binding parameters for both low and highly bound analytes. The accuracy and precision of SPME binding studies permit comparison of binding proteins from different species, determination of the effect of structure on binding, and partitioning behavior of a given compound or series of compounds. One aspect that makes SPME particularly attractive in these studies is its capability to sample directly biological fluid or tissue of interest either in vivo or ex vivo. This greatly simplifies the protocols, especially when dealing with tissue samples, as steps such as pulverization, homogenization, and repeated solvent extraction are integrated into one and replaced by a simple and much faster SPME workflow. This makes SPME a very useful alternative tool in various absorption and distribution studies in pharmacology, toxicology, and beyond.

10.7 Comparison of SPME to Other Binding Methods

A recent study compared the determination of binding parameters using SPME versus fluorescence and NMR for carbamazepine binding to human serum albumin. Both fluorescence and SPME successfully identified carbamazepine binding to high-affinity site with excellent agreement in K_a values despite vastly different analytical approaches: K_a of $3.00 \times 10^4\ M^{-1}$ for fluorescence and $3.38 \times 10^4\ M^{-1}$ for SPME. However, SPME data also clearly indicated a presence of second low-affinity binding site with K_a of $7.50 \times 10^3\ M^{-1}$ which could not be observed using fluorescence measurement. This study used high drug to protein ratio to enable fluorescence measurements. However, SPME also outperformed fluorescence and NMR spectroscopic studies by its capability to measure binding under more physiologically relevant conditions where % PPB binding of carbamazepine in mouse plasma was determined as 96.1 ± 0.6% across the range of drug concentrations tested from 0.04 to 9.3 μM. This example shows that excellent precision of % PPB determination can be achieved using SPME methods when experiment is designed appropriately and all parameters are well-controlled. Chalumot et al. [34] used SPME to investigate binding stoichiometry and constants for inclusion complexes that can form between aromatic compounds and β-cyclodextrin. For 1:2 stoichiometry, they also used classical modified Benesi–Hildebrand method which is based on measuring change in fluorescence or absorbance in the presence and absence of cyclodextrin and confirmed presence of 1:2 binding stoichiometry that was initially detected by SPME. The two methods also showed good agreement in binding constants at high cyclodextrin concentrations. The authors ascribe the disagreement at other concentrations by the inability to obtain reproducible binding data using Benesi–Hildebrand method in cases of weak and strong interactions. Overall, the authors conclude that SPME is a viable and simple method highly applicable to study guest–host molecule interactions.

In addition to binding parameters, determination of free concentrations in various samples is also extremely important and useful. SPME can provide accurate determination of this concentration as discussed throughout this chapter. Table 10.3 compares the use of pre-equilibrium SPME, equilibrium SPME, and microdialysis for the determination of free concentrations of various pharmaceuticals and other common groundwater contaminants that can bioaccumulate in fish muscle [9]. The results show that all three methods can yield equivalent free concentrations within the experimental error, as long as an appropriate experimental design for each approach is utilized. However, the results also highlight that SPME outperforms microdialysis for highly hydrophobic drugs where solubility issues and adsorptive losses/non-specific binding contribute to poor performance of microdialysis as observed for fluoxetine. Bojko et al. [35] also compared the performance of SPME to ultrafiltration for tranexamic acid, drug with very low degree of protein binding (3% PPB) and found equivalent free concentrations, with no significant difference between the two methods.

Table 10.3 Free analyte concentrations in rainbow trout dorsal–epaxial muscle (ng/g) following 8 days of exposure to the analyte mixture (n = 9) obtained using pre-equilibrium SPME, equilibrium SPME and microdialysis

Analyte	Pre-equilibrium SPME (20 min)	Equilibrium SPME (18 h)	Microdialysis
Atrazine	0.12 ± 0.02	0.13 ± 0.01	0.12 ± 0.03
Atorvastatin	ND	ND	ND
Bisphenol A	ND	ND	ND
Carbamazepine	0.25 ± 0.05	0.22 ± 0.05	0.27 ± 0.04
Diclofenac	1.25 ± 0.4	1.18 ± 0.3	1.20 ± 0.4
Fluoxetine	6.48 ± 0.3	6.95 ± 0.8	NR
Gemfibrozil	0.15 ± 0.05	0.12 ± 0.04	0.16 ± 0.05
Ibuprofen	0.14 ± 0.05	0.13 ± 0.03	0.12 ± 0.02
Naproxen	ND	ND	ND

ND not detected, *NR* not reported due to unacceptable precision

10.8 Determination of Free Concentrations: In Vivo and Ex Vivo

Figure 10.5 shows example time profiles of receptor concentration, total ligand concentration, and free ligand concentration. Although total ligand concentration shows first order kinetics with rapid decrease of chlorhexidine concentration within

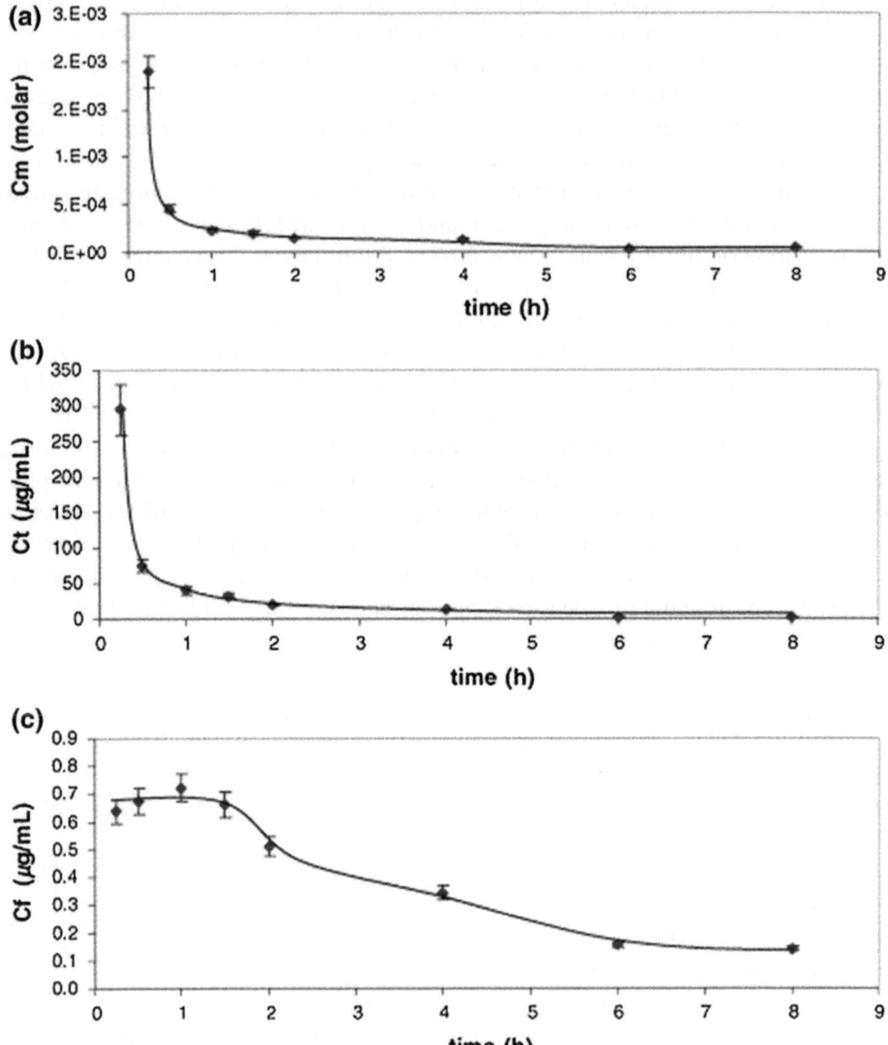

Fig. 10.5 Variation of concentration in time during the pharmacokinetic study (mouthrinsing with 1.0 mg/mL chlorhexidine base): **a** binding matrix concentration (molar), **b** total concentration of chlorhexidine (µg/mL), and **c** free concentration of chlorhexidine (µg/mL). Reprinted with the permission from Ref. [5]. Copyright 2005 Elsevier

the first hour of mouth-rinsing, free concentration of chlorhexidine shows different kinetics with clear plateau for the first two hours within dosing [5]. This example illustrates that it is not sufficient to monitor only total concentration of a given analyte, and that knowing free concentration can help explain the observed biological effect under study.

Birjandi et al. evaluated the effect of varying concentration of human serum albumin in the range from 5 to 100 g/L on the extraction efficiency of polyunsaturated fatty acids by SPME. They found constant extraction efficiencies in the normal range of human serum albumin, but significant differences at very low (<20 g/L) and very high (>70 g/L) albumin concentrations [23]. This should be considered in experimental design to ensure accuracy of total concentrations if they are of interest in a given study. The concentration of binding proteins can vary greatly from individual to individual and can depend on age, pregnancy, and disease status of the individual in question. Therefore, the common measurement of total concentrations only is not sufficient. Drug–drug interactions can also be mediated through binding of the same protein and contribute to adverse events. Historically, one of the main issues with performing free concentration determinations of highly bound drugs was insufficient sensitivity. However, with the new developments in analytical instrumentation, and new techniques that incorporate enrichment and sample clean-up in a single step such as SPME, the measurement of free concentrations becomes feasible for a wide variety of drugs. The future work for the determination of protein binding and free concentrations should move away from the determination of binding in single samples and begin to consider inter-individual differences in the extent of binding. Solid-phase microextraction is uniquely positioned to contribute to this area. Such efforts are uniquely important in personalized medicine to set-up individualized dosage regimens and to understand fate and toxicity of various exogenous chemicals, collectively known as exposome.

10.9 Conclusions

In summary, SPME is a valuable tool for binding studies. The main advantages of SPME for this application include short analysis time, ability to directly study complex samples such as tissue and whole blood and the ability to perform binding studies at any concentrations. Also, during SPME method development, the time required to reach binding equilibrium is evaluated, which ensures good accuracy of the resulting data without relying on the assumptions of slow or fast equilibration. In addition, these types of SPME studies can be automated in 96-well plate format to increase sample throughput. The flexibility of SPME format also affords unique opportunities to obtain various binding parameters using small sample volumes and the information obtained using SPME can be highly complementary to other techniques. The main disadvantages of SPME include time required to develop SPME method and verify assumptions such as fouling or analyte uptake, depending on the experimental design. SPME will also not perform very well in cases of extremely low binding, where it may not be possible to distinguish amount of analyte extracted within experimental error in presence of absence of receptor. SPME will also not perform well if using extremely high ligand concentrations and if there is saturation of coating which is particular danger for very thin coatings as well as adsorptive coatings. Finally, depending on the format of SPME employed, the amounts extracted can be very small, which in turn necessitates use of highly sensitive analytical instrumentation.

References

1. Zhang F, Xue J, Shao J et al (2012) Compilation of 222 drugs' plasma protein binding data and guidance for study designs. Drug Discov Today 17(9–10):475–485. doi:10.1016/j.drudis.2011.12.018
2. Vuignier K, Schappler J, Veuthey J et al (2010) Drug-protein binding: a critical review of analytical tools. Anal Bioanal Chem 398(1):53–66. doi:10.1007/s00216-010-3737-1
3. Illamola SM, Hirt D, Tréluyer JM et al (2015) Challenges regarding analysis of unbound fraction of highly bound protein antiretroviral drugs in several biological matrices: lack of harmonisation and guidelines. Drug Discov Today 20(4):466–474. doi:10.1016/j.drudis.2014.11.010
4. Musteata FM, Pawliszyn J (2005) Study of ligand–receptor binding using SPME: investigation of receptor, free, and total ligand concentrations. J Proteome Res 4(3):789–800. doi:10.1021/pr049768z
5. Musteata FM, Pawliszyn J (2005) Assay of stability, free and total concentration of chlorhexidine in saliva by solid phase microextraction. J Pharm Biomed Anal 37(5):1015–1024. http://dx.doi.org/10.1016/j.jpba.2004.09.055
6. Vuckovic D, Pawliszyn J (2009) Automated study of ligand–receptor binding using solid-phase microextraction. J Pharm Biomed Anal 50(4):550–555. doi:10.1016/j.jpba.2008.08.023
7. Vuckovic D, Cudjoe E, Musteata FM et al (2010) Automated solid-phase microextraction and thin-film microextraction for high-throughput analysis of biological fluids and ligand–receptor binding studies. Nat Protoc 5:140–161. doi:10.1038/nprot.2009.180
8. Heringa MB, Hermens JLM (2003) Measurement of free concentrations using negligible depletion-solid phase microextraction (nd-SPME). TrAC Trends Anal Chem 22(10):575–587. http://dx.doi.org/10.1016/S0165-9936(03)01006-9
9. Zhang X, Oakes KD, Hoque ME et al (2011) Pre-equilibrium solid-phase microextraction of free analyte in complex samples: correction for mass transfer variation from protein binding and matrix tortuosity. Anal Chem 83(9):3365–3370. doi:10.1021/ac2004899
10. Peltenburg H, Bosman IJ, Hermens JLM (2015) Sensitive determination of plasma protein binding of cationic drugs using mixed-mode solid-phase microextraction. J Pharm Biomed Anal 115:534–542. doi:10.1016/j.jpba.2015.08.002
11. Heringa MB, Hogevonder C, Busser F et al (2006) Measurement of the free concentration of octylphenol in biological samples with negligible depletion-solid phase microextraction (nd-SPME): analysis of matrix effects. J Chromatogr B 834(1–2):35–41. http://dx.doi.org/10.1016/j.jchromb.2006.02.009
12. Oomen AG, Mayer P, Tolls J (2000) Nonequilibrium solid-phase microextraction for determination of the freely dissolved concentration of hydrophobic organic compounds: matrix effects and limitations. Anal Chem 72(13):2802–2808. doi:10.1021/ac9914162
13. Hulme EC, Trevethick MA (2010) Ligand binding assays at equilibrium: validation and interpretation. Br J Pharmacol 161(6):1219–1237. doi:10.1111/j.1476-5381.2009.00604.x

14. Alam MN, Ricardez-Sandoval L, Pawliszyn J (2015) Numerical modeling of solid-phase microextraction: binding matrix effect on equilibrium time. Anal Chem 87(19):9846–9854. doi:10.1021/acs.analchem.5b02239
15. Broeders JJW, Blaauboer BJ, Hermens JLM (2011) Development of a negligible depletion-solid phase microextraction method to determine the free concentration of chlorpromazine in aqueous samples containing albumin. J Chromatogr A 1218(47):8529–8535. doi:10.1016/j.chroma.2011.09.064
16. Bojko B, Vuckovic D, Pawliszyn J (2012) Comparison of solid phase microextraction versus spectroscopic techniques for binding studies of carbamazepine. J Pharm Biomed Anal 66:91–99. doi:10.1016/j.jpba.2012.03.005
17. Musteata FM, Pawliszyn J (2006) Determination of free concentration of paclitaxel in liposome formulations. J Pharm Pharm Sci 9(2):231–237
18. Nilsson LB (2013) The bioanalytical challenge of determining unbound concentration and protein binding for drugs. Bioanalysis 5(24):3033–3050. doi:10.4155/bio.13.274
19. Musteata FM, Pawliszyn J, Qian MG et al (2006) Determination of drug plasma protein binding by solid phase microextraction. J Pharm Sci 95(8):1712–1722. http://0-dx.doi.org.mercury.concordia.ca/10.1002/jps.20558
20. Musteata FM (2012) Ligand–receptor binding and determination of free concentrations. In: Pawliszyn J (ed) Anonymous handbook of solid phase microextraction. Elsevier, pp 383–397. doi:10.1016/B978-0-12-416017-0.00011-5
21. Buscher B, Laakso S, Mascher H et al (2014) Bioanalysis for plasma protein binding studies in drug discovery and drug development: views and recommendations of the European bioanalysis forum. Bioanalysis 6(5):673–682. doi:10.4155/bio.13.338
22. Bessonneau V, Zhan Y, De Lannoy IAM et al (2015) In vivo solid-phase microextraction liquid chromatography-tandem mass spectrometry for monitoring blood eicosanoids time profile after lipopolysaccharide-induced inflammation in Sprague-Dawley rats. J Chromatogr A 1424:134–138. doi:10.1016/j.chroma.2015.10.067
23. Birjandi AP, Mirnaghi FS, Bojko B et al (2014) Application of solid phase microextraction for quantitation of polyunsaturated fatty acids in biological fluids. Anal Chem 86(24):12022–12029. doi:10.1021/ac502627w
24. Wu Q, Wu D, Guan Y (2013) In vivo fast equilibrium microextraction by stable and biocompatible nanofiber membrane sandwiched in microfluidic device. Anal Chem 85(23):11524–11531. doi:10.1021/ac4028438
25. Wu Q, Wu D, Guan Y (2013) Fast equilibrium micro-extraction from biological fluids with biocompatible core–sheath electrospun nanofibers. Anal Chem 85(12):5924–5932. doi:10.1021/ac4006974
26. Vuckovic D, Gien B, de Lannoy I et al (2011) In vivo solid-phase microextraction for single rodent pharmacokinetics studies of carbamazepine and carbamazpine-10,11-epoxide in mice. J Chromatogr A 1218:3367–3375. doi:10.1016/j.chroma.2010.07.060
27. Artola-Garciano E, Vaes WHJ, Hermens JLM (2000) Validation of negligible depletion solid-phase microextraction as a tool to determine tissue/blood partition coefficients for semivolatile and nonvolatile organic chemicals. Toxicol Appl Pharmacol 166:138–144. doi:10.1006/taap.2000.8969
28. Heringa MB, Pastor D, Algra J et al (2002) Negligible depletion solid-phase microextraction with radiolabeled analytes to study free concentrations and protein binding: an example with [3H]estradiol. Anal Chem 74:5993–5997. doi:10.1021/ac0204552
29. Zhou F, Zhao M, Su G et al (2014) Binding of aroma compounds with myofibrillar proteins modified by a hydroxyl-radical-induced oxidative system. J Agric Food Chem 62(39):9544–9552. doi:10.1021/jf502540p
30. Kühn J, Zhu X-, Considine T et al (2007) Binding of 2-nonanone and milk proteins in aqueous model systems. J Agric Food Chem 55(9):3599–3604. doi:10.1021/jf063517o
31. Prosen H, Fingler S, Zupančič-Kralj L et al (2007) Partitioning of selected environmental pollutants into organic matter as determined by solid-phase microextraction. Chemosphere 66(8):1580–1589. doi:10.1016/j.chemosphere.2006.08.006

32. Bandow N, Altenburger R, Brack W (2010) Application of nd-SPME to determine freely dissolved concentrations in the presence of green algae and algae-water partition coefficients. Chemosphere 79(11):1070–1076. doi:10.1016/j.chemosphere.2010.03.021
33. Alix A, Collot D, Nénon J- et al (2001) Measurement of insecticide uptake and effective fraction in a beneficial insect using solid-phase microextraction. Anal Chem 73(13):3107–3111. doi:10.1021/ac0013148
34. Chalumot G, Yao C, Pino V et al (2009) Determining the stoichiometry and binding constants of inclusion complexes formed between aromatic compounds and β-cyclodextrin by solid-phase microextraction coupled to high-performance liquid chromatography. J Chromatogr A 1216(27):5242–5248. doi:10.1016/j.chroma.2009.05.017
35. Bojko B, Vuckovic D, Cudjoe E et al (2011) Determination of tranexamic acid concentration by solid phase microextraction and liquid chromatography-tandem mass spectrometry: first step to in vivo analysis. J Chromatogr B Anal Tech Biomed Life Sci 879(32):3781–3787. doi:10.1016/j.jchromb.2011.08.003

MIX
Papier aus verantwortungsvollen Quellen
Paper from responsible sources
FSC® C105338

If you have any concerns about our products,
you can contact us on
ProductSafety@springernature.com

In case Publisher is established outside the EU,
the EU authorized representative is:
Springer Nature Customer Service Center GmbH
Europaplatz 3, 69115 Heidelberg, Germany

Printed by Libri Plureos GmbH
in Hamburg, Germany